The Visible One Million (1,000,000)

Your Personal Life List to Big Numbers

Dale W. Mitchell

The Visible One Million (1,000,000)

Your Personal Life List to Big Numbers

Dale W. Mitchell

This is the personal Number Life List of:

_____

Commenced on:

_____

The Visible One Million
(1,000,000)
Your Personal Life List to Big Numbers

ISBN-13:
978-1720916741

ISBN-10:
1720916748

First Edition

DEDICATIONS

To my One in a Million person, my lovely wife, Valerie.
She is the sum of all good things.

To my Two in a Million persons: those astonishing brothers, Boston and Trenton Reyes.
They are a vector of all good things.

To the good people of the Greater Corvallis Watershed Area who have generously taken us in.
They are the π of Love's Circle.

Nine hundred thousand, nine hundred and ninety-eight hugs to all, plus two more for our pets!

Keep Oregon Weird.

Table of Contents

Prologue

"How to Make 1,000,000 Dollars

Sell a $200 product to 5,000 people.
Sell a $500 product to 2,000 people.
Sell a $1,000 product to 1,000 people.
Sell a $2,000 product to 500 people.
Sell a $4,000 product to 250 people.

5,000 people pay $17 dollars a month for 12 months.
2,000 people pay $42 dollars a month for 12 months.
1,000 people pay $83 dollars a month for 12 months.
500 people pay $167 dollars a month for 12 months.
250 people pay $333 dollars a month for 12 months."

Quora.com

How to Be a Millionaire

Welcome. This book has a deceptively simple objective: to show you the size of the number One Million (1,000,000). To start with I am sure most of you already have a pretty good idea of the size of one million *whatevers*. You can do the math many ways and, by running the figures yourself, you have gained a decent 'internal' feel for the magnitude you are reaching. The quote above is dependent on multiple paths to reach its purported one million-dollar mark. There are also several books on scaling things up, or comparing, perhaps, the height of the Eifel Tower to that of Mount Everest. I have even own a cute little paperback that repeats the phrase, *"I'm sorry"*, one million times[1]. My wife, Valerie, likes that one. I own several of these compendiums and I am intrigued by them. Perhaps you already own one or more of these books, as well.

So, what is different here—or the burning question, "Why buy this book?"

Difference abounds. To begin with, it does not convey its information by scaling or comparison. Instead, I line up 1,000,000 little lozenges (◊) and let them fall where they may and thereby tell my tale. I have stuck to a very stripped-down and unvarying setup of 2,000 lozenges per page, resulting in a 500 page 'block' of raw, full-contact, in-your-face, counting. The one million I present herein, is, therefore, not a mental estimate, however good you might be at this skill. It is exact.

Another method I am using to convey what one million really is like, is the shear bulk of this volume. For this reason, I am keeping the other bookish matters—like this Prologue, short. Heft this book and you are hefting the number 1,000,000. Thumb through its pages and thousands speed by. If you visualize each lozenge as a year, that's centuries you are time hopping through. If you would rather picture my ◊◊◊s as hippies, you will reach the approximate attendance at the 1969 Woodstock Concert (around 40,000 souls) on page 53. If dollars are your thing, *Man*, you re gon'a need at least *seven* of these books to buy that pad today in Bel Air, California. *Bummer.* (Hint, this latter 'multi-book' function is ideal for disgruntled Congresspersons thundering about how expensive to their taxpayers would be that latest pork barrel boondoggle from the opposing Party.)

---

[1]   See Sources, p. 539 for the titles of the relevant books.

Here is another key difference. I have added fill-in spaces so that you can write in whatever you want about any number. All you do is color-in-in a lozenge (♦) to mark the number you are favoring and then write your note about it in the space provided. If you want to say more about that number (or pick more than one number from the same 'lozenge lineup'), than there are also more lines provided at the bottom of each page. Now you can see a deeper functionally. This book is not just a debutante ball for the number one million. It is also a talent show for, say, the number 444,037 or for 1236, or even 3. If you like *any* unit below one million, this book highlights the lot, and all are equally important. In this work, *The Visible One Million*, you also hold *The Visible Number of Your Choice*! Seldom have numbers been made so personal before!

As a bonus, there is a **Special Section** highlighting the numbers of everyday life. So, should you be especially fond of any number from 1 to 99, you will find it set out in all its glory. (These smaller numbers also get repeated in the **Main Listings**, so that block of 500 pages retains its visually-complete representation of a million.) For those of you who are passionate about Zero (0) (you know who you are . . .), that most august Number of Enigma has graciously consented to make an 'appearance', as well.

The philosophy of this book is wrapped up in making numbers personal, whether your usage is serious or humorous. Start seeing the numbers around you.

Enjoy!

*****

O h, by the way, I promised to explain a way to use this book to make yourself a millionaire. With *The Visible One Million*, it a snap! Just get a big box or other suitable storage place. Drop a dollar bill into said big box or other suitable storage place. Mark off the first lozenge in the Main Listings. Then repeat. Do not spend any of these dollar bills or allow others access to your box or other suitable storage place. After you have added your final dollar bill to your box or other suitable storage place and darkened the last 'checkbox' in this book; you will suddenly and delightfully find that you are a millionaire! Your stack of dollar bills should be about 43 inches (or three feet, seven inches) tall. Of course, it will help if your box or other suitable storage place is rodent and mold proof and that you check for things like this rather frequently . . .So, if your stash does not stack as high as about 43 inches, you might have a problem. In that case, you can consider starting over with a new copy of this book and a better box or other suitable storage place— and this time, not telling your weird Uncle Herkey that you have restarted your Do It Yourself millionaire project.

Hard work pays off, they say!

Introduction

"One million memories, ten thousand inside jokes, one hundred shared secrets, one reason: Best Friends."

From the internet, *Anonymous*

How to Use This Book

This is not a difficult book to use. It does not even have an index. For the most part, simple looking at the thing, combined with some necessary page manipulations, will disclose all its inner secrets. Numbers are arraigned in two formats.

The first is a short **Special Section** focused on the popular daily numbers of **1** to **99**, plus that ever-intriguing enigma, 0. Each number has a lozenge (◊) provided as a 'checkbox' to allow you to mark it you wish. All you need to do is color-in (or not, your choice) said lozenge (♦) to signify a number as your own. If you want to write something about your chosen number, space has been provided for a short note. This section starts on page 21.

The **Main Listings** follow the Special Section and are considerably longer—500 pages longer. This has all the numbers (symbolized as lozenges, rather than written out) from number **1** to number **1,000,000**. Each line ends with a number, such as 1,550—see example below—that is the written-out *last* number (◊) of its line. Each page of these Main Listings has 2,000 lozenges, with a double-underline visually separating each of the two included *Thousands*. As with the previous section, there are lozenges waiting to be colored in (or not) and a space at the end of each 'lozenge lineup' to write a short note about any number(s) you chose. If you should need more writing space, six more lines are provided at the bottom of each page. This section starts on page 33.

Below, I present a bottom portion of the first page of the of the Main Listings with a few notes:

◇◇◇◇◇ ◇◇◇◇◇ ◇◇◇◇◇ ◇◇◇◇◇ ◇◇◇◇◇ ◇◇◇◇◇ ◇◇◇◇◇ ◇◇◇◇◇ ◇◇◇◇◇ ◇◇◇◇◇	1,050
◇◇◇◇◇ ◇◇◇◇◇ ◇◇◇◇◇ ◇◇◇◇◇ ◇◇◇◇◇ ◇◇◇◇◇ ◇◇◇◇◇ ◇◇◇◇◇ ◇◇◇◇◇ ◇◇◇◇◇	1,100
◇◇◇◇◇ ◇◇◇◇◇ ◇◇◇◇◇ ◇◇◇◇◇ ◇◇◇◇◇ ◇◇◇◇◇ ◇◇◇◇◇ ◇◇◇◇◇ ◇◇◇◇◇ ◇◇◇◇◇	1,150
◇◇◇◇◇ ◇◇◇◇◇ ◇◇◇◇◇ ◇◇◇◇◇ ◇◇◇◇◇ ◇◇◇◇◇ ◇◇◇◇◇ ◇◇◇◇◇ ◇◇◇◇◇ ◇◇◇◇◇	1,200
◇◇◇◇◇ ◇◇◇◇◇ ◇◇◇◇◇ ◇◇◇◇◇ ◇◇◇◇◇ ◇◇◇◇◇ ◇◇◇◇◇ ◇◇◇◇◇ ◇◇◇◇◇ ◇◇◇◇◇	1,250
◇◇◇◇◇ ◇◇◇◇◇ ◇◇◇◇◇ ◇◇◇◇◇ ◇◇◇◇◇ ◇◇◇◇◇ ◇◇◇◇◇ ◇◇◇◇◇ ◇◇◇◇◇ ◇◇◇◇◇	1,300
◇◇◇◇◇ ◇◇◇◇◇ ◇◇◇◇◇ ◇◇◇◇◇ ◇◇◇◇◇ ◇◇◇◇◇ ◇◇◇◇◇ ◇◇◇◇◇ ◇◇◇◇◇ ◇◇◇◇◇	1,350
◇◇◇◇◇ ◇◇◇◇◇ ◇◇◇◇◇ ◇◇◇◇◇ ◇◇◇◇◇ ◇◇◇◇◇ ◇◇◇◇◇ ◇◇◇◇◇ ◇◇◇◇◇ ◇◇◇◇◇	1,400
◇◇◇◇◇ ◇◇◇◇◇ ◇◇◇◇◇ ◇◇◇◇◇ ◇◇◇◇◇ ◇◇◇◇◇ ◇◇◇◇◇ ◇◇◇◇◇ ◇◇◇◇◇ ◇◇◇◇◇	1,450
◇◇◇◇◇ ◇◇◇◇◇ ◇◇◇◇◇ ◇◇◇◇◇ ◇◇◇◇◇ ◇◇◇◇◇ ◇◇◇◇◇ ◇◇◇◇◇ ◇◇◇◇◇ ◇◇◇◇◇	1,500
◇◇◇◇◇ ◇◇◇◇◇ ◇◇◇◇◇ ◇◇◇◇◇ ◇◇◇◇◇ ◇◇◇◇◇ ◇◇◇◇◇ ◇◇◇◇◇ ◇◇◇◇◇ ◇◇◇◇◇	1,550
◇◇◇◇◇ ◇◇◇◇◇ ◇◇◇◇◇ ◇◇◇◇◇ ◇◇◇◇◇ ◇◇◇◇◇ ◇◇◇◇◇ ◇◇◇◇◇ ◇◇◇◇◇ ◇◇◇◇◇	1,600
◇◇◇◇◇ ◇◇◇◇◇ ◇◇◇◇◇ ◇◇◇◇◇ ◇◇◇◇◇ ◇◇◇◇◇ ◇◇◇◇◇ ◇◇◇◇◇ ◇◇◇◇◇ ◇◇◇◇◇	1,650
◇◇◇◇◇ ◇◇◇◇◇ ◇◇◇◇◇ ◇◇◇◇◇ ◇◇◇◇◇ ◇◇◇◇◇ ◇◇◇◇◇ ◇◇◇◇◇ ◇◇◇◇◇ ◇◇◇◇◇	1,700
◇◇◇◇◇ ◇◇◇◇◇ ◇◇◇◇◇ ◇◇◇◇◇ ◇◇◇◇◇ ◇◇◇◇◇ ◇◇◇◇◇ ◇◇◇◇◇ ◇◇◇◇◇ ◇◇◇◇◇	1,750
◇◇◇◇◇ ◇◇◇◇◇ ◇◇◇◇◇ ◇◇◇◇◇ ◇◇◇◇◇ ◇◇◇◇◇ ◇◇◇◇◇ ◇◇◇◇◇ ◇◇◇◇◇ ◇◇◇◇◇	1,800
◇◇◇◇◇ ◇◇◇◇◇ ◇◇◇◇◇ ◇◇◇◇◇ ◇◇◇◇◇ ◇◇◇◇◇ ◇◇◇◇◇ ◇◇◇◇◇ ◇◇◇◇◇ ◇◇◇◇◇	1,850
◇◇◇◇◇ ◇◇◇◇◇ ◇◇◇◇◇ ◇◇◇◇◇ ◇◇◇◇◇ ◇◇◇◇◇ ◇◇◇◇◇ ◇◇◇◇◇ ◇◇◇◇◇ ◇◇◇◇◇	1,900
◇◇◇◇◇ ◇◇◇◇◆ ◇◇◇◇◇ ◇◇◇◇◆ ◇◇◇◇◇ ◇◇◇◇◇ ◇◇◇◇◇ ◇◇◇◇◇ ◇◇◇◇◇ ◇◇◇◇◇	1,950 *1920, Women can vote*
◇◇◇◇◇ ◇◇◇◇◇ ◇◇◇◇◇ ◇◇◇◇◇ ◇◇◇◇◇ ◇◇◇◇◇ ◇◇◇◇◇ ◇◇◇◇◇ ◇◇◇◇◇ ◇◇◇◇◇	**2,000**

1910	*My father is born!*

You are now free to make use of this book. I hope you like it, even if that usage results in this volume taking a long, noisy, page-fluttering, arc through the lower atmosphere . . . or, conversely, for those chest-clutching, bouncing up-and-down squeals of pure rapture its very existence keeps pumping through you, Dear Reader.

"Freedom is the oxygen of the soul."

Moshe Dayan

The Lozenge (or 'Checkbox')

Marking this in makes for a quick way to scan the entire book for all your encounters. Normally in a checklist or life list the checkbox is just that: a box; but I feel this shape is more edgy and evocative. You can modify it to convey specialized information by marking in just the bottom half or the converse; if, say, you want to track whether a particular number was also seen by your significant other *or* that it means something else to that person. Leaving it blank might express yet another personal notion.

Freedom, not my plans for your fate, is what is symbolized by the Lozenge.

Bigger Numbers

"Now they know how many holes it takes to fill the Albert Hall."

John Lennon & Paul Cartney, *A Day in the Life*, The Beatles

Because a million gets us thinking about big numbers, I have added a short section on numbers in multiples larger than the subject of this book. Theoretically you could use multiple copies of this book to explore aspects of these bigger-than-one-million quantities. (I will not mind, please buy as many copies of my book as you wish.)

Note: we just cannot simply forever use a smooth multiplication to produce larger-than-life amounts that still remain in 'chunks' convenient for human comprehension. No one seems to want their 'Trillion' made by multiplying a trillion trillions! Because of this, Americans (and now most of the world) get to 'their' higher numbers by multiplying by thousands (Short Scale) while the traditional British and Continental systems multiply 'their' big numbers in blocks of millions (Long Scale). (By the way, both these systems actually started in France.) This is a very significant controversy, both from the standpoint of simple logic (what should be more basic and necessary to math and the hearts of mathematicians than whole-number counting?) and the looming giants of world finance. Presently, there seems little hope or ground for agreement (but, see Russ Rowlett's proposed solution below).

The data for the charts below is mostly modified from articles in *Wikipedia*.

(The abbreviation *SI* refers to the French, *Système International d'Unités*, who watch over these sorts of things.)

The Numbers to One Million (1,000,000)

	Agreed Upon Modern Usage			Number of zeros you will need
	Up to the millions, these numbers are agreed upon.			

		SI Symbol	*SI Prefix*	
1	One			1^0
10	Ten	da-	deca-	10^1
100	Hundred	h-	hecto-	10^2
1,000	Thousand	k-	kilo-	10^3
10,000	Ten Thousand	*myria-* non-SI		10^4
100,000	Hundred Thousand			10^5
1,000,000	**Million**	M-	Mega-	10^6

Chart of the Great High Number Schism

A Brief Comparison of the *Short Scale* vs. *Long Scale* Controversy

Short Scale				Long Scale—Traditional British (& *European*)	
	SI Symbol	*SI Prefix*		*SI Prefix*	
Billion	G-	*Giga-*	10^9	Giga-	Thousand Million (*Milliard*)
Trillion	T-	*Tera-*	10^{12}	Tera-	**Billion**
Quadrillion	P-	*Peta-*	10^{15}	Peta-	Thousand Billion (*Billiard*)
Quintillion	E-	*Exa-*	10^{18}	Exa-	**Trillion**
Sextillion	Z-	*Zetta-*	10^{21}	Zetta-	Thousand Trillion (*Trilliard*)
Septillion	Y-	*Yotta-*	10^{24}	Yotta-	**Quadrillion**
Octillion	X-		10^{27}		Thousand Quadrillion (*Quadrilliard*)
Nonillion	W-		10^{30}		**Quintillion**
Decillion	V-		10^{33}		Thousand Quintillion
Undecillion			10^{36}		**Sextillion**
Duodecillion			10^{39}		Thousand Sextillion
Tredecillion			10^{42}		**Septillion**
Quattuordecillion			10^{45}		Thousand Septillion
Quinquadecillion			10^{48}		**Octillion**
Sedecillion			10^{51}		Thousand Octillion
Septendecillion			10^{54}		**Nonillion**
Octodecillion			10^{57}		Thousand Nonillion
Novendecillion			10^{60}		**Decillion**
Vigintillion			10^{63}		Thousand Decillion
Unvigintillion			10^{66}		**Undecillion**
Duovigintillion			10^{69}		Thousand Undecillion
Tresvigintillion			10^{72}		**Duodecillion**
Quattuorvigintillion			10^{75}		Thousand Duodecillion
Quinquavigintillion			10^{78}		**Tredecillion**
Sesvigintillion			10^{81}		Thousand Tredecillion
Septemvigintillion			10^{84}		**Quattuordecillion**
Octovigintillion			10^{87}		Thousand Quattuordecillion
Novemvigintillion			10^{90}		**Quindecillion**
Trigintillion			10^{93}		Thousand Quindecillion
Untrigintillion			10^{96}		**Sedecillion**
Duotrigintillion			10^{99}		Thousand Sedecillion

Googol is a famous, but wholly informal, 'Big Number'.

"Googol" (incorrectly "Google") 10^{100}

Also *Ten Duotrigintillion*   Also *Ten Sexdecilliard*

And *lots* more.

I might add, when we get to numbers in the range of a *Googol* (10^{100}), we have reached "a quantity that surpasses even the number of hydrogen atoms in the observable universe." (*livescience.com*). You can decide for yourself if there is a useful life style for the even more Brobdingnagian numbers that follow?

An Alternative Big Number System

Finally, there is a good idea for making an end run around the split above. A short scale system based on Greek, rather than Latin, might be more acceptable to all. This was proposed in 2001 by Russ Rowlett, Director of the Center for Mathematics and Science Education at the University of North Carolina at Chapel Hill.

I present the chart below.

Russ Rowlett's System of Big, Fat, Greek, Numbers

Standard Short Scale		Rowlett's Short Scale
Billion	10^9	**Gillion**
Trillion	10^{12}	**Tetrillion**
Quadrillion	10^{15}	**Pentillion**
Quintillion	10^{18}	**Hexillion**
Sextillion	10^{21}	**Heptillion**
Septillion	10^{24}	**Oktillion**
Octillion	10^{27}	**Ennillion**
Nonillion	10^{30}	**Dekillion**
Decillion	10^{33}	Hendekillion
Undecillion	10^{36}	Dodekillion
Duodecillion	10^{39}	Trisdekillion
Tredecillion	10^{42}	Tetradekillion
Quattuordecillion	10^{45}	Pentadekillion
Quinquadecillion	10^{48}	Hexadekillion
Sedecillion	10^{51}	Heptadekillion
Septendecillion	10^{54}	Oktadekillion
Octodecillion	10^{57}	Enneadekillion
Novendecillion	10^{60}	**Icosillion**
Vigintillion	10^{63}	Icosihenillion
Unvigintillion	10^{66}	Icosidillion
Duovigintillion	10^{69}	Icositrillion
Tresvigintillion	10^{72}	Icositetrillion
Quattuorvigintillion	10^{75}	Icosipentillion
Quinquavigintillion	10^{78}	Icosihexillion
Sesvigintillion	10^{81}	Icosiheptillion
Septemvigintillion	10^{84}	Icosioktillion
Octovigintillion	10^{87}	Icosiennillion
Novemvigintillion	10^{90}	**Triacontillion**

Littler Numbers

"Sometimes we underestimate the influence of little things."

Charles W. Chestnutt

It is easy to see how this book's illustrated one million can be stretched to realms of ever more ponderous numbers. For most purposes, all you need do is acquire more copies of this book. Surprisingly, the opposite is also possible (not implying that you can acquire less than one copy of this book, obviously!). You can use your one copy of *The Visible One Million* to get a feel for what it might mean to be very, very, small: *one millionth* of your size, or the size or other quality of anything you want. To do this takes imagination, because visualizing this shrinkage is not a completely straightforward matter.

For simplicity, we will use a generic six-foot-tall 'you' as our example. Using a part of the illustration of the first page of the Main Listings below, start with the first (leftmost) lozenge (normally, this book's Number One) in the top row and imagine that lozenge (◊) stands for a generic six-foot 'you', hands-to-hips, the wind in your hair, tall and proud.

◊	The Number 1 to the Number 2,000	
Start		Notes
◊◊◊◊◊ ◊◊◊◊◊ ◊◊◊◊◊ ◊◊◊◊◊ ◊◊◊◊◊ ◊◊◊◊◊ ◊◊◊◊◊ ◊◊◊◊◊ ◊◊◊◊◊ ◊◊◊◊◊	50	
◊◊◊◊◊ ◊◊◊◊◊ ◊◊◊◊◊ ◊◊◊◊◊ ◊◊◊◊◊ ◊◊◊◊◊ ◊◊◊◊◊ ◊◊◊◊◊ ◊◊◊◊◊ ◊◊◊◊◊	100	

The next lozenge, counting right, is 'you'; now three feet of, still-confident, tallness. 'You' have shrunk one half your height. At the third ◊, 'your' stature is now halved again and 'you' stand at 18 inches (or 1 ½ feet) in 'your' stockings. It will take *three* of 'you' to equal 'your' old six-foot frame. (How 'you' will get two other scaled-down copies of 'yourself' is your business—*and* 'you' are getting a tad worried about the whereabouts of that ferret.) Let's cut to the chase. Looking at the lozenge at the end of this line (50) means that 'you' will now have been shortened *fifty times*. Welcome to one twelfth of an inch. 'You' can ride large beetles (or be eaten by them)! Moreover, you' must chase down, assemble, and line up fifty of 'you all', if 'you' want to make a line of six feet. (Like herding cats, isn't it?) It goes like this through the whole of the Main Listings. The realities of scaling one million times smaller (or, to illustrate other qualities, *quieter*, or *more transparent*) reaches kingdoms astonishingly distant and strange. Your 'height' is a mind-boggling 0.000072 of an inch and you live in a land of the very teensy! From six feet, the journey of a million halves is sobering. Most nucleated cells are bigger than you.

Yet . . . viruses, molecules, atoms and other such really tiny things, are LOTS smaller still . . .

Below, I present a list of some of the realms of small numbers. The abbreviation *SI* again refers to the French, *Système International d'Unités*. Have fun!

United States, Canadian and modern British usage				Number of zeros you will need
Short Scale Counting $(1{,}000 \times 1{,}000)^N$				

SI Symbol	SI Prefix			
"The Unit"	1		One	-1^0
d	deci-	-10	Tenth	-10^1
c	centi-	-100	Hundredth	-10^2
m	milli-	-1,000	Thousandth	-10^3
	dimi-	-10,000	Ten Thousandth	-10^4
		-100,000	Hundred Thousandth	-10^5
μ	micro-	-1,000,000	**Millionth**	-10^6
n	nano-		**Billionth**	-10^9
p	pico-		**Trillionth**	-10^{12}
f	femto-		**Quadrillionth**	-10^{15}
a	atto-		**Quintillionth**	-10^{18}
z	zepto-		**Sextillionth**	-10^{21}
y	yocto-		**Septillionth**	-10^{24}
x	xona-		**Octillionth**	-10^{27}
			Nonillionth	-10^{30}
			Decillionth	-10^{33}

45

12

8

88

79

1   10

0

96

A Special Section to the First Number Realms

54

41   5

42

| The number 1 to the number 99—with a surprise cameo appearance by non-other than Zero! |

61

2

18

9

11   33   32

23

63   3   47

99   46

58

20   53

78   27

The Unique Number 0

The Number Zero!

◊ | 0 |

The Number 1 to the Number 9

The Realm of One-digit Numbers!

◊ | 1 |
◊ | 2 |
◊ | 3 |
◊ | 4 |
◊ | 5 |
◊ | 6 |
◊ | 7 |
◊ | 8 |
◊ | 9 |

The Number 10 to the Number 19

The Realm of Two-digit Numbers!

◊ | **10**
◊ | **11**
◊ | **12**
◊ | **13**
◊ | **14**
◊ | **15**
◊ | **16**
◊ | **17**
◊ | **18**
◊ | **19**

The Number 20 to the Number 29

◊ | 20 |
◊ | 21 |
◊ | 22 |
◊ | 23 |
◊ | 24 |
◊ | 25 |
◊ | 26 |
◊ | 27 |
◊ | 28 |
◊ | 29 |

The Number 30 to the Number 39

◊ | **30**
◊ | **31**
◊ | **32**
◊ | **33**
◊ | **34**
◊ | **35**
◊ | **36**
◊ | **37**
◊ | **38**
◊ | **39**

The Number 40 to the Number 49

◊ | **40**
◊ | **41**
◊ | **42**
◊ | **43**
◊ | **44**
◊ | **45**
◊ | **46**
◊ | **47**
◊ | **48**
◊ | **49**

The Number 50 to the Number 59

◊ | **50**
◊ | **51**
◊ | **52**
◊ | **53**
◊ | **54**
◊ | **55**
◊ | **56**
◊ | **57**
◊ | **58**
◊ | **59**

The Number 60 to the Number 69

◊ | 60
◊ | 61
◊ | 62
◊ | 63
◊ | 64
◊ | 65
◊ | 66
◊ | 67
◊ | 68
◊ | 69

The Number 70 to the Number 79

◊ | **70** |
◊ | **71** |
◊ | **72** |
◊ | **73** |
◊ | **74** |
◊ | **75** |
◊ | **76** |
◊ | **77** |
◊ | **78** |
◊ | **79** |

The Number 80 to the Number 89

◊ **80**
◊ **81**
◊ **82**
◊ **83**
◊ **84**
◊ **85**
◊ **86**
◊ **87**
◊ **88**
◊ **89**

The Number 90 to the Number 99

◊ | **90** | _____
◊ | **91** | _____
◊ | **92** | _____
◊ | **93** | _____
◊ | **94** | _____
◊ | **95** | _____
◊ | **96** | _____
◊ | **97** | _____
◊ | **98** | _____
◊ | **99** | _____

1   102,674   100

        8,839   998,458
                            444   683,025

666

                                                                12

              75   23,856

550,665   1804

              53,065

                                                        396,591

The Main Listings: 1 to 1,000,000

                                                        5,515

              144   2

476,903   42

                                                        10,003

        1,066

                                                1,000

              47,431

246   4,808   55

                                                        70,723

111   2,676

                            77,981

              18   338,562

                                                        12

                            1,946   666

                            1,953

        80   7

◊ The Number 1 to the Number 2,000

Start

Notes

50	
100	
150	
200	
250	
300	
350	
400	
450	
500	
550	
600	
650	
700	
750	
800	
850	
900	
950	
1,000	
1,050	
1,100	
1,150	
1,200	
1,250	
1,300	
1,350	
1,400	
1,450	
1,500	
1,550	
1,600	
1,650	
1,700	
1,750	
1,800	
1,850	
1,900	
1,950	
2,000	

The Number 2,001 to the Number 4,000

Start

Notes

	2,050
	2,100
	2,150
	2,200
	2,250
	2,300
	2,350
	2,400
	2,450
	2,500
	2,550
	2,600
	2,650
	2,700
	2,750
	2,800
	2,850
	2,900
	2,950
	3,000
	3,050
	3,100
	3,150
	3,200
	3,250
	3,300
	3,350
	3,400
	3,450
	3,500
	3,550
	3,600
	3,650
	3,700
	3,750
	3,800
	3,850
	3,900
	3,950
	4,000

The Number 4,001 to the Number 6,000

◊

Start		Notes

Number
4,050
4,100
4,150
4,200
4,250
4,300
4,350
4,400
4,450
4,500
4,550
4,600
4,650
4,700
4,750
4,800
4,850
4,900
4,950
5,000
5,050
5,100
5,150
5,200
5,250
5,300
5,350
5,400
5,450
5,500
5,550
5,600
5,650
5,700
5,750
5,800
5,850
5,900
5,950
6,000

◊ The Number 6,001 to the Number 8,000

Start   **Notes**

	6,050
	6,100
	6,150
	6,200
	6,250
	6,300
	6,350
	6,400
	6,450
	6,500
	6,550
	6,600
	6,650
	6,700
	6,750
	6,800
	6,850
	6,900
	6,950
	7,000
	7,050
	7,100
	7,150
	7,200
	7,250
	7,300
	7,350
	7,400
	7,450
	7,500
	7,550
	7,600
	7,650
	7,700
	7,750
	7,800
	7,850
	7,900
	7,950
	8,000

The Number 8,001 to the Number 10,000

Start	Notes
	8,050
	8,100
	8,150
	8,200
	8,250
	8,300
	8,350
	8,400
	8,450
	8,500
	8,550
	8,600
	8,650
	8,700
	8,750
	8,800
	8,850
	8,900
	8,950
	9,000
	9,050
	9,100
	9,150
	9,200
	9,250
	9,300
	9,350
	9,400
	9,450
	9,500
	9,550
	9,600
	9,650
	9,700
	9,750
	9,800
	9,850
	9,900
	9,950
	10,000

The Number 10,001 to the Number 12,000

Start   Notes

◇◇◇◇◇ ◇◇◇◇◇ ◇◇◇◇◇ ◇◇◇◇◇ ◇◇◇◇◇ ◇◇◇◇◇ ◇◇◇◇◇ ◇◇◇◇◇ ◇◇◇◇◇ ◇◇◇◇◇	10,050
◇◇◇◇◇ ◇◇◇◇◇ ◇◇◇◇◇ ◇◇◇◇◇ ◇◇◇◇◇ ◇◇◇◇◇ ◇◇◇◇◇ ◇◇◇◇◇ ◇◇◇◇◇ ◇◇◇◇◇	10,100
◇◇◇◇◇ ◇◇◇◇◇ ◇◇◇◇◇ ◇◇◇◇◇ ◇◇◇◇◇ ◇◇◇◇◇ ◇◇◇◇◇ ◇◇◇◇◇ ◇◇◇◇◇ ◇◇◇◇◇	10,150
◇◇◇◇◇ ◇◇◇◇◇ ◇◇◇◇◇ ◇◇◇◇◇ ◇◇◇◇◇ ◇◇◇◇◇ ◇◇◇◇◇ ◇◇◇◇◇ ◇◇◇◇◇ ◇◇◇◇◇	10,200
◇◇◇◇◇ ◇◇◇◇◇ ◇◇◇◇◇ ◇◇◇◇◇ ◇◇◇◇◇ ◇◇◇◇◇ ◇◇◇◇◇ ◇◇◇◇◇ ◇◇◇◇◇ ◇◇◇◇◇	10,250
◇◇◇◇◇ ◇◇◇◇◇ ◇◇◇◇◇ ◇◇◇◇◇ ◇◇◇◇◇ ◇◇◇◇◇ ◇◇◇◇◇ ◇◇◇◇◇ ◇◇◇◇◇ ◇◇◇◇◇	10,300
◇◇◇◇◇ ◇◇◇◇◇ ◇◇◇◇◇ ◇◇◇◇◇ ◇◇◇◇◇ ◇◇◇◇◇ ◇◇◇◇◇ ◇◇◇◇◇ ◇◇◇◇◇ ◇◇◇◇◇	10,350
◇◇◇◇◇ ◇◇◇◇◇ ◇◇◇◇◇ ◇◇◇◇◇ ◇◇◇◇◇ ◇◇◇◇◇ ◇◇◇◇◇ ◇◇◇◇◇ ◇◇◇◇◇ ◇◇◇◇◇	10,400
◇◇◇◇◇ ◇◇◇◇◇ ◇◇◇◇◇ ◇◇◇◇◇ ◇◇◇◇◇ ◇◇◇◇◇ ◇◇◇◇◇ ◇◇◇◇◇ ◇◇◇◇◇ ◇◇◇◇◇	10,450
◇◇◇◇◇ ◇◇◇◇◇ ◇◇◇◇◇ ◇◇◇◇◇ ◇◇◇◇◇ ◇◇◇◇◇ ◇◇◇◇◇ ◇◇◇◇◇ ◇◇◇◇◇ ◇◇◇◇◇	10,500
◇◇◇◇◇ ◇◇◇◇◇ ◇◇◇◇◇ ◇◇◇◇◇ ◇◇◇◇◇ ◇◇◇◇◇ ◇◇◇◇◇ ◇◇◇◇◇ ◇◇◇◇◇ ◇◇◇◇◇	10,550
◇◇◇◇◇ ◇◇◇◇◇ ◇◇◇◇◇ ◇◇◇◇◇ ◇◇◇◇◇ ◇◇◇◇◇ ◇◇◇◇◇ ◇◇◇◇◇ ◇◇◇◇◇ ◇◇◇◇◇	10,600
◇◇◇◇◇ ◇◇◇◇◇ ◇◇◇◇◇ ◇◇◇◇◇ ◇◇◇◇◇ ◇◇◇◇◇ ◇◇◇◇◇ ◇◇◇◇◇ ◇◇◇◇◇ ◇◇◇◇◇	10,650
◇◇◇◇◇ ◇◇◇◇◇ ◇◇◇◇◇ ◇◇◇◇◇ ◇◇◇◇◇ ◇◇◇◇◇ ◇◇◇◇◇ ◇◇◇◇◇ ◇◇◇◇◇ ◇◇◇◇◇	10,700
◇◇◇◇◇ ◇◇◇◇◇ ◇◇◇◇◇ ◇◇◇◇◇ ◇◇◇◇◇ ◇◇◇◇◇ ◇◇◇◇◇ ◇◇◇◇◇ ◇◇◇◇◇ ◇◇◇◇◇	10,750
◇◇◇◇◇ ◇◇◇◇◇ ◇◇◇◇◇ ◇◇◇◇◇ ◇◇◇◇◇ ◇◇◇◇◇ ◇◇◇◇◇ ◇◇◇◇◇ ◇◇◇◇◇ ◇◇◇◇◇	10,800
◇◇◇◇◇ ◇◇◇◇◇ ◇◇◇◇◇ ◇◇◇◇◇ ◇◇◇◇◇ ◇◇◇◇◇ ◇◇◇◇◇ ◇◇◇◇◇ ◇◇◇◇◇ ◇◇◇◇◇	10,850
◇◇◇◇◇ ◇◇◇◇◇ ◇◇◇◇◇ ◇◇◇◇◇ ◇◇◇◇◇ ◇◇◇◇◇ ◇◇◇◇◇ ◇◇◇◇◇ ◇◇◇◇◇ ◇◇◇◇◇	10,900
◇◇◇◇◇ ◇◇◇◇◇ ◇◇◇◇◇ ◇◇◇◇◇ ◇◇◇◇◇ ◇◇◇◇◇ ◇◇◇◇◇ ◇◇◇◇◇ ◇◇◇◇◇ ◇◇◇◇◇	10,950
◇◇◇◇◇ ◇◇◇◇◇ ◇◇◇◇◇ ◇◇◇◇◇ ◇◇◇◇◇ ◇◇◇◇◇ ◇◇◇◇◇ ◇◇◇◇◇ ◇◇◇◇◇ ◇◇◇◇◇	**11,000**
◇◇◇◇◇ ◇◇◇◇◇ ◇◇◇◇◇ ◇◇◇◇◇ ◇◇◇◇◇ ◇◇◇◇◇ ◇◇◇◇◇ ◇◇◇◇◇ ◇◇◇◇◇ ◇◇◇◇◇	11,050
◇◇◇◇◇ ◇◇◇◇◇ ◇◇◇◇◇ ◇◇◇◇◇ ◇◇◇◇◇ ◇◇◇◇◇ ◇◇◇◇◇ ◇◇◇◇◇ ◇◇◇◇◇ ◇◇◇◇◇	11,100
◇◇◇◇◇ ◇◇◇◇◇ ◇◇◇◇◇ ◇◇◇◇◇ ◇◇◇◇◇ ◇◇◇◇◇ ◇◇◇◇◇ ◇◇◇◇◇ ◇◇◇◇◇ ◇◇◇◇◇	11,150
◇◇◇◇◇ ◇◇◇◇◇ ◇◇◇◇◇ ◇◇◇◇◇ ◇◇◇◇◇ ◇◇◇◇◇ ◇◇◇◇◇ ◇◇◇◇◇ ◇◇◇◇◇ ◇◇◇◇◇	11,200
◇◇◇◇◇ ◇◇◇◇◇ ◇◇◇◇◇ ◇◇◇◇◇ ◇◇◇◇◇ ◇◇◇◇◇ ◇◇◇◇◇ ◇◇◇◇◇ ◇◇◇◇◇ ◇◇◇◇◇	11,250
◇◇◇◇◇ ◇◇◇◇◇ ◇◇◇◇◇ ◇◇◇◇◇ ◇◇◇◇◇ ◇◇◇◇◇ ◇◇◇◇◇ ◇◇◇◇◇ ◇◇◇◇◇ ◇◇◇◇◇	11,300
◇◇◇◇◇ ◇◇◇◇◇ ◇◇◇◇◇ ◇◇◇◇◇ ◇◇◇◇◇ ◇◇◇◇◇ ◇◇◇◇◇ ◇◇◇◇◇ ◇◇◇◇◇ ◇◇◇◇◇	11,350
◇◇◇◇◇ ◇◇◇◇◇ ◇◇◇◇◇ ◇◇◇◇◇ ◇◇◇◇◇ ◇◇◇◇◇ ◇◇◇◇◇ ◇◇◇◇◇ ◇◇◇◇◇ ◇◇◇◇◇	11,400
◇◇◇◇◇ ◇◇◇◇◇ ◇◇◇◇◇ ◇◇◇◇◇ ◇◇◇◇◇ ◇◇◇◇◇ ◇◇◇◇◇ ◇◇◇◇◇ ◇◇◇◇◇ ◇◇◇◇◇	11,450
◇◇◇◇◇ ◇◇◇◇◇ ◇◇◇◇◇ ◇◇◇◇◇ ◇◇◇◇◇ ◇◇◇◇◇ ◇◇◇◇◇ ◇◇◇◇◇ ◇◇◇◇◇ ◇◇◇◇◇	11,500
◇◇◇◇◇ ◇◇◇◇◇ ◇◇◇◇◇ ◇◇◇◇◇ ◇◇◇◇◇ ◇◇◇◇◇ ◇◇◇◇◇ ◇◇◇◇◇ ◇◇◇◇◇ ◇◇◇◇◇	11,550
◇◇◇◇◇ ◇◇◇◇◇ ◇◇◇◇◇ ◇◇◇◇◇ ◇◇◇◇◇ ◇◇◇◇◇ ◇◇◇◇◇ ◇◇◇◇◇ ◇◇◇◇◇ ◇◇◇◇◇	11,600
◇◇◇◇◇ ◇◇◇◇◇ ◇◇◇◇◇ ◇◇◇◇◇ ◇◇◇◇◇ ◇◇◇◇◇ ◇◇◇◇◇ ◇◇◇◇◇ ◇◇◇◇◇ ◇◇◇◇◇	11,650
◇◇◇◇◇ ◇◇◇◇◇ ◇◇◇◇◇ ◇◇◇◇◇ ◇◇◇◇◇ ◇◇◇◇◇ ◇◇◇◇◇ ◇◇◇◇◇ ◇◇◇◇◇ ◇◇◇◇◇	11,700
◇◇◇◇◇ ◇◇◇◇◇ ◇◇◇◇◇ ◇◇◇◇◇ ◇◇◇◇◇ ◇◇◇◇◇ ◇◇◇◇◇ ◇◇◇◇◇ ◇◇◇◇◇ ◇◇◇◇◇	11,750
◇◇◇◇◇ ◇◇◇◇◇ ◇◇◇◇◇ ◇◇◇◇◇ ◇◇◇◇◇ ◇◇◇◇◇ ◇◇◇◇◇ ◇◇◇◇◇ ◇◇◇◇◇ ◇◇◇◇◇	11,800
◇◇◇◇◇ ◇◇◇◇◇ ◇◇◇◇◇ ◇◇◇◇◇ ◇◇◇◇◇ ◇◇◇◇◇ ◇◇◇◇◇ ◇◇◇◇◇ ◇◇◇◇◇ ◇◇◇◇◇	11,850
◇◇◇◇◇ ◇◇◇◇◇ ◇◇◇◇◇ ◇◇◇◇◇ ◇◇◇◇◇ ◇◇◇◇◇ ◇◇◇◇◇ ◇◇◇◇◇ ◇◇◇◇◇ ◇◇◇◇◇	11,900
◇◇◇◇◇ ◇◇◇◇◇ ◇◇◇◇◇ ◇◇◇◇◇ ◇◇◇◇◇ ◇◇◇◇◇ ◇◇◇◇◇ ◇◇◇◇◇ ◇◇◇◇◇ ◇◇◇◇◇	11,950
◇◇◇◇◇ ◇◇◇◇◇ ◇◇◇◇◇ ◇◇◇◇◇ ◇◇◇◇◇ ◇◇◇◇◇ ◇◇◇◇◇ ◇◇◇◇◇ ◇◇◇◇◇ ◇◇◇◇◇	**12,000**

◇ The Number 12,001 to the Number 14,000

Start	End #	Notes
	12,050	
	12,100	
	12,150	
	12,200	
	12,250	
	12,300	
	12,350	
	12,400	
	12,450	
	12,500	
	12,550	
	12,600	
	12,650	
	12,700	
	12,750	
	12,800	
	12,850	
	12,900	
	12,950	
	13,000	
	13,050	
	13,100	
	13,150	
	13,200	
	13,250	
	13,300	
	13,350	
	13,400	
	13,450	
	13,500	
	13,550	
	13,600	
	13,650	
	13,700	
	13,750	
	13,800	
	13,850	
	13,900	
	13,950	
	14,000	

The Number 14,001 to the Number 16,000

Start

Notes

14,050	
14,100	
14,150	
14,200	
14,250	
14,300	
14,350	
14,400	
14,450	
14,500	
14,550	
14,600	
14,650	
14,700	
14,750	
14,800	
14,850	
14,900	
14,950	
15,000	
15,050	
15,100	
15,150	
15,200	
15,250	
15,300	
15,350	
15,400	
15,450	
15,500	
15,550	
15,600	
15,650	
15,700	
15,750	
15,800	
15,850	
15,900	
15,950	
16,000	

The Number 16,001 to the Number 18,000

Start

Notes

	16,050
	16,100
	16,150
	16,200
	16,250
	16,300
	16,350
	16,400
	16,450
	16,500
	16,550
	16,600
	16,650
	16,700
	16,750
	16,800
	16,850
	16,900
	16,950
	17,000
	17,050
	17,100
	17,150
	17,200
	17,250
	17,300
	17,350
	17,400
	17,450
	17,500
	17,550
	17,600
	17,650
	17,700
	17,750
	17,800
	17,850
	17,900
	17,950
	18,000

The Number 18,001 to the Number 20,000

Start

Notes

	18,050
	18,100
	18,150
	18,200
	18,250
	18,300
	18,350
	18,400
	18,450
	18,500
	18,550
	18,600
	18,650
	18,700
	18,750
	18,800
	18,850
	18,900
	18,950
	19,000
	19,050
	19,100
	19,150
	19,200
	19,250
	19,300
	19,350
	19,400
	19,450
	19,500
	19,550
	19,600
	19,650
	19,700
	19,750
	19,800
	19,850
	19,900
	19,950
	20,000

The Number 20,001 to the Number 22,000

Start

Notes

	20,050
	20,100
	20,150
	20,200
	20,250
	20,300
	20,350
	20,400
	20,450
	20,500
	20,550
	20,600
	20,650
	20,700
	20,750
	20,800
	20,850
	20,900
	20,950
	21,000
	21,050
	21,100
	21,150
	21,200
	21,250
	21,300
	21,350
	21,400
	21,450
	21,500
	21,550
	21,600
	21,650
	21,700
	21,750
	21,800
	21,850
	21,900
	21,950
	22,000

The Number 22,001 to the Number 24,000

Start

Notes

	22,050
	22,100
	22,150
	22,200
	22,250
	22,300
	22,350
	22,400
	22,450
	22,500
	22,550
	22,600
	22,650
	22,700
	22,750
	22,800
	22,850
	22,900
	22,950
	23,000
	23,050
	23,100
	23,150
	23,200
	23,250
	23,300
	23,350
	23,400
	23,450
	23,500
	23,550
	23,600
	23,650
	23,700
	23,750
	23,800
	23,850
	23,900
	23,950
	24,000

◇ The Number 24,001 to the Number 26,000

Start   Notes

	24,050
	24,100
	24,150
	24,200
	24,250
	24,300
	24,350
	24,400
	24,450
	24,500
	24,550
	24,600
	24,650
	24,700
	24,750
	24,800
	24,850
	24,900
	24,950
	25,000
	25,050
	25,100
	25,150
	25,200
	25,250
	25,300
	25,350
	25,400
	25,450
	25,500
	25,550
	25,600
	25,650
	25,700
	25,750
	25,800
	25,850
	25,900
	25,950
	26,000

◊ The Number 26,001 to the Number 28,000

Start		Notes
	26,050	
	26,100	
	26,150	
	26,200	
	26,250	
	26,300	
	26,350	
	26,400	
	26,450	
	26,500	
	26,550	
	26,600	
	26,650	
	26,700	
	26,750	
	26,800	
	26,850	
	26,900	
	26,950	
	27,000	
	27,050	
	27,100	
	27,150	
	27,200	
	27,250	
	27,300	
	27,350	
	27,400	
	27,450	
	27,500	
	27,550	
	27,600	
	27,650	
	27,700	
	27,750	
	27,800	
	27,850	
	27,900	
	27,950	
	28,000	

The Number 28,001 to the Number 30,000

Start   **Notes**

◇◇◇◇◇ ◇◇◇◇◇ ◇◇◇◇◇ ◇◇◇◇◇ ◇◇◇◇◇ ◇◇◇◇◇ ◇◇◇◇◇ ◇◇◇◇◇ ◇◇◇◇◇ ◇◇◇◇◇	28,050
◇◇◇◇◇ ◇◇◇◇◇ ◇◇◇◇◇ ◇◇◇◇◇ ◇◇◇◇◇ ◇◇◇◇◇ ◇◇◇◇◇ ◇◇◇◇◇ ◇◇◇◇◇ ◇◇◇◇◇	28,100
◇◇◇◇◇ ◇◇◇◇◇ ◇◇◇◇◇ ◇◇◇◇◇ ◇◇◇◇◇ ◇◇◇◇◇ ◇◇◇◇◇ ◇◇◇◇◇ ◇◇◇◇◇ ◇◇◇◇◇	28,150
◇◇◇◇◇ ◇◇◇◇◇ ◇◇◇◇◇ ◇◇◇◇◇ ◇◇◇◇◇ ◇◇◇◇◇ ◇◇◇◇◇ ◇◇◇◇◇ ◇◇◇◇◇ ◇◇◇◇◇	28,200
◇◇◇◇◇ ◇◇◇◇◇ ◇◇◇◇◇ ◇◇◇◇◇ ◇◇◇◇◇ ◇◇◇◇◇ ◇◇◇◇◇ ◇◇◇◇◇ ◇◇◇◇◇ ◇◇◇◇◇	28,250
◇◇◇◇◇ ◇◇◇◇◇ ◇◇◇◇◇ ◇◇◇◇◇ ◇◇◇◇◇ ◇◇◇◇◇ ◇◇◇◇◇ ◇◇◇◇◇ ◇◇◇◇◇ ◇◇◇◇◇	28,300
◇◇◇◇◇ ◇◇◇◇◇ ◇◇◇◇◇ ◇◇◇◇◇ ◇◇◇◇◇ ◇◇◇◇◇ ◇◇◇◇◇ ◇◇◇◇◇ ◇◇◇◇◇ ◇◇◇◇◇	28,350
◇◇◇◇◇ ◇◇◇◇◇ ◇◇◇◇◇ ◇◇◇◇◇ ◇◇◇◇◇ ◇◇◇◇◇ ◇◇◇◇◇ ◇◇◇◇◇ ◇◇◇◇◇ ◇◇◇◇◇	28,400
◇◇◇◇◇ ◇◇◇◇◇ ◇◇◇◇◇ ◇◇◇◇◇ ◇◇◇◇◇ ◇◇◇◇◇ ◇◇◇◇◇ ◇◇◇◇◇ ◇◇◇◇◇ ◇◇◇◇◇	28,450
◇◇◇◇◇ ◇◇◇◇◇ ◇◇◇◇◇ ◇◇◇◇◇ ◇◇◇◇◇ ◇◇◇◇◇ ◇◇◇◇◇ ◇◇◇◇◇ ◇◇◇◇◇ ◇◇◇◇◇	28,500
◇◇◇◇◇ ◇◇◇◇◇ ◇◇◇◇◇ ◇◇◇◇◇ ◇◇◇◇◇ ◇◇◇◇◇ ◇◇◇◇◇ ◇◇◇◇◇ ◇◇◇◇◇ ◇◇◇◇◇	28,550
◇◇◇◇◇ ◇◇◇◇◇ ◇◇◇◇◇ ◇◇◇◇◇ ◇◇◇◇◇ ◇◇◇◇◇ ◇◇◇◇◇ ◇◇◇◇◇ ◇◇◇◇◇ ◇◇◇◇◇	28,600
◇◇◇◇◇ ◇◇◇◇◇ ◇◇◇◇◇ ◇◇◇◇◇ ◇◇◇◇◇ ◇◇◇◇◇ ◇◇◇◇◇ ◇◇◇◇◇ ◇◇◇◇◇ ◇◇◇◇◇	28,650
◇◇◇◇◇ ◇◇◇◇◇ ◇◇◇◇◇ ◇◇◇◇◇ ◇◇◇◇◇ ◇◇◇◇◇ ◇◇◇◇◇ ◇◇◇◇◇ ◇◇◇◇◇ ◇◇◇◇◇	28,700
◇◇◇◇◇ ◇◇◇◇◇ ◇◇◇◇◇ ◇◇◇◇◇ ◇◇◇◇◇ ◇◇◇◇◇ ◇◇◇◇◇ ◇◇◇◇◇ ◇◇◇◇◇ ◇◇◇◇◇	28,750
◇◇◇◇◇ ◇◇◇◇◇ ◇◇◇◇◇ ◇◇◇◇◇ ◇◇◇◇◇ ◇◇◇◇◇ ◇◇◇◇◇ ◇◇◇◇◇ ◇◇◇◇◇ ◇◇◇◇◇	28,800
◇◇◇◇◇ ◇◇◇◇◇ ◇◇◇◇◇ ◇◇◇◇◇ ◇◇◇◇◇ ◇◇◇◇◇ ◇◇◇◇◇ ◇◇◇◇◇ ◇◇◇◇◇ ◇◇◇◇◇	28,850
◇◇◇◇◇ ◇◇◇◇◇ ◇◇◇◇◇ ◇◇◇◇◇ ◇◇◇◇◇ ◇◇◇◇◇ ◇◇◇◇◇ ◇◇◇◇◇ ◇◇◇◇◇ ◇◇◇◇◇	28,900
◇◇◇◇◇ ◇◇◇◇◇ ◇◇◇◇◇ ◇◇◇◇◇ ◇◇◇◇◇ ◇◇◇◇◇ ◇◇◇◇◇ ◇◇◇◇◇ ◇◇◇◇◇ ◇◇◇◇◇	28,950
◇◇◇◇◇ ◇◇◇◇◇ ◇◇◇◇◇ ◇◇◇◇◇ ◇◇◇◇◇ ◇◇◇◇◇ ◇◇◇◇◇ ◇◇◇◇◇ ◇◇◇◇◇ ◇◇◇◇◇	**29,000**
◇◇◇◇◇ ◇◇◇◇◇ ◇◇◇◇◇ ◇◇◇◇◇ ◇◇◇◇◇ ◇◇◇◇◇ ◇◇◇◇◇ ◇◇◇◇◇ ◇◇◇◇◇ ◇◇◇◇◇	29,050
◇◇◇◇◇ ◇◇◇◇◇ ◇◇◇◇◇ ◇◇◇◇◇ ◇◇◇◇◇ ◇◇◇◇◇ ◇◇◇◇◇ ◇◇◇◇◇ ◇◇◇◇◇ ◇◇◇◇◇	29,100
◇◇◇◇◇ ◇◇◇◇◇ ◇◇◇◇◇ ◇◇◇◇◇ ◇◇◇◇◇ ◇◇◇◇◇ ◇◇◇◇◇ ◇◇◇◇◇ ◇◇◇◇◇ ◇◇◇◇◇	29,150
◇◇◇◇◇ ◇◇◇◇◇ ◇◇◇◇◇ ◇◇◇◇◇ ◇◇◇◇◇ ◇◇◇◇◇ ◇◇◇◇◇ ◇◇◇◇◇ ◇◇◇◇◇ ◇◇◇◇◇	29,200
◇◇◇◇◇ ◇◇◇◇◇ ◇◇◇◇◇ ◇◇◇◇◇ ◇◇◇◇◇ ◇◇◇◇◇ ◇◇◇◇◇ ◇◇◇◇◇ ◇◇◇◇◇ ◇◇◇◇◇	29,250
◇◇◇◇◇ ◇◇◇◇◇ ◇◇◇◇◇ ◇◇◇◇◇ ◇◇◇◇◇ ◇◇◇◇◇ ◇◇◇◇◇ ◇◇◇◇◇ ◇◇◇◇◇ ◇◇◇◇◇	29,300
◇◇◇◇◇ ◇◇◇◇◇ ◇◇◇◇◇ ◇◇◇◇◇ ◇◇◇◇◇ ◇◇◇◇◇ ◇◇◇◇◇ ◇◇◇◇◇ ◇◇◇◇◇ ◇◇◇◇◇	29,350
◇◇◇◇◇ ◇◇◇◇◇ ◇◇◇◇◇ ◇◇◇◇◇ ◇◇◇◇◇ ◇◇◇◇◇ ◇◇◇◇◇ ◇◇◇◇◇ ◇◇◇◇◇ ◇◇◇◇◇	29,400
◇◇◇◇◇ ◇◇◇◇◇ ◇◇◇◇◇ ◇◇◇◇◇ ◇◇◇◇◇ ◇◇◇◇◇ ◇◇◇◇◇ ◇◇◇◇◇ ◇◇◇◇◇ ◇◇◇◇◇	29,450
◇◇◇◇◇ ◇◇◇◇◇ ◇◇◇◇◇ ◇◇◇◇◇ ◇◇◇◇◇ ◇◇◇◇◇ ◇◇◇◇◇ ◇◇◇◇◇ ◇◇◇◇◇ ◇◇◇◇◇	29,500
◇◇◇◇◇ ◇◇◇◇◇ ◇◇◇◇◇ ◇◇◇◇◇ ◇◇◇◇◇ ◇◇◇◇◇ ◇◇◇◇◇ ◇◇◇◇◇ ◇◇◇◇◇ ◇◇◇◇◇	29,550
◇◇◇◇◇ ◇◇◇◇◇ ◇◇◇◇◇ ◇◇◇◇◇ ◇◇◇◇◇ ◇◇◇◇◇ ◇◇◇◇◇ ◇◇◇◇◇ ◇◇◇◇◇ ◇◇◇◇◇	29,600
◇◇◇◇◇ ◇◇◇◇◇ ◇◇◇◇◇ ◇◇◇◇◇ ◇◇◇◇◇ ◇◇◇◇◇ ◇◇◇◇◇ ◇◇◇◇◇ ◇◇◇◇◇ ◇◇◇◇◇	29,650
◇◇◇◇◇ ◇◇◇◇◇ ◇◇◇◇◇ ◇◇◇◇◇ ◇◇◇◇◇ ◇◇◇◇◇ ◇◇◇◇◇ ◇◇◇◇◇ ◇◇◇◇◇ ◇◇◇◇◇	29,700
◇◇◇◇◇ ◇◇◇◇◇ ◇◇◇◇◇ ◇◇◇◇◇ ◇◇◇◇◇ ◇◇◇◇◇ ◇◇◇◇◇ ◇◇◇◇◇ ◇◇◇◇◇ ◇◇◇◇◇	29,750
◇◇◇◇◇ ◇◇◇◇◇ ◇◇◇◇◇ ◇◇◇◇◇ ◇◇◇◇◇ ◇◇◇◇◇ ◇◇◇◇◇ ◇◇◇◇◇ ◇◇◇◇◇ ◇◇◇◇◇	29,800
◇◇◇◇◇ ◇◇◇◇◇ ◇◇◇◇◇ ◇◇◇◇◇ ◇◇◇◇◇ ◇◇◇◇◇ ◇◇◇◇◇ ◇◇◇◇◇ ◇◇◇◇◇ ◇◇◇◇◇	29,850
◇◇◇◇◇ ◇◇◇◇◇ ◇◇◇◇◇ ◇◇◇◇◇ ◇◇◇◇◇ ◇◇◇◇◇ ◇◇◇◇◇ ◇◇◇◇◇ ◇◇◇◇◇ ◇◇◇◇◇	29,900
◇◇◇◇◇ ◇◇◇◇◇ ◇◇◇◇◇ ◇◇◇◇◇ ◇◇◇◇◇ ◇◇◇◇◇ ◇◇◇◇◇ ◇◇◇◇◇ ◇◇◇◇◇ ◇◇◇◇◇	29,950
◇◇◇◇◇ ◇◇◇◇◇ ◇◇◇◇◇ ◇◇◇◇◇ ◇◇◇◇◇ ◇◇◇◇◇ ◇◇◇◇◇ ◇◇◇◇◇ ◇◇◇◇◇ ◇◇◇◇◇	**30,000**

◊ The Number 30,001 to the Number 32,000

Start   **Notes**

Numbers	Notes
30,050	
30,100	
30,150	
30,200	
30,250	
30,300	
30,350	
30,400	
30,450	
30,500	
30,550	
30,600	
30,650	
30,700	
30,750	
30,800	
30,850	
30,900	
30,950	
31,000	
31,050	
31,100	
31,150	
31,200	
31,250	
31,300	
31,350	
31,400	
31,450	
31,500	
31,550	
31,600	
31,650	
31,700	
31,750	
31,800	
31,850	
31,900	
31,950	
32,000	

The Number 32,001 to the Number 34,000

Start

	Notes
32,050	
32,100	
32,150	
32,200	
32,250	
32,300	
32,350	
32,400	
32,450	
32,500	
32,550	
32,600	
32,650	
32,700	
32,750	
32,800	
32,850	
32,900	
32,950	
33,000	
33,050	
33,100	
33,150	
33,200	
33,250	
33,300	
33,350	
33,400	
33,450	
33,500	
33,550	
33,600	
33,650	
33,700	
33,750	
33,800	
33,850	
33,900	
33,950	
34,000	

The Number 34,001 to the Number 36,000

Start		Notes
	34,050	
	34,100	
	34,150	
	34,200	
	34,250	
	34,300	
	34,350	
	34,400	
	34,450	
	34,500	
	34,550	
	34,600	
	34,650	
	34,700	
	34,750	
	34,800	
	34,850	
	34,900	
	34,950	
	35,000	
	35,050	
	35,100	
	35,150	
	35,200	
	35,250	
	35,300	
	35,350	
	35,400	
	35,450	
	35,500	
	35,550	
	35,600	
	35,650	
	35,700	
	35,750	
	35,800	
	35,850	
	35,900	
	35,950	
	36,000	

The Number 36,001 to the Number 38,000

Start		Notes

	36,050	
	36,100	
	36,150	
	36,200	
	36,250	
	36,300	
	36,350	
	36,400	
	36,450	
	36,500	
	36,550	
	36,600	
	36,650	
	36,700	
	36,750	
	36,800	
	36,850	
	36,900	
	36,950	
	37,000	
	37,050	
	37,100	
	37,150	
	37,200	
	37,250	
	37,300	
	37,350	
	37,400	
	37,450	
	37,500	
	37,550	
	37,600	
	37,650	
	37,700	
	37,750	
	37,800	
	37,850	
	37,900	
	37,950	
	38,000	

The Number 38,001 to the Number 40,000

Start   **Notes**

	38,050
	38,100
	38,150
	38,200
	38,250
	38,300
	38,350
	38,400
	38,450
	38,500
	38,550
	38,600
	38,650
	38,700
	38,750
	38,800
	38,850
	38,900
	38,950
	39,000
	39,050
	39,100
	39,150
	39,200
	39,250
	39,300
	39,350
	39,400
	39,450
	39,500
	39,550
	39,600
	39,650
	39,700
	39,750
	39,800
	39,850
	39,900
	39,950
	40,000

The Number 40,001 to the Number 42,000

Start

Notes

	40,050
	40,100
	40,150
	40,200
	40,250
	40,300
	40,350
	40,400
	40,450
	40,500
	40,550
	40,600
	40,650
	40,700
	40,750
	40,800
	40,850
	40,900
	40,950
	41,000
	41,050
	41,100
	41,150
	41,200
	41,250
	41,300
	41,350
	41,400
	41,450
	41,500
	41,550
	41,600
	41,650
	41,700
	41,750
	41,800
	41,850
	41,900
	41,950
	42,000

The Number 42,001 to the Number 44,000

Start | Notes

42,050
42,100
42,150
42,200
42,250
42,300
42,350
42,400
42,450
42,500
42,550
42,600
42,650
42,700
42,750
42,800
42,850
42,900
42,950
43,000
43,050
43,100
43,150
43,200
43,250
43,300
43,350
43,400
43,450
43,500
43,550
43,600
43,650
43,700
43,750
43,800
43,850
43,900
43,950
44,000

The Number 44,001 to the Number 46,000

Start

	Notes
44,050	
44,100	
44,150	
44,200	
44,250	
44,300	
44,350	
44,400	
44,450	
44,500	
44,550	
44,600	
44,650	
44,700	
44,750	
44,800	
44,850	
44,900	
44,950	
45,000	
45,050	
45,100	
45,150	
45,200	
45,250	
45,300	
45,350	
45,400	
45,450	
45,500	
45,550	
45,600	
45,650	
45,700	
45,750	
45,800	
45,850	
45,900	
45,950	
46,000	

The Number 46,001 to the Number 48,000

Start   Notes

	Number	Notes
	46,050	
	46,100	
	46,150	
	46,200	
	46,250	
	46,300	
	46,350	
	46,400	
	46,450	
	46,500	
	46,550	
	46,600	
	46,650	
	46,700	
	46,750	
	46,800	
	46,850	
	46,900	
	46,950	
	47,000	
	47,050	
	47,100	
	47,150	
	47,200	
	47,250	
	47,300	
	47,350	
	47,400	
	47,450	
	47,500	
	47,550	
	47,600	
	47,650	
	47,700	
	47,750	
	47,800	
	47,850	
	47,900	
	47,950	
	48,000	

The Number 48,001 to the Number 50,000

Start   Notes

	48,050
	48,100
	48,150
	48,200
	48,250
	48,300
	48,350
	48,400
	48,450
	48,500
	48,550
	48,600
	48,650
	48,700
	48,750
	48,800
	48,850
	48,900
	48,950
	49,000
	49,050
	49,100
	49,150
	49,200
	49,250
	49,300
	49,350
	49,400
	49,450
	49,500
	49,550
	49,600
	49,650
	49,700
	49,750
	49,800
	49,850
	49,900
	49,950
	50,000

The Number 50,001 to the Number 52,000

Start   Notes

	50,050
	50,100
	50,150
	50,200
	50,250
	50,300
	50,350
	50,400
	50,450
	50,500
	50,550
	50,600
	50,650
	50,700
	50,750
	50,800
	50,850
	50,900
	50,950
	51,000
	51,050
	51,100
	51,150
	51,200
	51,250
	51,300
	51,350
	51,400
	51,450
	51,500
	51,550
	51,600
	51,650
	51,700
	51,750
	51,800
	51,850
	51,900
	51,950
	52,000

The Number 52,001 to the Number 54,000

Start

Notes

52,050	
52,100	
52,150	
52,200	
52,250	
52,300	
52,350	
52,400	
52,450	
52,500	
52,550	
52,600	
52,650	
52,700	
52,750	
52,800	
52,850	
52,900	
52,950	
53,000	
53,050	
53,100	
53,150	
53,200	
53,250	
53,300	
53,350	
53,400	
53,450	
53,500	
53,550	
53,600	
53,650	
53,700	
53,750	
53,800	
53,850	
53,900	
53,950	
54,000	

The Number 54,001 to the Number 56,000

Start

	Notes
	54,050
	54,100
	54,150
	54,200
	54,250
	54,300
	54,350
	54,400
	54,450
	54,500
	54,550
	54,600
	54,650
	54,700
	54,750
	54,800
	54,850
	54,900
	54,950
	55,000
	55,050
	55,100
	55,150
	55,200
	55,250
	55,300
	55,350
	55,400
	55,450
	55,500
	55,550
	55,600
	55,650
	55,700
	55,750
	55,800
	55,850
	55,900
	55,950
	56,000

The Number 56,001 to the Number 58,000

Start	Notes
56,050	
56,100	
56,150	
56,200	
56,250	
56,300	
56,350	
56,400	
56,450	
56,500	
56,550	
56,600	
56,650	
56,700	
56,750	
56,800	
56,850	
56,900	
56,950	
57,000	
57,050	
57,100	
57,150	
57,200	
57,250	
57,300	
57,350	
57,400	
57,450	
57,500	
57,550	
57,600	
57,650	
57,700	
57,750	
57,800	
57,850	
57,900	
57,950	
58,000	

The Number 58,001 to the Number 60,000

Start	Notes
58,050	
58,100	
58,150	
58,200	
58,250	
58,300	
58,350	
58,400	
58,450	
58,500	
58,550	
58,600	
58,650	
58,700	
58,750	
58,800	
58,850	
58,900	
58,950	
59,000	
59,050	
59,100	
59,150	
59,200	
59,250	
59,300	
59,350	
59,400	
59,450	
59,500	
59,550	
59,600	
59,650	
59,700	
59,750	
59,800	
59,850	
59,900	
59,950	
60,000	

◇ The Number 60,001 to the Number 62,000

Start	Notes
60,050	
60,100	
60,150	
60,200	
60,250	
60,300	
60,350	
60,400	
60,450	
60,500	
60,550	
60,600	
60,650	
60,700	
60,750	
60,800	
60,850	
60,900	
60,950	
61,000	
61,050	
61,100	
61,150	
61,200	
61,250	
61,300	
61,350	
61,400	
61,450	
61,500	
61,550	
61,600	
61,650	
61,700	
61,750	
61,800	
61,850	
61,900	
61,950	
62,000	

The Number 62,001 to the Number 64,000

Start

Notes

	62,050
	62,100
	62,150
	62,200
	62,250
	62,300
	62,350
	62,400
	62,450
	62,500
	62,550
	62,600
	62,650
	62,700
	62,750
	62,800
	62,850
	62,900
	62,950
	63,000
	63,050
	63,100
	63,150
	63,200
	63,250
	63,300
	63,350
	63,400
	63,450
	63,500
	63,550
	63,600
	63,650
	63,700
	63,750
	63,800
	63,850
	63,900
	63,950
	64,000

The Number 64,001 to the Number 66,000

Start

Notes

	64,050
	64,100
	64,150
	64,200
	64,250
	64,300
	64,350
	64,400
	64,450
	64,500
	64,550
	64,600
	64,650
	64,700
	64,750
	64,800
	64,850
	64,900
	64,950
	65,000
	65,050
	65,100
	65,150
	65,200
	65,250
	65,300
	65,350
	65,400
	65,450
	65,500
	65,550
	65,600
	65,650
	65,700
	65,750
	65,800
	65,850
	65,900
	65,950
	66,000

The Number 66,001 to the Number 68,000

Start   Notes

	66,050
	66,100
	66,150
	66,200
	66,250
	66,300
	66,350
	66,400
	66,450
	66,500
	66,550
	66,600
	66,650
	66,700
	66,750
	66,800
	66,850
	66,900
	66,950
	67,000
	67,050
	67,100
	67,150
	67,200
	67,250
	67,300
	67,350
	67,400
	67,450
	67,500
	67,550
	67,600
	67,650
	67,700
	67,750
	67,800
	67,850
	67,900
	67,950
	68,000

The Number 68,001 to the Number 70,000

Start

	Notes
68,050	
68,100	
68,150	
68,200	
68,250	
68,300	
68,350	
68,400	
68,450	
68,500	
68,550	
68,600	
68,650	
68,700	
68,750	
68,800	
68,850	
68,900	
68,950	
69,000	
69,050	
69,100	
69,150	
69,200	
69,250	
69,300	
69,350	
69,400	
69,450	
69,500	
69,550	
69,600	
69,650	
69,700	
69,750	
69,800	
69,850	
69,900	
69,950	
70,000	

The Number 70,001 to the Number 72,000

Start

Notes

	70,050
	70,100
	70,150
	70,200
	70,250
	70,300
	70,350
	70,400
	70,450
	70,500
	70,550
	70,600
	70,650
	70,700
	70,750
	70,800
	70,850
	70,900
	70,950
	71,000
	71,050
	71,100
	71,150
	71,200
	71,250
	71,300
	71,350
	71,400
	71,450
	71,500
	71,550
	71,600
	71,650
	71,700
	71,750
	71,800
	71,850
	71,900
	71,950
	72,000

The Number 72,001 to the Number 74,000

Start

	Notes
72,050	
72,100	
72,150	
72,200	
72,250	
72,300	
72,350	
72,400	
72,450	
72,500	
72,550	
72,600	
72,650	
72,700	
72,750	
72,800	
72,850	
72,900	
72,950	
73,000	
73,050	
73,100	
73,150	
73,200	
73,250	
73,300	
73,350	
73,400	
73,450	
73,500	
73,550	
73,600	
73,650	
73,700	
73,750	
73,800	
73,850	
73,900	
73,950	
74,000	

The Number 74,001 to the Number 76,000

Start   Notes

	74,050
	74,100
	74,150
	74,200
	74,250
	74,300
	74,350
	74,400
	74,450
	74,500
	74,550
	74,600
	74,650
	74,700
	74,750
	74,800
	74,850
	74,900
	74,950
	75,000
	75,050
	75,100
	75,150
	75,200
	75,250
	75,300
	75,350
	75,400
	75,450
	75,500
	75,550
	75,600
	75,650
	75,700
	75,750
	75,800
	75,850
	75,900
	75,950
	76,000

The Number 76,001 to the Number 78,000

Start	Notes
76,050
76,100
76,150
76,200
76,250
76,300
76,350
76,400
76,450
76,500
76,550
76,600
76,650
76,700
76,750
76,800
76,850
76,900
76,950
77,000
77,050
77,100
77,150
77,200
77,250
77,300
77,350
77,400
77,450
77,500
77,550
77,600
77,650
77,700
77,750
77,800
77,850
77,900
77,950
78,000

The Number 78,001 to the Number 80,000

Start | | | Notes

	Number	Notes
	78,050	
	78,100	
	78,150	
	78,200	
	78,250	
	78,300	
	78,350	
	78,400	
	78,450	
	78,500	
	78,550	
	78,600	
	78,650	
	78,700	
	78,750	
	78,800	
	78,850	
	78,900	
	78,950	
	79,000	
	79,050	
	79,100	
	79,150	
	79,200	
	79,250	
	79,300	
	79,350	
	79,400	
	79,450	
	79,500	
	79,550	
	79,600	
	79,650	
	79,700	
	79,750	
	79,800	
	79,850	
	79,900	
	79,950	
	80,000	

The Number 80,001 to the Number 82,000

Start

Notes

	80,050
	80,100
	80,150
	80,200
	80,250
	80,300
	80,350
	80,400
	80,450
	80,500
	80,550
	80,600
	80,650
	80,700
	80,750
	80,800
	80,850
	80,900
	80,950
	81,000
	81,050
	81,100
	81,150
	81,200
	81,250
	81,300
	81,350
	81,400
	81,450
	81,500
	81,550
	81,600
	81,650
	81,700
	81,750
	81,800
	81,850
	81,900
	81,950
	82,000

The Number 82,001 to the Number 84,000

Start		Notes
	82,050	
	82,100	
	82,150	
	82,200	
	82,250	
	82,300	
	82,350	
	82,400	
	82,450	
	82,500	
	82,550	
	82,600	
	82,650	
	82,700	
	82,750	
	82,800	
	82,850	
	82,900	
	82,950	
	83,000	
	83,050	
	83,100	
	83,150	
	83,200	
	83,250	
	83,300	
	83,350	
	83,400	
	83,450	
	83,500	
	83,550	
	83,600	
	83,650	
	83,700	
	83,750	
	83,800	
	83,850	
	83,900	
	83,950	
	84,000	

The Number 84,001 to the Number 86,000

Start   Notes

	Number	Notes
◇◇◇◇◇ ◇◇◇◇◇ ◇◇◇◇◇ ◇◇◇◇◇ ◇◇◇◇◇ ◇◇◇◇◇ ◇◇◇◇◇ ◇◇◇◇◇ ◇◇◇◇◇ ◇◇◇◇◇	84,050	
◇◇◇◇◇ ◇◇◇◇◇ ◇◇◇◇◇ ◇◇◇◇◇ ◇◇◇◇◇ ◇◇◇◇◇ ◇◇◇◇◇ ◇◇◇◇◇ ◇◇◇◇◇ ◇◇◇◇◇	84,100	
◇◇◇◇◇ ◇◇◇◇◇ ◇◇◇◇◇ ◇◇◇◇◇ ◇◇◇◇◇ ◇◇◇◇◇ ◇◇◇◇◇ ◇◇◇◇◇ ◇◇◇◇◇ ◇◇◇◇◇	84,150	
◇◇◇◇◇ ◇◇◇◇◇ ◇◇◇◇◇ ◇◇◇◇◇ ◇◇◇◇◇ ◇◇◇◇◇ ◇◇◇◇◇ ◇◇◇◇◇ ◇◇◇◇◇ ◇◇◇◇◇	84,200	
◇◇◇◇◇ ◇◇◇◇◇ ◇◇◇◇◇ ◇◇◇◇◇ ◇◇◇◇◇ ◇◇◇◇◇ ◇◇◇◇◇ ◇◇◇◇◇ ◇◇◇◇◇ ◇◇◇◇◇	84,250	
◇◇◇◇◇ ◇◇◇◇◇ ◇◇◇◇◇ ◇◇◇◇◇ ◇◇◇◇◇ ◇◇◇◇◇ ◇◇◇◇◇ ◇◇◇◇◇ ◇◇◇◇◇ ◇◇◇◇◇	84,300	
◇◇◇◇◇ ◇◇◇◇◇ ◇◇◇◇◇ ◇◇◇◇◇ ◇◇◇◇◇ ◇◇◇◇◇ ◇◇◇◇◇ ◇◇◇◇◇ ◇◇◇◇◇ ◇◇◇◇◇	84,350	
◇◇◇◇◇ ◇◇◇◇◇ ◇◇◇◇◇ ◇◇◇◇◇ ◇◇◇◇◇ ◇◇◇◇◇ ◇◇◇◇◇ ◇◇◇◇◇ ◇◇◇◇◇ ◇◇◇◇◇	84,400	
◇◇◇◇◇ ◇◇◇◇◇ ◇◇◇◇◇ ◇◇◇◇◇ ◇◇◇◇◇ ◇◇◇◇◇ ◇◇◇◇◇ ◇◇◇◇◇ ◇◇◇◇◇ ◇◇◇◇◇	84,450	
◇◇◇◇◇ ◇◇◇◇◇ ◇◇◇◇◇ ◇◇◇◇◇ ◇◇◇◇◇ ◇◇◇◇◇ ◇◇◇◇◇ ◇◇◇◇◇ ◇◇◇◇◇ ◇◇◇◇◇	84,500	
◇◇◇◇◇ ◇◇◇◇◇ ◇◇◇◇◇ ◇◇◇◇◇ ◇◇◇◇◇ ◇◇◇◇◇ ◇◇◇◇◇ ◇◇◇◇◇ ◇◇◇◇◇ ◇◇◇◇◇	84,550	
◇◇◇◇◇ ◇◇◇◇◇ ◇◇◇◇◇ ◇◇◇◇◇ ◇◇◇◇◇ ◇◇◇◇◇ ◇◇◇◇◇ ◇◇◇◇◇ ◇◇◇◇◇ ◇◇◇◇◇	84,600	
◇◇◇◇◇ ◇◇◇◇◇ ◇◇◇◇◇ ◇◇◇◇◇ ◇◇◇◇◇ ◇◇◇◇◇ ◇◇◇◇◇ ◇◇◇◇◇ ◇◇◇◇◇ ◇◇◇◇◇	84,650	
◇◇◇◇◇ ◇◇◇◇◇ ◇◇◇◇◇ ◇◇◇◇◇ ◇◇◇◇◇ ◇◇◇◇◇ ◇◇◇◇◇ ◇◇◇◇◇ ◇◇◇◇◇ ◇◇◇◇◇	84,700	
◇◇◇◇◇ ◇◇◇◇◇ ◇◇◇◇◇ ◇◇◇◇◇ ◇◇◇◇◇ ◇◇◇◇◇ ◇◇◇◇◇ ◇◇◇◇◇ ◇◇◇◇◇ ◇◇◇◇◇	84,750	
◇◇◇◇◇ ◇◇◇◇◇ ◇◇◇◇◇ ◇◇◇◇◇ ◇◇◇◇◇ ◇◇◇◇◇ ◇◇◇◇◇ ◇◇◇◇◇ ◇◇◇◇◇ ◇◇◇◇◇	84,800	
◇◇◇◇◇ ◇◇◇◇◇ ◇◇◇◇◇ ◇◇◇◇◇ ◇◇◇◇◇ ◇◇◇◇◇ ◇◇◇◇◇ ◇◇◇◇◇ ◇◇◇◇◇ ◇◇◇◇◇	84,850	
◇◇◇◇◇ ◇◇◇◇◇ ◇◇◇◇◇ ◇◇◇◇◇ ◇◇◇◇◇ ◇◇◇◇◇ ◇◇◇◇◇ ◇◇◇◇◇ ◇◇◇◇◇ ◇◇◇◇◇	84,900	
◇◇◇◇◇ ◇◇◇◇◇ ◇◇◇◇◇ ◇◇◇◇◇ ◇◇◇◇◇ ◇◇◇◇◇ ◇◇◇◇◇ ◇◇◇◇◇ ◇◇◇◇◇ ◇◇◇◇◇	84,950	
◇◇◇◇◇ ◇◇◇◇◇ ◇◇◇◇◇ ◇◇◇◇◇ ◇◇◇◇◇ ◇◇◇◇◇ ◇◇◇◇◇ ◇◇◇◇◇ ◇◇◇◇◇ ◇◇◇◇◇	**85,000**	
◇◇◇◇◇ ◇◇◇◇◇ ◇◇◇◇◇ ◇◇◇◇◇ ◇◇◇◇◇ ◇◇◇◇◇ ◇◇◇◇◇ ◇◇◇◇◇ ◇◇◇◇◇ ◇◇◇◇◇	85,050	
◇◇◇◇◇ ◇◇◇◇◇ ◇◇◇◇◇ ◇◇◇◇◇ ◇◇◇◇◇ ◇◇◇◇◇ ◇◇◇◇◇ ◇◇◇◇◇ ◇◇◇◇◇ ◇◇◇◇◇	85,100	
◇◇◇◇◇ ◇◇◇◇◇ ◇◇◇◇◇ ◇◇◇◇◇ ◇◇◇◇◇ ◇◇◇◇◇ ◇◇◇◇◇ ◇◇◇◇◇ ◇◇◇◇◇ ◇◇◇◇◇	85,150	
◇◇◇◇◇ ◇◇◇◇◇ ◇◇◇◇◇ ◇◇◇◇◇ ◇◇◇◇◇ ◇◇◇◇◇ ◇◇◇◇◇ ◇◇◇◇◇ ◇◇◇◇◇ ◇◇◇◇◇	85,200	
◇◇◇◇◇ ◇◇◇◇◇ ◇◇◇◇◇ ◇◇◇◇◇ ◇◇◇◇◇ ◇◇◇◇◇ ◇◇◇◇◇ ◇◇◇◇◇ ◇◇◇◇◇ ◇◇◇◇◇	85,250	
◇◇◇◇◇ ◇◇◇◇◇ ◇◇◇◇◇ ◇◇◇◇◇ ◇◇◇◇◇ ◇◇◇◇◇ ◇◇◇◇◇ ◇◇◇◇◇ ◇◇◇◇◇ ◇◇◇◇◇	85,300	
◇◇◇◇◇ ◇◇◇◇◇ ◇◇◇◇◇ ◇◇◇◇◇ ◇◇◇◇◇ ◇◇◇◇◇ ◇◇◇◇◇ ◇◇◇◇◇ ◇◇◇◇◇ ◇◇◇◇◇	85,350	
◇◇◇◇◇ ◇◇◇◇◇ ◇◇◇◇◇ ◇◇◇◇◇ ◇◇◇◇◇ ◇◇◇◇◇ ◇◇◇◇◇ ◇◇◇◇◇ ◇◇◇◇◇ ◇◇◇◇◇	85,400	
◇◇◇◇◇ ◇◇◇◇◇ ◇◇◇◇◇ ◇◇◇◇◇ ◇◇◇◇◇ ◇◇◇◇◇ ◇◇◇◇◇ ◇◇◇◇◇ ◇◇◇◇◇ ◇◇◇◇◇	85,450	
◇◇◇◇◇ ◇◇◇◇◇ ◇◇◇◇◇ ◇◇◇◇◇ ◇◇◇◇◇ ◇◇◇◇◇ ◇◇◇◇◇ ◇◇◇◇◇ ◇◇◇◇◇ ◇◇◇◇◇	85,500	
◇◇◇◇◇ ◇◇◇◇◇ ◇◇◇◇◇ ◇◇◇◇◇ ◇◇◇◇◇ ◇◇◇◇◇ ◇◇◇◇◇ ◇◇◇◇◇ ◇◇◇◇◇ ◇◇◇◇◇	85,550	
◇◇◇◇◇ ◇◇◇◇◇ ◇◇◇◇◇ ◇◇◇◇◇ ◇◇◇◇◇ ◇◇◇◇◇ ◇◇◇◇◇ ◇◇◇◇◇ ◇◇◇◇◇ ◇◇◇◇◇	85,600	
◇◇◇◇◇ ◇◇◇◇◇ ◇◇◇◇◇ ◇◇◇◇◇ ◇◇◇◇◇ ◇◇◇◇◇ ◇◇◇◇◇ ◇◇◇◇◇ ◇◇◇◇◇ ◇◇◇◇◇	85,650	
◇◇◇◇◇ ◇◇◇◇◇ ◇◇◇◇◇ ◇◇◇◇◇ ◇◇◇◇◇ ◇◇◇◇◇ ◇◇◇◇◇ ◇◇◇◇◇ ◇◇◇◇◇ ◇◇◇◇◇	85,700	
◇◇◇◇◇ ◇◇◇◇◇ ◇◇◇◇◇ ◇◇◇◇◇ ◇◇◇◇◇ ◇◇◇◇◇ ◇◇◇◇◇ ◇◇◇◇◇ ◇◇◇◇◇ ◇◇◇◇◇	85,750	
◇◇◇◇◇ ◇◇◇◇◇ ◇◇◇◇◇ ◇◇◇◇◇ ◇◇◇◇◇ ◇◇◇◇◇ ◇◇◇◇◇ ◇◇◇◇◇ ◇◇◇◇◇ ◇◇◇◇◇	85,800	
◇◇◇◇◇ ◇◇◇◇◇ ◇◇◇◇◇ ◇◇◇◇◇ ◇◇◇◇◇ ◇◇◇◇◇ ◇◇◇◇◇ ◇◇◇◇◇ ◇◇◇◇◇ ◇◇◇◇◇	85,850	
◇◇◇◇◇ ◇◇◇◇◇ ◇◇◇◇◇ ◇◇◇◇◇ ◇◇◇◇◇ ◇◇◇◇◇ ◇◇◇◇◇ ◇◇◇◇◇ ◇◇◇◇◇ ◇◇◇◇◇	85,900	
◇◇◇◇◇ ◇◇◇◇◇ ◇◇◇◇◇ ◇◇◇◇◇ ◇◇◇◇◇ ◇◇◇◇◇ ◇◇◇◇◇ ◇◇◇◇◇ ◇◇◇◇◇ ◇◇◇◇◇	85,950	
◇◇◇◇◇ ◇◇◇◇◇ ◇◇◇◇◇ ◇◇◇◇◇ ◇◇◇◇◇ ◇◇◇◇◇ ◇◇◇◇◇ ◇◇◇◇◇ ◇◇◇◇◇ ◇◇◇◇◇	**86,000**	

The Number 86,001 to the Number 88,000

Start   **Notes**

	86,050
	86,100
	86,150
	86,200
	86,250
	86,300
	86,350
	86,400
	86,450
	86,500
	86,550
	86,600
	86,650
	86,700
	86,750
	86,800
	86,850
	86,900
	86,950
	87,000
	87,050
	87,100
	87,150
	87,200
	87,250
	87,300
	87,350
	87,400
	87,450
	87,500
	87,550
	87,600
	87,650
	87,700
	87,750
	87,800
	87,850
	87,900
	87,950
	88,000

The Number 88,001 to the Number 90,000

Start   Notes

◊◊◊◊◊ ◊◊◊◊◊ ◊◊◊◊◊ ◊◊◊◊◊ ◊◊◊◊◊ ◊◊◊◊◊ ◊◊◊◊◊ ◊◊◊◊◊ ◊◊◊◊◊ ◊◊◊◊◊	88,050
◊◊◊◊◊ ◊◊◊◊◊ ◊◊◊◊◊ ◊◊◊◊◊ ◊◊◊◊◊ ◊◊◊◊◊ ◊◊◊◊◊ ◊◊◊◊◊ ◊◊◊◊◊ ◊◊◊◊◊	88,100
◊◊◊◊◊ ◊◊◊◊◊ ◊◊◊◊◊ ◊◊◊◊◊ ◊◊◊◊◊ ◊◊◊◊◊ ◊◊◊◊◊ ◊◊◊◊◊ ◊◊◊◊◊ ◊◊◊◊◊	88,150
◊◊◊◊◊ ◊◊◊◊◊ ◊◊◊◊◊ ◊◊◊◊◊ ◊◊◊◊◊ ◊◊◊◊◊ ◊◊◊◊◊ ◊◊◊◊◊ ◊◊◊◊◊ ◊◊◊◊◊	88,200
◊◊◊◊◊ ◊◊◊◊◊ ◊◊◊◊◊ ◊◊◊◊◊ ◊◊◊◊◊ ◊◊◊◊◊ ◊◊◊◊◊ ◊◊◊◊◊ ◊◊◊◊◊ ◊◊◊◊◊	88,250
◊◊◊◊◊ ◊◊◊◊◊ ◊◊◊◊◊ ◊◊◊◊◊ ◊◊◊◊◊ ◊◊◊◊◊ ◊◊◊◊◊ ◊◊◊◊◊ ◊◊◊◊◊ ◊◊◊◊◊	88,300
◊◊◊◊◊ ◊◊◊◊◊ ◊◊◊◊◊ ◊◊◊◊◊ ◊◊◊◊◊ ◊◊◊◊◊ ◊◊◊◊◊ ◊◊◊◊◊ ◊◊◊◊◊ ◊◊◊◊◊	88,350
◊◊◊◊◊ ◊◊◊◊◊ ◊◊◊◊◊ ◊◊◊◊◊ ◊◊◊◊◊ ◊◊◊◊◊ ◊◊◊◊◊ ◊◊◊◊◊ ◊◊◊◊◊ ◊◊◊◊◊	88,400
◊◊◊◊◊ ◊◊◊◊◊ ◊◊◊◊◊ ◊◊◊◊◊ ◊◊◊◊◊ ◊◊◊◊◊ ◊◊◊◊◊ ◊◊◊◊◊ ◊◊◊◊◊ ◊◊◊◊◊	88,450
◊◊◊◊◊ ◊◊◊◊◊ ◊◊◊◊◊ ◊◊◊◊◊ ◊◊◊◊◊ ◊◊◊◊◊ ◊◊◊◊◊ ◊◊◊◊◊ ◊◊◊◊◊ ◊◊◊◊◊	88,500
◊◊◊◊◊ ◊◊◊◊◊ ◊◊◊◊◊ ◊◊◊◊◊ ◊◊◊◊◊ ◊◊◊◊◊ ◊◊◊◊◊ ◊◊◊◊◊ ◊◊◊◊◊ ◊◊◊◊◊	88,550
◊◊◊◊◊ ◊◊◊◊◊ ◊◊◊◊◊ ◊◊◊◊◊ ◊◊◊◊◊ ◊◊◊◊◊ ◊◊◊◊◊ ◊◊◊◊◊ ◊◊◊◊◊ ◊◊◊◊◊	88,600
◊◊◊◊◊ ◊◊◊◊◊ ◊◊◊◊◊ ◊◊◊◊◊ ◊◊◊◊◊ ◊◊◊◊◊ ◊◊◊◊◊ ◊◊◊◊◊ ◊◊◊◊◊ ◊◊◊◊◊	88,650
◊◊◊◊◊ ◊◊◊◊◊ ◊◊◊◊◊ ◊◊◊◊◊ ◊◊◊◊◊ ◊◊◊◊◊ ◊◊◊◊◊ ◊◊◊◊◊ ◊◊◊◊◊ ◊◊◊◊◊	88,700
◊◊◊◊◊ ◊◊◊◊◊ ◊◊◊◊◊ ◊◊◊◊◊ ◊◊◊◊◊ ◊◊◊◊◊ ◊◊◊◊◊ ◊◊◊◊◊ ◊◊◊◊◊ ◊◊◊◊◊	88,750
◊◊◊◊◊ ◊◊◊◊◊ ◊◊◊◊◊ ◊◊◊◊◊ ◊◊◊◊◊ ◊◊◊◊◊ ◊◊◊◊◊ ◊◊◊◊◊ ◊◊◊◊◊ ◊◊◊◊◊	88,800
◊◊◊◊◊ ◊◊◊◊◊ ◊◊◊◊◊ ◊◊◊◊◊ ◊◊◊◊◊ ◊◊◊◊◊ ◊◊◊◊◊ ◊◊◊◊◊ ◊◊◊◊◊ ◊◊◊◊◊	88,850
◊◊◊◊◊ ◊◊◊◊◊ ◊◊◊◊◊ ◊◊◊◊◊ ◊◊◊◊◊ ◊◊◊◊◊ ◊◊◊◊◊ ◊◊◊◊◊ ◊◊◊◊◊ ◊◊◊◊◊	88,900
◊◊◊◊◊ ◊◊◊◊◊ ◊◊◊◊◊ ◊◊◊◊◊ ◊◊◊◊◊ ◊◊◊◊◊ ◊◊◊◊◊ ◊◊◊◊◊ ◊◊◊◊◊ ◊◊◊◊◊	88,950
◊◊◊◊◊ ◊◊◊◊◊ ◊◊◊◊◊ ◊◊◊◊◊ ◊◊◊◊◊ ◊◊◊◊◊ ◊◊◊◊◊ ◊◊◊◊◊ ◊◊◊◊◊ ◊◊◊◊◊	**89,000**
◊◊◊◊◊ ◊◊◊◊◊ ◊◊◊◊◊ ◊◊◊◊◊ ◊◊◊◊◊ ◊◊◊◊◊ ◊◊◊◊◊ ◊◊◊◊◊ ◊◊◊◊◊ ◊◊◊◊◊	89,050
◊◊◊◊◊ ◊◊◊◊◊ ◊◊◊◊◊ ◊◊◊◊◊ ◊◊◊◊◊ ◊◊◊◊◊ ◊◊◊◊◊ ◊◊◊◊◊ ◊◊◊◊◊ ◊◊◊◊◊	89,100
◊◊◊◊◊ ◊◊◊◊◊ ◊◊◊◊◊ ◊◊◊◊◊ ◊◊◊◊◊ ◊◊◊◊◊ ◊◊◊◊◊ ◊◊◊◊◊ ◊◊◊◊◊ ◊◊◊◊◊	89,150
◊◊◊◊◊ ◊◊◊◊◊ ◊◊◊◊◊ ◊◊◊◊◊ ◊◊◊◊◊ ◊◊◊◊◊ ◊◊◊◊◊ ◊◊◊◊◊ ◊◊◊◊◊ ◊◊◊◊◊	89,200
◊◊◊◊◊ ◊◊◊◊◊ ◊◊◊◊◊ ◊◊◊◊◊ ◊◊◊◊◊ ◊◊◊◊◊ ◊◊◊◊◊ ◊◊◊◊◊ ◊◊◊◊◊ ◊◊◊◊◊	89,250
◊◊◊◊◊ ◊◊◊◊◊ ◊◊◊◊◊ ◊◊◊◊◊ ◊◊◊◊◊ ◊◊◊◊◊ ◊◊◊◊◊ ◊◊◊◊◊ ◊◊◊◊◊ ◊◊◊◊◊	89,300
◊◊◊◊◊ ◊◊◊◊◊ ◊◊◊◊◊ ◊◊◊◊◊ ◊◊◊◊◊ ◊◊◊◊◊ ◊◊◊◊◊ ◊◊◊◊◊ ◊◊◊◊◊ ◊◊◊◊◊	89,350
◊◊◊◊◊ ◊◊◊◊◊ ◊◊◊◊◊ ◊◊◊◊◊ ◊◊◊◊◊ ◊◊◊◊◊ ◊◊◊◊◊ ◊◊◊◊◊ ◊◊◊◊◊ ◊◊◊◊◊	89,400
◊◊◊◊◊ ◊◊◊◊◊ ◊◊◊◊◊ ◊◊◊◊◊ ◊◊◊◊◊ ◊◊◊◊◊ ◊◊◊◊◊ ◊◊◊◊◊ ◊◊◊◊◊ ◊◊◊◊◊	89,450
◊◊◊◊◊ ◊◊◊◊◊ ◊◊◊◊◊ ◊◊◊◊◊ ◊◊◊◊◊ ◊◊◊◊◊ ◊◊◊◊◊ ◊◊◊◊◊ ◊◊◊◊◊ ◊◊◊◊◊	89,500
◊◊◊◊◊ ◊◊◊◊◊ ◊◊◊◊◊ ◊◊◊◊◊ ◊◊◊◊◊ ◊◊◊◊◊ ◊◊◊◊◊ ◊◊◊◊◊ ◊◊◊◊◊ ◊◊◊◊◊	89,550
◊◊◊◊◊ ◊◊◊◊◊ ◊◊◊◊◊ ◊◊◊◊◊ ◊◊◊◊◊ ◊◊◊◊◊ ◊◊◊◊◊ ◊◊◊◊◊ ◊◊◊◊◊ ◊◊◊◊◊	89,600
◊◊◊◊◊ ◊◊◊◊◊ ◊◊◊◊◊ ◊◊◊◊◊ ◊◊◊◊◊ ◊◊◊◊◊ ◊◊◊◊◊ ◊◊◊◊◊ ◊◊◊◊◊ ◊◊◊◊◊	89,650
◊◊◊◊◊ ◊◊◊◊◊ ◊◊◊◊◊ ◊◊◊◊◊ ◊◊◊◊◊ ◊◊◊◊◊ ◊◊◊◊◊ ◊◊◊◊◊ ◊◊◊◊◊ ◊◊◊◊◊	89,700
◊◊◊◊◊ ◊◊◊◊◊ ◊◊◊◊◊ ◊◊◊◊◊ ◊◊◊◊◊ ◊◊◊◊◊ ◊◊◊◊◊ ◊◊◊◊◊ ◊◊◊◊◊ ◊◊◊◊◊	89,750
◊◊◊◊◊ ◊◊◊◊◊ ◊◊◊◊◊ ◊◊◊◊◊ ◊◊◊◊◊ ◊◊◊◊◊ ◊◊◊◊◊ ◊◊◊◊◊ ◊◊◊◊◊ ◊◊◊◊◊	89,800
◊◊◊◊◊ ◊◊◊◊◊ ◊◊◊◊◊ ◊◊◊◊◊ ◊◊◊◊◊ ◊◊◊◊◊ ◊◊◊◊◊ ◊◊◊◊◊ ◊◊◊◊◊ ◊◊◊◊◊	89,850
◊◊◊◊◊ ◊◊◊◊◊ ◊◊◊◊◊ ◊◊◊◊◊ ◊◊◊◊◊ ◊◊◊◊◊ ◊◊◊◊◊ ◊◊◊◊◊ ◊◊◊◊◊ ◊◊◊◊◊	89,900
◊◊◊◊◊ ◊◊◊◊◊ ◊◊◊◊◊ ◊◊◊◊◊ ◊◊◊◊◊ ◊◊◊◊◊ ◊◊◊◊◊ ◊◊◊◊◊ ◊◊◊◊◊ ◊◊◊◊◊	89,950
◊◊◊◊◊ ◊◊◊◊◊ ◊◊◊◊◊ ◊◊◊◊◊ ◊◊◊◊◊ ◊◊◊◊◊ ◊◊◊◊◊ ◊◊◊◊◊ ◊◊◊◊◊ ◊◊◊◊◊	**90,000**

The Number 90,001 to the Number 92,000

Start										Notes
										90,050
										90,100
										90,150
										90,200
										90,250
										90,300
										90,350
										90,400
										90,450
										90,500
										90,550
										90,600
										90,650
										90,700
										90,750
										90,800
										90,850
										90,900
										90,950
										91,000
										91,050
										91,100
										91,150
										91,200
										91,250
										91,300
										91,350
										91,400
										91,450
										91,500
										91,550
										91,600
										91,650
										91,700
										91,750
										91,800
										91,850
										91,900
										91,950
										92,000

The Number 92,001 to the Number 94,000

Start

Notes

92,050	
92,100	
92,150	
92,200	
92,250	
92,300	
92,350	
92,400	
92,450	
92,500	
92,550	
92,600	
92,650	
92,700	
92,750	
92,800	
92,850	
92,900	
92,950	
93,000	
93,050	
93,100	
93,150	
93,200	
93,250	
93,300	
93,350	
93,400	
93,450	
93,500	
93,550	
93,600	
93,650	
93,700	
93,750	
93,800	
93,850	
93,900	
93,950	
94,000	

The Number 94,001 to the Number 96,000

Start		Notes
	94,050	
	94,100	
	94,150	
	94,200	
	94,250	
	94,300	
	94,350	
	94,400	
	94,450	
	94,500	
	94,550	
	94,600	
	94,650	
	94,700	
	94,750	
	94,800	
	94,850	
	94,900	
	94,950	
	95,000	
	95,050	
	95,100	
	95,150	
	95,200	
	95,250	
	95,300	
	95,350	
	95,400	
	95,450	
	95,500	
	95,550	
	95,600	
	95,650	
	95,700	
	95,750	
	95,800	
	95,850	
	95,900	
	95,950	
	96,000	

The Number 96,001 to the Number 98,000

Start

Notes

	96,050
	96,100
	96,150
	96,200
	96,250
	96,300
	96,350
	96,400
	96,450
	96,500
	96,550
	96,600
	96,650
	96,700
	96,750
	96,800
	96,850
	96,900
	96,950
	97,000
	97,050
	97,100
	97,150
	97,200
	97,250
	97,300
	97,350
	97,400
	97,450
	97,500
	97,550
	97,600
	97,650
	97,700
	97,750
	97,800
	97,850
	97,900
	97,950
	98,000

◇ The Number 98,001 to the Number 100,000

Start		Notes
98,050
98,100
98,150
98,200
98,250
98,300
98,350
98,400
98,450
98,500
98,550
98,600
98,650
98,700
98,750
98,800
98,850
98,900
98,950
99,000
99,050
99,100
99,150
99,200
99,250
99,300
99,350
99,400
99,450
99,500
99,550
99,600
99,650
99,700
99,750
99,800
99,850
99,900
99,950
100,000

◊ The Number 100,001 to the Number 102,000

Start	End #	Notes
	100,050	
	100,100	
	100,150	
	100,200	
	100,250	
	100,300	
	100,350	
	100,400	
	100,450	
	100,500	
	100,550	
	100,600	
	100,650	
	100,700	
	100,750	
	100,800	
	100,850	
	100,900	
	100,950	
	101,000	
	101,050	
	101,100	
	101,150	
	101,200	
	101,250	
	101,300	
	101,350	
	101,400	
	101,450	
	101,500	
	101,550	
	101,600	
	101,650	
	101,700	
	101,750	
	101,800	
	101,850	
	101,900	
	101,950	
	102,000	

The Number 102,001 to the Number 104,000

Start

Notes

	102,050
	102,100
	102,150
	102,200
	102,250
	102,300
	102,350
	102,400
	102,450
	102,500
	102,550
	102,600
	102,650
	102,700
	102,750
	102,800
	102,850
	102,900
	102,950
	103,000
	103,050
	103,100
	103,150
	103,200
	103,250
	103,300
	103,350
	103,400
	103,450
	103,500
	103,550
	103,600
	103,650
	103,700
	103,750
	103,800
	103,850
	103,900
	103,950
	104,000

The Number 104,001 to the Number 106,000

Start

Notes

	104,050
	104,100
	104,150
	104,200
	104,250
	104,300
	104,350
	104,400
	104,450
	104,500
	104,550
	104,600
	104,650
	104,700
	104,750
	104,800
	104,850
	104,900
	104,950
	105,000
	105,050
	105,100
	105,150
	105,200
	105,250
	105,300
	105,350
	105,400
	105,450
	105,500
	105,550
	105,600
	105,650
	105,700
	105,750
	105,800
	105,850
	105,900
	105,950
	106,000

The Number 106,001 to the Number 108,000

Start

Notes

	106,050
	106,100
	106,150
	106,200
	106,250
	106,300
	106,350
	106,400
	106,450
	106,500
	106,550
	106,600
	106,650
	106,700
	106,750
	106,800
	106,850
	106,900
	106,950
	107,000
	107,050
	107,100
	107,150
	107,200
	107,250
	107,300
	107,350
	107,400
	107,450
	107,500
	107,550
	107,600
	107,650
	107,700
	107,750
	107,800
	107,850
	107,900
	107,950
	108,000

The Number 108,001 to the Number 110,000

Start   Notes

	108,050
	108,100
	108,150
	108,200
	108,250
	108,300
	108,350
	108,400
	108,450
	108,500
	108,550
	108,600
	108,650
	108,700
	108,750
	108,800
	108,850
	108,900
	108,950
	109,000
	109,050
	109,100
	109,150
	109,200
	109,250
	109,300
	109,350
	109,400
	109,450
	109,500
	109,550
	109,600
	109,650
	109,700
	109,750
	109,800
	109,850
	109,900
	109,950
	110,000

The Number 110,001 to The Number 112,000

Start		Notes
110,050
110,100
110,150
110,200
110,250
110,300
110,350
110,400
110,450
110,500
110,550
110,600
110,650
110,700
110,750
110,800
110,850
110,900
110,950
111,000
111,050
111,100
111,150
111,200
111,250
111,300
111,350
111,400
111,450
111,500
111,550
111,600
111,650
111,700
111,750
111,800
111,850
111,900
111,950
112,000

The Number 112,001 to the Number 114,000

Start

Notes

	112,050	
	112,100	
	112,150	
	112,200	
	112,250	
	112,300	
	112,350	
	112,400	
	112,450	
	112,500	
	112,550	
	112,600	
	112,650	
	112,700	
	112,750	
	112,800	
	112,850	
	112,900	
	112,950	
	113,000	
	113,050	
	113,100	
	113,150	
	113,200	
	113,250	
	113,300	
	113,350	
	113,400	
	113,450	
	113,500	
	113,550	
	113,600	
	113,650	
	113,700	
	113,750	
	113,800	
	113,850	
	113,900	
	113,950	
	114,000	

The Number 114,001 to the Number 116,000

Start											Notes
										114,050	
										114,100	
										114,150	
										114,200	
										114,250	
										114,300	
										114,350	
										114,400	
										114,450	
										114,500	
										114,550	
										114,600	
										114,650	
										114,700	
										114,750	
										114,800	
										114,850	
										114,900	
										114,950	
										115,000	
										115,050	
										115,100	
										115,150	
										115,200	
										115,250	
										115,300	
										115,350	
										115,400	
										115,450	
										115,500	
										115,550	
										115,600	
										115,650	
										115,700	
										115,750	
										115,800	
										115,850	
										115,900	
										115,950	
										116,000	

The Number 116,001 to the Number 118,000

Start		Notes

	116,050
	116,100
	116,150
	116,200
	116,250
	116,300
	116,350
	116,400
	116,450
	116,500
	116,550
	116,600
	116,650
	116,700
	116,750
	116,800
	116,850
	116,900
	116,950
	117,000
	117,050
	117,100
	117,150
	117,200
	117,250
	117,300
	117,350
	117,400
	117,450
	117,500
	117,550
	117,600
	117,650
	117,700
	117,750
	117,800
	117,850
	117,900
	117,950
	118,000

The Number 118,001 to the Number 120,000

Start   **Notes**

	118,050
	118,100
	118,150
	118,200
	118,250
	118,300
	118,350
	118,400
	118,450
	118,500
	118,550
	118,600
	118,650
	118,700
	118,750
	118,800
	118,850
	118,900
	118,950
	119,000
	119,050
	119,100
	119,150
	119,200
	119,250
	119,300
	119,350
	119,400
	119,450
	119,500
	119,550
	119,600
	119,650
	119,700
	119,750
	119,800
	119,850
	119,900
	119,950
	120,000

The Number 120,001 to the Number 122,000

Start

	Notes
120,050	
120,100	
120,150	
120,200	
120,250	
120,300	
120,350	
120,400	
120,450	
120,500	
120,550	
120,600	
120,650	
120,700	
120,750	
120,800	
120,850	
120,900	
120,950	
121,000	
121,050	
121,100	
121,150	
121,200	
121,250	
121,300	
121,350	
121,400	
121,450	
121,500	
121,550	
121,600	
121,650	
121,700	
121,750	
121,800	
121,850	
121,900	
121,950	
122,000	

The Number 122,001 to the Number 124,000

Start	Notes
◊◊◊◊◊ ◊◊◊◊◊ ◊◊◊◊◊ ◊◊◊◊◊ ◊◊◊◊◊ ◊◊◊◊◊ ◊◊◊◊◊ ◊◊◊◊◊ ◊◊◊◊◊ ◊◊◊◊◊	122,050
◊◊◊◊◊ ◊◊◊◊◊ ◊◊◊◊◊ ◊◊◊◊◊ ◊◊◊◊◊ ◊◊◊◊◊ ◊◊◊◊◊ ◊◊◊◊◊ ◊◊◊◊◊ ◊◊◊◊◊	122,100
◊◊◊◊◊ ◊◊◊◊◊ ◊◊◊◊◊ ◊◊◊◊◊ ◊◊◊◊◊ ◊◊◊◊◊ ◊◊◊◊◊ ◊◊◊◊◊ ◊◊◊◊◊ ◊◊◊◊◊	122,150
◊◊◊◊◊ ◊◊◊◊◊ ◊◊◊◊◊ ◊◊◊◊◊ ◊◊◊◊◊ ◊◊◊◊◊ ◊◊◊◊◊ ◊◊◊◊◊ ◊◊◊◊◊ ◊◊◊◊◊	122,200
◊◊◊◊◊ ◊◊◊◊◊ ◊◊◊◊◊ ◊◊◊◊◊ ◊◊◊◊◊ ◊◊◊◊◊ ◊◊◊◊◊ ◊◊◊◊◊ ◊◊◊◊◊ ◊◊◊◊◊	122,250
◊◊◊◊◊ ◊◊◊◊◊ ◊◊◊◊◊ ◊◊◊◊◊ ◊◊◊◊◊ ◊◊◊◊◊ ◊◊◊◊◊ ◊◊◊◊◊ ◊◊◊◊◊ ◊◊◊◊◊	122,300
◊◊◊◊◊ ◊◊◊◊◊ ◊◊◊◊◊ ◊◊◊◊◊ ◊◊◊◊◊ ◊◊◊◊◊ ◊◊◊◊◊ ◊◊◊◊◊ ◊◊◊◊◊ ◊◊◊◊◊	122,350
◊◊◊◊◊ ◊◊◊◊◊ ◊◊◊◊◊ ◊◊◊◊◊ ◊◊◊◊◊ ◊◊◊◊◊ ◊◊◊◊◊ ◊◊◊◊◊ ◊◊◊◊◊ ◊◊◊◊◊	122,400
◊◊◊◊◊ ◊◊◊◊◊ ◊◊◊◊◊ ◊◊◊◊◊ ◊◊◊◊◊ ◊◊◊◊◊ ◊◊◊◊◊ ◊◊◊◊◊ ◊◊◊◊◊ ◊◊◊◊◊	122,450
◊◊◊◊◊ ◊◊◊◊◊ ◊◊◊◊◊ ◊◊◊◊◊ ◊◊◊◊◊ ◊◊◊◊◊ ◊◊◊◊◊ ◊◊◊◊◊ ◊◊◊◊◊ ◊◊◊◊◊	122,500
◊◊◊◊◊ ◊◊◊◊◊ ◊◊◊◊◊ ◊◊◊◊◊ ◊◊◊◊◊ ◊◊◊◊◊ ◊◊◊◊◊ ◊◊◊◊◊ ◊◊◊◊◊ ◊◊◊◊◊	122,550
◊◊◊◊◊ ◊◊◊◊◊ ◊◊◊◊◊ ◊◊◊◊◊ ◊◊◊◊◊ ◊◊◊◊◊ ◊◊◊◊◊ ◊◊◊◊◊ ◊◊◊◊◊ ◊◊◊◊◊	122,600
◊◊◊◊◊ ◊◊◊◊◊ ◊◊◊◊◊ ◊◊◊◊◊ ◊◊◊◊◊ ◊◊◊◊◊ ◊◊◊◊◊ ◊◊◊◊◊ ◊◊◊◊◊ ◊◊◊◊◊	122,650
◊◊◊◊◊ ◊◊◊◊◊ ◊◊◊◊◊ ◊◊◊◊◊ ◊◊◊◊◊ ◊◊◊◊◊ ◊◊◊◊◊ ◊◊◊◊◊ ◊◊◊◊◊ ◊◊◊◊◊	122,700
◊◊◊◊◊ ◊◊◊◊◊ ◊◊◊◊◊ ◊◊◊◊◊ ◊◊◊◊◊ ◊◊◊◊◊ ◊◊◊◊◊ ◊◊◊◊◊ ◊◊◊◊◊ ◊◊◊◊◊	122,750
◊◊◊◊◊ ◊◊◊◊◊ ◊◊◊◊◊ ◊◊◊◊◊ ◊◊◊◊◊ ◊◊◊◊◊ ◊◊◊◊◊ ◊◊◊◊◊ ◊◊◊◊◊ ◊◊◊◊◊	122,800
◊◊◊◊◊ ◊◊◊◊◊ ◊◊◊◊◊ ◊◊◊◊◊ ◊◊◊◊◊ ◊◊◊◊◊ ◊◊◊◊◊ ◊◊◊◊◊ ◊◊◊◊◊ ◊◊◊◊◊	122,850
◊◊◊◊◊ ◊◊◊◊◊ ◊◊◊◊◊ ◊◊◊◊◊ ◊◊◊◊◊ ◊◊◊◊◊ ◊◊◊◊◊ ◊◊◊◊◊ ◊◊◊◊◊ ◊◊◊◊◊	122,900
◊◊◊◊◊ ◊◊◊◊◊ ◊◊◊◊◊ ◊◊◊◊◊ ◊◊◊◊◊ ◊◊◊◊◊ ◊◊◊◊◊ ◊◊◊◊◊ ◊◊◊◊◊ ◊◊◊◊◊	122,950
◊◊◊◊◊ ◊◊◊◊◊ ◊◊◊◊◊ ◊◊◊◊◊ ◊◊◊◊◊ ◊◊◊◊◊ ◊◊◊◊◊ ◊◊◊◊◊ ◊◊◊◊◊ ◊◊◊◊◊	**123,000**
◊◊◊◊◊ ◊◊◊◊◊ ◊◊◊◊◊ ◊◊◊◊◊ ◊◊◊◊◊ ◊◊◊◊◊ ◊◊◊◊◊ ◊◊◊◊◊ ◊◊◊◊◊ ◊◊◊◊◊	123,050
◊◊◊◊◊ ◊◊◊◊◊ ◊◊◊◊◊ ◊◊◊◊◊ ◊◊◊◊◊ ◊◊◊◊◊ ◊◊◊◊◊ ◊◊◊◊◊ ◊◊◊◊◊ ◊◊◊◊◊	123,100
◊◊◊◊◊ ◊◊◊◊◊ ◊◊◊◊◊ ◊◊◊◊◊ ◊◊◊◊◊ ◊◊◊◊◊ ◊◊◊◊◊ ◊◊◊◊◊ ◊◊◊◊◊ ◊◊◊◊◊	123,150
◊◊◊◊◊ ◊◊◊◊◊ ◊◊◊◊◊ ◊◊◊◊◊ ◊◊◊◊◊ ◊◊◊◊◊ ◊◊◊◊◊ ◊◊◊◊◊ ◊◊◊◊◊ ◊◊◊◊◊	123,200
◊◊◊◊◊ ◊◊◊◊◊ ◊◊◊◊◊ ◊◊◊◊◊ ◊◊◊◊◊ ◊◊◊◊◊ ◊◊◊◊◊ ◊◊◊◊◊ ◊◊◊◊◊ ◊◊◊◊◊	123,250
◊◊◊◊◊ ◊◊◊◊◊ ◊◊◊◊◊ ◊◊◊◊◊ ◊◊◊◊◊ ◊◊◊◊◊ ◊◊◊◊◊ ◊◊◊◊◊ ◊◊◊◊◊ ◊◊◊◊◊	123,300
◊◊◊◊◊ ◊◊◊◊◊ ◊◊◊◊◊ ◊◊◊◊◊ ◊◊◊◊◊ ◊◊◊◊◊ ◊◊◊◊◊ ◊◊◊◊◊ ◊◊◊◊◊ ◊◊◊◊◊	123,350
◊◊◊◊◊ ◊◊◊◊◊ ◊◊◊◊◊ ◊◊◊◊◊ ◊◊◊◊◊ ◊◊◊◊◊ ◊◊◊◊◊ ◊◊◊◊◊ ◊◊◊◊◊ ◊◊◊◊◊	123,400
◊◊◊◊◊ ◊◊◊◊◊ ◊◊◊◊◊ ◊◊◊◊◊ ◊◊◊◊◊ ◊◊◊◊◊ ◊◊◊◊◊ ◊◊◊◊◊ ◊◊◊◊◊ ◊◊◊◊◊	123,450
◊◊◊◊◊ ◊◊◊◊◊ ◊◊◊◊◊ ◊◊◊◊◊ ◊◊◊◊◊ ◊◊◊◊◊ ◊◊◊◊◊ ◊◊◊◊◊ ◊◊◊◊◊ ◊◊◊◊◊	123,500
◊◊◊◊◊ ◊◊◊◊◊ ◊◊◊◊◊ ◊◊◊◊◊ ◊◊◊◊◊ ◊◊◊◊◊ ◊◊◊◊◊ ◊◊◊◊◊ ◊◊◊◊◊ ◊◊◊◊◊	123,550
◊◊◊◊◊ ◊◊◊◊◊ ◊◊◊◊◊ ◊◊◊◊◊ ◊◊◊◊◊ ◊◊◊◊◊ ◊◊◊◊◊ ◊◊◊◊◊ ◊◊◊◊◊ ◊◊◊◊◊	123,600
◊◊◊◊◊ ◊◊◊◊◊ ◊◊◊◊◊ ◊◊◊◊◊ ◊◊◊◊◊ ◊◊◊◊◊ ◊◊◊◊◊ ◊◊◊◊◊ ◊◊◊◊◊ ◊◊◊◊◊	123,650
◊◊◊◊◊ ◊◊◊◊◊ ◊◊◊◊◊ ◊◊◊◊◊ ◊◊◊◊◊ ◊◊◊◊◊ ◊◊◊◊◊ ◊◊◊◊◊ ◊◊◊◊◊ ◊◊◊◊◊	123,700
◊◊◊◊◊ ◊◊◊◊◊ ◊◊◊◊◊ ◊◊◊◊◊ ◊◊◊◊◊ ◊◊◊◊◊ ◊◊◊◊◊ ◊◊◊◊◊ ◊◊◊◊◊ ◊◊◊◊◊	123,750
◊◊◊◊◊ ◊◊◊◊◊ ◊◊◊◊◊ ◊◊◊◊◊ ◊◊◊◊◊ ◊◊◊◊◊ ◊◊◊◊◊ ◊◊◊◊◊ ◊◊◊◊◊ ◊◊◊◊◊	123,800
◊◊◊◊◊ ◊◊◊◊◊ ◊◊◊◊◊ ◊◊◊◊◊ ◊◊◊◊◊ ◊◊◊◊◊ ◊◊◊◊◊ ◊◊◊◊◊ ◊◊◊◊◊ ◊◊◊◊◊	123,850
◊◊◊◊◊ ◊◊◊◊◊ ◊◊◊◊◊ ◊◊◊◊◊ ◊◊◊◊◊ ◊◊◊◊◊ ◊◊◊◊◊ ◊◊◊◊◊ ◊◊◊◊◊ ◊◊◊◊◊	123,900
◊◊◊◊◊ ◊◊◊◊◊ ◊◊◊◊◊ ◊◊◊◊◊ ◊◊◊◊◊ ◊◊◊◊◊ ◊◊◊◊◊ ◊◊◊◊◊ ◊◊◊◊◊ ◊◊◊◊◊	123,950
◊◊◊◊◊ ◊◊◊◊◊ ◊◊◊◊◊ ◊◊◊◊◊ ◊◊◊◊◊ ◊◊◊◊◊ ◊◊◊◊◊ ◊◊◊◊◊ ◊◊◊◊◊ ◊◊◊◊◊	**124,000**

The Number 124,001 to the Number 126,000

Start   **Notes**

◊◊◊◊◊ ◊◊◊◊◊ ◊◊◊◊◊ ◊◊◊◊◊ ◊◊◊◊◊ ◊◊◊◊◊ ◊◊◊◊◊ ◊◊◊◊◊ ◊◊◊◊◊ ◊◊◊◊◊	124,050
◊◊◊◊◊ ◊◊◊◊◊ ◊◊◊◊◊ ◊◊◊◊◊ ◊◊◊◊◊ ◊◊◊◊◊ ◊◊◊◊◊ ◊◊◊◊◊ ◊◊◊◊◊ ◊◊◊◊◊	124,100
◊◊◊◊◊ ◊◊◊◊◊ ◊◊◊◊◊ ◊◊◊◊◊ ◊◊◊◊◊ ◊◊◊◊◊ ◊◊◊◊◊ ◊◊◊◊◊ ◊◊◊◊◊ ◊◊◊◊◊	124,150
◊◊◊◊◊ ◊◊◊◊◊ ◊◊◊◊◊ ◊◊◊◊◊ ◊◊◊◊◊ ◊◊◊◊◊ ◊◊◊◊◊ ◊◊◊◊◊ ◊◊◊◊◊ ◊◊◊◊◊	124,200
◊◊◊◊◊ ◊◊◊◊◊ ◊◊◊◊◊ ◊◊◊◊◊ ◊◊◊◊◊ ◊◊◊◊◊ ◊◊◊◊◊ ◊◊◊◊◊ ◊◊◊◊◊ ◊◊◊◊◊	124,250
◊◊◊◊◊ ◊◊◊◊◊ ◊◊◊◊◊ ◊◊◊◊◊ ◊◊◊◊◊ ◊◊◊◊◊ ◊◊◊◊◊ ◊◊◊◊◊ ◊◊◊◊◊ ◊◊◊◊◊	124,300
◊◊◊◊◊ ◊◊◊◊◊ ◊◊◊◊◊ ◊◊◊◊◊ ◊◊◊◊◊ ◊◊◊◊◊ ◊◊◊◊◊ ◊◊◊◊◊ ◊◊◊◊◊ ◊◊◊◊◊	124,350
◊◊◊◊◊ ◊◊◊◊◊ ◊◊◊◊◊ ◊◊◊◊◊ ◊◊◊◊◊ ◊◊◊◊◊ ◊◊◊◊◊ ◊◊◊◊◊ ◊◊◊◊◊ ◊◊◊◊◊	124,400
◊◊◊◊◊ ◊◊◊◊◊ ◊◊◊◊◊ ◊◊◊◊◊ ◊◊◊◊◊ ◊◊◊◊◊ ◊◊◊◊◊ ◊◊◊◊◊ ◊◊◊◊◊ ◊◊◊◊◊	124,450
◊◊◊◊◊ ◊◊◊◊◊ ◊◊◊◊◊ ◊◊◊◊◊ ◊◊◊◊◊ ◊◊◊◊◊ ◊◊◊◊◊ ◊◊◊◊◊ ◊◊◊◊◊ ◊◊◊◊◊	124,500
◊◊◊◊◊ ◊◊◊◊◊ ◊◊◊◊◊ ◊◊◊◊◊ ◊◊◊◊◊ ◊◊◊◊◊ ◊◊◊◊◊ ◊◊◊◊◊ ◊◊◊◊◊ ◊◊◊◊◊	124,550
◊◊◊◊◊ ◊◊◊◊◊ ◊◊◊◊◊ ◊◊◊◊◊ ◊◊◊◊◊ ◊◊◊◊◊ ◊◊◊◊◊ ◊◊◊◊◊ ◊◊◊◊◊ ◊◊◊◊◊	124,600
◊◊◊◊◊ ◊◊◊◊◊ ◊◊◊◊◊ ◊◊◊◊◊ ◊◊◊◊◊ ◊◊◊◊◊ ◊◊◊◊◊ ◊◊◊◊◊ ◊◊◊◊◊ ◊◊◊◊◊	124,650
◊◊◊◊◊ ◊◊◊◊◊ ◊◊◊◊◊ ◊◊◊◊◊ ◊◊◊◊◊ ◊◊◊◊◊ ◊◊◊◊◊ ◊◊◊◊◊ ◊◊◊◊◊ ◊◊◊◊◊	124,700
◊◊◊◊◊ ◊◊◊◊◊ ◊◊◊◊◊ ◊◊◊◊◊ ◊◊◊◊◊ ◊◊◊◊◊ ◊◊◊◊◊ ◊◊◊◊◊ ◊◊◊◊◊ ◊◊◊◊◊	124,750
◊◊◊◊◊ ◊◊◊◊◊ ◊◊◊◊◊ ◊◊◊◊◊ ◊◊◊◊◊ ◊◊◊◊◊ ◊◊◊◊◊ ◊◊◊◊◊ ◊◊◊◊◊ ◊◊◊◊◊	124,800
◊◊◊◊◊ ◊◊◊◊◊ ◊◊◊◊◊ ◊◊◊◊◊ ◊◊◊◊◊ ◊◊◊◊◊ ◊◊◊◊◊ ◊◊◊◊◊ ◊◊◊◊◊ ◊◊◊◊◊	124,850
◊◊◊◊◊ ◊◊◊◊◊ ◊◊◊◊◊ ◊◊◊◊◊ ◊◊◊◊◊ ◊◊◊◊◊ ◊◊◊◊◊ ◊◊◊◊◊ ◊◊◊◊◊ ◊◊◊◊◊	124,900
◊◊◊◊◊ ◊◊◊◊◊ ◊◊◊◊◊ ◊◊◊◊◊ ◊◊◊◊◊ ◊◊◊◊◊ ◊◊◊◊◊ ◊◊◊◊◊ ◊◊◊◊◊ ◊◊◊◊◊	124,950
◊◊◊◊◊ ◊◊◊◊◊ ◊◊◊◊◊ ◊◊◊◊◊ ◊◊◊◊◊ ◊◊◊◊◊ ◊◊◊◊◊ ◊◊◊◊◊ ◊◊◊◊◊ ◊◊◊◊◊	**125,000**
◊◊◊◊◊ ◊◊◊◊◊ ◊◊◊◊◊ ◊◊◊◊◊ ◊◊◊◊◊ ◊◊◊◊◊ ◊◊◊◊◊ ◊◊◊◊◊ ◊◊◊◊◊ ◊◊◊◊◊	125,050
◊◊◊◊◊ ◊◊◊◊◊ ◊◊◊◊◊ ◊◊◊◊◊ ◊◊◊◊◊ ◊◊◊◊◊ ◊◊◊◊◊ ◊◊◊◊◊ ◊◊◊◊◊ ◊◊◊◊◊	125,100
◊◊◊◊◊ ◊◊◊◊◊ ◊◊◊◊◊ ◊◊◊◊◊ ◊◊◊◊◊ ◊◊◊◊◊ ◊◊◊◊◊ ◊◊◊◊◊ ◊◊◊◊◊ ◊◊◊◊◊	125,150
◊◊◊◊◊ ◊◊◊◊◊ ◊◊◊◊◊ ◊◊◊◊◊ ◊◊◊◊◊ ◊◊◊◊◊ ◊◊◊◊◊ ◊◊◊◊◊ ◊◊◊◊◊ ◊◊◊◊◊	125,200
◊◊◊◊◊ ◊◊◊◊◊ ◊◊◊◊◊ ◊◊◊◊◊ ◊◊◊◊◊ ◊◊◊◊◊ ◊◊◊◊◊ ◊◊◊◊◊ ◊◊◊◊◊ ◊◊◊◊◊	125,250
◊◊◊◊◊ ◊◊◊◊◊ ◊◊◊◊◊ ◊◊◊◊◊ ◊◊◊◊◊ ◊◊◊◊◊ ◊◊◊◊◊ ◊◊◊◊◊ ◊◊◊◊◊ ◊◊◊◊◊	125,300
◊◊◊◊◊ ◊◊◊◊◊ ◊◊◊◊◊ ◊◊◊◊◊ ◊◊◊◊◊ ◊◊◊◊◊ ◊◊◊◊◊ ◊◊◊◊◊ ◊◊◊◊◊ ◊◊◊◊◊	125,350
◊◊◊◊◊ ◊◊◊◊◊ ◊◊◊◊◊ ◊◊◊◊◊ ◊◊◊◊◊ ◊◊◊◊◊ ◊◊◊◊◊ ◊◊◊◊◊ ◊◊◊◊◊ ◊◊◊◊◊	125,400
◊◊◊◊◊ ◊◊◊◊◊ ◊◊◊◊◊ ◊◊◊◊◊ ◊◊◊◊◊ ◊◊◊◊◊ ◊◊◊◊◊ ◊◊◊◊◊ ◊◊◊◊◊ ◊◊◊◊◊	125,450
◊◊◊◊◊ ◊◊◊◊◊ ◊◊◊◊◊ ◊◊◊◊◊ ◊◊◊◊◊ ◊◊◊◊◊ ◊◊◊◊◊ ◊◊◊◊◊ ◊◊◊◊◊ ◊◊◊◊◊	125,500
◊◊◊◊◊ ◊◊◊◊◊ ◊◊◊◊◊ ◊◊◊◊◊ ◊◊◊◊◊ ◊◊◊◊◊ ◊◊◊◊◊ ◊◊◊◊◊ ◊◊◊◊◊ ◊◊◊◊◊	125,550
◊◊◊◊◊ ◊◊◊◊◊ ◊◊◊◊◊ ◊◊◊◊◊ ◊◊◊◊◊ ◊◊◊◊◊ ◊◊◊◊◊ ◊◊◊◊◊ ◊◊◊◊◊ ◊◊◊◊◊	125,600
◊◊◊◊◊ ◊◊◊◊◊ ◊◊◊◊◊ ◊◊◊◊◊ ◊◊◊◊◊ ◊◊◊◊◊ ◊◊◊◊◊ ◊◊◊◊◊ ◊◊◊◊◊ ◊◊◊◊◊	125,650
◊◊◊◊◊ ◊◊◊◊◊ ◊◊◊◊◊ ◊◊◊◊◊ ◊◊◊◊◊ ◊◊◊◊◊ ◊◊◊◊◊ ◊◊◊◊◊ ◊◊◊◊◊ ◊◊◊◊◊	125,700
◊◊◊◊◊ ◊◊◊◊◊ ◊◊◊◊◊ ◊◊◊◊◊ ◊◊◊◊◊ ◊◊◊◊◊ ◊◊◊◊◊ ◊◊◊◊◊ ◊◊◊◊◊ ◊◊◊◊◊	125,750
◊◊◊◊◊ ◊◊◊◊◊ ◊◊◊◊◊ ◊◊◊◊◊ ◊◊◊◊◊ ◊◊◊◊◊ ◊◊◊◊◊ ◊◊◊◊◊ ◊◊◊◊◊ ◊◊◊◊◊	125,800
◊◊◊◊◊ ◊◊◊◊◊ ◊◊◊◊◊ ◊◊◊◊◊ ◊◊◊◊◊ ◊◊◊◊◊ ◊◊◊◊◊ ◊◊◊◊◊ ◊◊◊◊◊ ◊◊◊◊◊	125,850
◊◊◊◊◊ ◊◊◊◊◊ ◊◊◊◊◊ ◊◊◊◊◊ ◊◊◊◊◊ ◊◊◊◊◊ ◊◊◊◊◊ ◊◊◊◊◊ ◊◊◊◊◊ ◊◊◊◊◊	125,900
◊◊◊◊◊ ◊◊◊◊◊ ◊◊◊◊◊ ◊◊◊◊◊ ◊◊◊◊◊ ◊◊◊◊◊ ◊◊◊◊◊ ◊◊◊◊◊ ◊◊◊◊◊ ◊◊◊◊◊	125,950
◊◊◊◊◊ ◊◊◊◊◊ ◊◊◊◊◊ ◊◊◊◊◊ ◊◊◊◊◊ ◊◊◊◊◊ ◊◊◊◊◊ ◊◊◊◊◊ ◊◊◊◊◊ ◊◊◊◊◊	**126,000**

The Number 126,001 to the Number 128,000

Start   Notes

	126,050
	126,100
	126,150
	126,200
	126,250
	126,300
	126,350
	126,400
	126,450
	126,500
	126,550
	126,600
	126,650
	126,700
	126,750
	126,800
	126,850
	126,900
	126,950
	127,000
	127,050
	127,100
	127,150
	127,200
	127,250
	127,300
	127,350
	127,400
	127,450
	127,500
	127,550
	127,600
	127,650
	127,700
	127,750
	127,800
	127,850
	127,900
	127,950
	128,000

The Number 128,001 to the Number 130,000

Start   Notes

	128,050
	128,100
	128,150
	128,200
	128,250
	128,300
	128,350
	128,400
	128,450
	128,500
	128,550
	128,600
	128,650
	128,700
	128,750
	128,800
	128,850
	128,900
	128,950
	129,000
	129,050
	129,100
	129,150
	129,200
	129,250
	129,300
	129,350
	129,400
	129,450
	129,500
	129,550
	129,600
	129,650
	129,700
	129,750
	129,800
	129,850
	129,900
	129,950
	130,000

The Number 130,001 to the Number 132,000

Start   Notes

	130,050
	130,100
	130,150
	130,200
	130,250
	130,300
	130,350
	130,400
	130,450
	130,500
	130,550
	130,600
	130,650
	130,700
	130,750
	130,800
	130,850
	130,900
	130,950
	131,000
	131,050
	131,100
	131,150
	131,200
	131,250
	131,300
	131,350
	131,400
	131,450
	131,500
	131,550
	131,600
	131,650
	131,700
	131,750
	131,800
	131,850
	131,900
	131,950
	132,000

The Number 132,001 to the Number 134,000

Start		Notes
	132,050	
	132,100	
	132,150	
	132,200	
	132,250	
	132,300	
	132,350	
	132,400	
	132,450	
	132,500	
	132,550	
	132,600	
	132,650	
	132,700	
	132,750	
	132,800	
	132,850	
	132,900	
	132,950	
	133,000	
	133,050	
	133,100	
	133,150	
	133,200	
	133,250	
	133,300	
	133,350	
	133,400	
	133,450	
	133,500	
	133,550	
	133,600	
	133,650	
	133,700	
	133,750	
	133,800	
	133,850	
	133,900	
	133,950	
	134,000	

The Number 134,001 to the Number 136,000

Start   **Notes**

134,050	
134,100	
134,150	
134,200	
134,250	
134,300	
134,350	
134,400	
134,450	
134,500	
134,550	
134,600	
134,650	
134,700	
134,750	
134,800	
134,850	
134,900	
134,950	
135,000	
135,050	
135,100	
135,150	
135,200	
135,250	
135,300	
135,350	
135,400	
135,450	
135,500	
135,550	
135,600	
135,650	
135,700	
135,750	
135,800	
135,850	
135,900	
135,950	
136,000	

The Number 136,001 to the Number 138,000

Start

	Number	Notes
	136,050	
	136,100	
	136,150	
	136,200	
	136,250	
	136,300	
	136,350	
	136,400	
	136,450	
	136,500	
	136,550	
	136,600	
	136,650	
	136,700	
	136,750	
	136,800	
	136,850	
	136,900	
	136,950	
	137,000	
	137,050	
	137,100	
	137,150	
	137,200	
	137,250	
	137,300	
	137,350	
	137,400	
	137,450	
	137,500	
	137,550	
	137,600	
	137,650	
	137,700	
	137,750	
	137,800	
	137,850	
	137,900	
	137,950	
	138,000	

The Number 138,001 to the Number 140,000

Start	Notes
◊◊◊◊◊ ◊◊◊◊◊ ◊◊◊◊◊ ◊◊◊◊◊ ◊◊◊◊◊ ◊◊◊◊◊ ◊◊◊◊◊ ◊◊◊◊◊ ◊◊◊◊◊ ◊◊◊◊◊	138,050
◊◊◊◊◊ ◊◊◊◊◊ ◊◊◊◊◊ ◊◊◊◊◊ ◊◊◊◊◊ ◊◊◊◊◊ ◊◊◊◊◊ ◊◊◊◊◊ ◊◊◊◊◊ ◊◊◊◊◊	138,100
◊◊◊◊◊ ◊◊◊◊◊ ◊◊◊◊◊ ◊◊◊◊◊ ◊◊◊◊◊ ◊◊◊◊◊ ◊◊◊◊◊ ◊◊◊◊◊ ◊◊◊◊◊ ◊◊◊◊◊	138,150
◊◊◊◊◊ ◊◊◊◊◊ ◊◊◊◊◊ ◊◊◊◊◊ ◊◊◊◊◊ ◊◊◊◊◊ ◊◊◊◊◊ ◊◊◊◊◊ ◊◊◊◊◊ ◊◊◊◊◊	138,200
◊◊◊◊◊ ◊◊◊◊◊ ◊◊◊◊◊ ◊◊◊◊◊ ◊◊◊◊◊ ◊◊◊◊◊ ◊◊◊◊◊ ◊◊◊◊◊ ◊◊◊◊◊ ◊◊◊◊◊	138,250
◊◊◊◊◊ ◊◊◊◊◊ ◊◊◊◊◊ ◊◊◊◊◊ ◊◊◊◊◊ ◊◊◊◊◊ ◊◊◊◊◊ ◊◊◊◊◊ ◊◊◊◊◊ ◊◊◊◊◊	138,300
◊◊◊◊◊ ◊◊◊◊◊ ◊◊◊◊◊ ◊◊◊◊◊ ◊◊◊◊◊ ◊◊◊◊◊ ◊◊◊◊◊ ◊◊◊◊◊ ◊◊◊◊◊ ◊◊◊◊◊	138,350
◊◊◊◊◊ ◊◊◊◊◊ ◊◊◊◊◊ ◊◊◊◊◊ ◊◊◊◊◊ ◊◊◊◊◊ ◊◊◊◊◊ ◊◊◊◊◊ ◊◊◊◊◊ ◊◊◊◊◊	138,400
◊◊◊◊◊ ◊◊◊◊◊ ◊◊◊◊◊ ◊◊◊◊◊ ◊◊◊◊◊ ◊◊◊◊◊ ◊◊◊◊◊ ◊◊◊◊◊ ◊◊◊◊◊ ◊◊◊◊◊	138,450
◊◊◊◊◊ ◊◊◊◊◊ ◊◊◊◊◊ ◊◊◊◊◊ ◊◊◊◊◊ ◊◊◊◊◊ ◊◊◊◊◊ ◊◊◊◊◊ ◊◊◊◊◊ ◊◊◊◊◊	138,500
◊◊◊◊◊ ◊◊◊◊◊ ◊◊◊◊◊ ◊◊◊◊◊ ◊◊◊◊◊ ◊◊◊◊◊ ◊◊◊◊◊ ◊◊◊◊◊ ◊◊◊◊◊ ◊◊◊◊◊	138,550
◊◊◊◊◊ ◊◊◊◊◊ ◊◊◊◊◊ ◊◊◊◊◊ ◊◊◊◊◊ ◊◊◊◊◊ ◊◊◊◊◊ ◊◊◊◊◊ ◊◊◊◊◊ ◊◊◊◊◊	138,600
◊◊◊◊◊ ◊◊◊◊◊ ◊◊◊◊◊ ◊◊◊◊◊ ◊◊◊◊◊ ◊◊◊◊◊ ◊◊◊◊◊ ◊◊◊◊◊ ◊◊◊◊◊ ◊◊◊◊◊	138,650
◊◊◊◊◊ ◊◊◊◊◊ ◊◊◊◊◊ ◊◊◊◊◊ ◊◊◊◊◊ ◊◊◊◊◊ ◊◊◊◊◊ ◊◊◊◊◊ ◊◊◊◊◊ ◊◊◊◊◊	138,700
◊◊◊◊◊ ◊◊◊◊◊ ◊◊◊◊◊ ◊◊◊◊◊ ◊◊◊◊◊ ◊◊◊◊◊ ◊◊◊◊◊ ◊◊◊◊◊ ◊◊◊◊◊ ◊◊◊◊◊	138,750
◊◊◊◊◊ ◊◊◊◊◊ ◊◊◊◊◊ ◊◊◊◊◊ ◊◊◊◊◊ ◊◊◊◊◊ ◊◊◊◊◊ ◊◊◊◊◊ ◊◊◊◊◊ ◊◊◊◊◊	138,800
◊◊◊◊◊ ◊◊◊◊◊ ◊◊◊◊◊ ◊◊◊◊◊ ◊◊◊◊◊ ◊◊◊◊◊ ◊◊◊◊◊ ◊◊◊◊◊ ◊◊◊◊◊ ◊◊◊◊◊	138,850
◊◊◊◊◊ ◊◊◊◊◊ ◊◊◊◊◊ ◊◊◊◊◊ ◊◊◊◊◊ ◊◊◊◊◊ ◊◊◊◊◊ ◊◊◊◊◊ ◊◊◊◊◊ ◊◊◊◊◊	138,900
◊◊◊◊◊ ◊◊◊◊◊ ◊◊◊◊◊ ◊◊◊◊◊ ◊◊◊◊◊ ◊◊◊◊◊ ◊◊◊◊◊ ◊◊◊◊◊ ◊◊◊◊◊ ◊◊◊◊◊	138,950
◊◊◊◊◊ ◊◊◊◊◊ ◊◊◊◊◊ ◊◊◊◊◊ ◊◊◊◊◊ ◊◊◊◊◊ ◊◊◊◊◊ ◊◊◊◊◊ ◊◊◊◊◊ ◊◊◊◊◊	**139,000**
◊◊◊◊◊ ◊◊◊◊◊ ◊◊◊◊◊ ◊◊◊◊◊ ◊◊◊◊◊ ◊◊◊◊◊ ◊◊◊◊◊ ◊◊◊◊◊ ◊◊◊◊◊ ◊◊◊◊◊	139,050
◊◊◊◊◊ ◊◊◊◊◊ ◊◊◊◊◊ ◊◊◊◊◊ ◊◊◊◊◊ ◊◊◊◊◊ ◊◊◊◊◊ ◊◊◊◊◊ ◊◊◊◊◊ ◊◊◊◊◊	139,100
◊◊◊◊◊ ◊◊◊◊◊ ◊◊◊◊◊ ◊◊◊◊◊ ◊◊◊◊◊ ◊◊◊◊◊ ◊◊◊◊◊ ◊◊◊◊◊ ◊◊◊◊◊ ◊◊◊◊◊	139,150
◊◊◊◊◊ ◊◊◊◊◊ ◊◊◊◊◊ ◊◊◊◊◊ ◊◊◊◊◊ ◊◊◊◊◊ ◊◊◊◊◊ ◊◊◊◊◊ ◊◊◊◊◊ ◊◊◊◊◊	139,200
◊◊◊◊◊ ◊◊◊◊◊ ◊◊◊◊◊ ◊◊◊◊◊ ◊◊◊◊◊ ◊◊◊◊◊ ◊◊◊◊◊ ◊◊◊◊◊ ◊◊◊◊◊ ◊◊◊◊◊	139,250
◊◊◊◊◊ ◊◊◊◊◊ ◊◊◊◊◊ ◊◊◊◊◊ ◊◊◊◊◊ ◊◊◊◊◊ ◊◊◊◊◊ ◊◊◊◊◊ ◊◊◊◊◊ ◊◊◊◊◊	139,300
◊◊◊◊◊ ◊◊◊◊◊ ◊◊◊◊◊ ◊◊◊◊◊ ◊◊◊◊◊ ◊◊◊◊◊ ◊◊◊◊◊ ◊◊◊◊◊ ◊◊◊◊◊ ◊◊◊◊◊	139,350
◊◊◊◊◊ ◊◊◊◊◊ ◊◊◊◊◊ ◊◊◊◊◊ ◊◊◊◊◊ ◊◊◊◊◊ ◊◊◊◊◊ ◊◊◊◊◊ ◊◊◊◊◊ ◊◊◊◊◊	139,400
◊◊◊◊◊ ◊◊◊◊◊ ◊◊◊◊◊ ◊◊◊◊◊ ◊◊◊◊◊ ◊◊◊◊◊ ◊◊◊◊◊ ◊◊◊◊◊ ◊◊◊◊◊ ◊◊◊◊◊	139,450
◊◊◊◊◊ ◊◊◊◊◊ ◊◊◊◊◊ ◊◊◊◊◊ ◊◊◊◊◊ ◊◊◊◊◊ ◊◊◊◊◊ ◊◊◊◊◊ ◊◊◊◊◊ ◊◊◊◊◊	139,500
◊◊◊◊◊ ◊◊◊◊◊ ◊◊◊◊◊ ◊◊◊◊◊ ◊◊◊◊◊ ◊◊◊◊◊ ◊◊◊◊◊ ◊◊◊◊◊ ◊◊◊◊◊ ◊◊◊◊◊	139,550
◊◊◊◊◊ ◊◊◊◊◊ ◊◊◊◊◊ ◊◊◊◊◊ ◊◊◊◊◊ ◊◊◊◊◊ ◊◊◊◊◊ ◊◊◊◊◊ ◊◊◊◊◊ ◊◊◊◊◊	139,600
◊◊◊◊◊ ◊◊◊◊◊ ◊◊◊◊◊ ◊◊◊◊◊ ◊◊◊◊◊ ◊◊◊◊◊ ◊◊◊◊◊ ◊◊◊◊◊ ◊◊◊◊◊ ◊◊◊◊◊	139,650
◊◊◊◊◊ ◊◊◊◊◊ ◊◊◊◊◊ ◊◊◊◊◊ ◊◊◊◊◊ ◊◊◊◊◊ ◊◊◊◊◊ ◊◊◊◊◊ ◊◊◊◊◊ ◊◊◊◊◊	139,700
◊◊◊◊◊ ◊◊◊◊◊ ◊◊◊◊◊ ◊◊◊◊◊ ◊◊◊◊◊ ◊◊◊◊◊ ◊◊◊◊◊ ◊◊◊◊◊ ◊◊◊◊◊ ◊◊◊◊◊	139,750
◊◊◊◊◊ ◊◊◊◊◊ ◊◊◊◊◊ ◊◊◊◊◊ ◊◊◊◊◊ ◊◊◊◊◊ ◊◊◊◊◊ ◊◊◊◊◊ ◊◊◊◊◊ ◊◊◊◊◊	139,800
◊◊◊◊◊ ◊◊◊◊◊ ◊◊◊◊◊ ◊◊◊◊◊ ◊◊◊◊◊ ◊◊◊◊◊ ◊◊◊◊◊ ◊◊◊◊◊ ◊◊◊◊◊ ◊◊◊◊◊	139,850
◊◊◊◊◊ ◊◊◊◊◊ ◊◊◊◊◊ ◊◊◊◊◊ ◊◊◊◊◊ ◊◊◊◊◊ ◊◊◊◊◊ ◊◊◊◊◊ ◊◊◊◊◊ ◊◊◊◊◊	139,900
◊◊◊◊◊ ◊◊◊◊◊ ◊◊◊◊◊ ◊◊◊◊◊ ◊◊◊◊◊ ◊◊◊◊◊ ◊◊◊◊◊ ◊◊◊◊◊ ◊◊◊◊◊ ◊◊◊◊◊	139,950
◊◊◊◊◊ ◊◊◊◊◊ ◊◊◊◊◊ ◊◊◊◊◊ ◊◊◊◊◊ ◊◊◊◊◊ ◊◊◊◊◊ ◊◊◊◊◊ ◊◊◊◊◊ ◊◊◊◊◊	**140,000**

The Number 140,001 to the Number 142,000

Start | | Notes

	140,050
	140,100
	140,150
	140,200
	140,250
	140,300
	140,350
	140,400
	140,450
	140,500
	140,550
	140,600
	140,650
	140,700
	140,750
	140,800
	140,850
	140,900
	140,950
	141,000
	141,050
	141,100
	141,150
	141,200
	141,250
	141,300
	141,350
	141,400
	141,450
	141,500
	141,550
	141,600
	141,650
	141,700
	141,750
	141,800
	141,850
	141,900
	141,950
	142,000

The Number 142,001 to the Number 144,000

Start	Notes
142,050	
142,100	
142,150	
142,200	
142,250	
142,300	
142,350	
142,400	
142,450	
142,500	
142,550	
142,600	
142,650	
142,700	
142,750	
142,800	
142,850	
142,900	
142,950	
143,000	
143,050	
143,100	
143,150	
143,200	
143,250	
143,300	
143,350	
143,400	
143,450	
143,500	
143,550	
143,600	
143,650	
143,700	
143,750	
143,800	
143,850	
143,900	
143,950	
144,000	

The Number 144,001 to the Number 146,000

Start   Notes

	144,050
	144,100
	144,150
	144,200
	144,250
	144,300
	144,350
	144,400
	144,450
	144,500
	144,550
	144,600
	144,650
	144,700
	144,750
	144,800
	144,850
	144,900
	144,950
	145,000
	145,050
	145,100
	145,150
	145,200
	145,250
	145,300
	145,350
	145,400
	145,450
	145,500
	145,550
	145,600
	145,650
	145,700
	145,750
	145,800
	145,850
	145,900
	145,950
	146,000

The Number 146,001 to the Number 148,000

Start   Notes

	146,050
	146,100
	146,150
	146,200
	146,250
	146,300
	146,350
	146,400
	146,450
	146,500
	146,550
	146,600
	146,650
	146,700
	146,750
	146,800
	146,850
	146,900
	146,950
	147,000
	147,050
	147,100
	147,150
	147,200
	147,250
	147,300
	147,350
	147,400
	147,450
	147,500
	147,550
	147,600
	147,650
	147,700
	147,750
	147,800
	147,850
	147,900
	147,950
	148,000

The Number 148,001 to the Number 150,000

Start											Notes
◇◇◇◇◇ ◇◇◇◇◇ ◇◇◇◇◇ ◇◇◇◇◇ ◇◇◇◇◇ ◇◇◇◇◇ ◇◇◇◇◇ ◇◇◇◇◇ ◇◇◇◇◇ ◇◇◇◇◇	148,050
◇◇◇◇◇ ◇◇◇◇◇ ◇◇◇◇◇ ◇◇◇◇◇ ◇◇◇◇◇ ◇◇◇◇◇ ◇◇◇◇◇ ◇◇◇◇◇ ◇◇◇◇◇ ◇◇◇◇◇	148,100
◇◇◇◇◇ ◇◇◇◇◇ ◇◇◇◇◇ ◇◇◇◇◇ ◇◇◇◇◇ ◇◇◇◇◇ ◇◇◇◇◇ ◇◇◇◇◇ ◇◇◇◇◇ ◇◇◇◇◇	148,150
◇◇◇◇◇ ◇◇◇◇◇ ◇◇◇◇◇ ◇◇◇◇◇ ◇◇◇◇◇ ◇◇◇◇◇ ◇◇◇◇◇ ◇◇◇◇◇ ◇◇◇◇◇ ◇◇◇◇◇	148,200
◇◇◇◇◇ ◇◇◇◇◇ ◇◇◇◇◇ ◇◇◇◇◇ ◇◇◇◇◇ ◇◇◇◇◇ ◇◇◇◇◇ ◇◇◇◇◇ ◇◇◇◇◇ ◇◇◇◇◇	148,250
◇◇◇◇◇ ◇◇◇◇◇ ◇◇◇◇◇ ◇◇◇◇◇ ◇◇◇◇◇ ◇◇◇◇◇ ◇◇◇◇◇ ◇◇◇◇◇ ◇◇◇◇◇ ◇◇◇◇◇	148,300
◇◇◇◇◇ ◇◇◇◇◇ ◇◇◇◇◇ ◇◇◇◇◇ ◇◇◇◇◇ ◇◇◇◇◇ ◇◇◇◇◇ ◇◇◇◇◇ ◇◇◇◇◇ ◇◇◇◇◇	148,350
◇◇◇◇◇ ◇◇◇◇◇ ◇◇◇◇◇ ◇◇◇◇◇ ◇◇◇◇◇ ◇◇◇◇◇ ◇◇◇◇◇ ◇◇◇◇◇ ◇◇◇◇◇ ◇◇◇◇◇	148,400
◇◇◇◇◇ ◇◇◇◇◇ ◇◇◇◇◇ ◇◇◇◇◇ ◇◇◇◇◇ ◇◇◇◇◇ ◇◇◇◇◇ ◇◇◇◇◇ ◇◇◇◇◇ ◇◇◇◇◇	148,450
◇◇◇◇◇ ◇◇◇◇◇ ◇◇◇◇◇ ◇◇◇◇◇ ◇◇◇◇◇ ◇◇◇◇◇ ◇◇◇◇◇ ◇◇◇◇◇ ◇◇◇◇◇ ◇◇◇◇◇	148,500
◇◇◇◇◇ ◇◇◇◇◇ ◇◇◇◇◇ ◇◇◇◇◇ ◇◇◇◇◇ ◇◇◇◇◇ ◇◇◇◇◇ ◇◇◇◇◇ ◇◇◇◇◇ ◇◇◇◇◇	148,550
◇◇◇◇◇ ◇◇◇◇◇ ◇◇◇◇◇ ◇◇◇◇◇ ◇◇◇◇◇ ◇◇◇◇◇ ◇◇◇◇◇ ◇◇◇◇◇ ◇◇◇◇◇ ◇◇◇◇◇	148,600
◇◇◇◇◇ ◇◇◇◇◇ ◇◇◇◇◇ ◇◇◇◇◇ ◇◇◇◇◇ ◇◇◇◇◇ ◇◇◇◇◇ ◇◇◇◇◇ ◇◇◇◇◇ ◇◇◇◇◇	148,650
◇◇◇◇◇ ◇◇◇◇◇ ◇◇◇◇◇ ◇◇◇◇◇ ◇◇◇◇◇ ◇◇◇◇◇ ◇◇◇◇◇ ◇◇◇◇◇ ◇◇◇◇◇ ◇◇◇◇◇	148,700
◇◇◇◇◇ ◇◇◇◇◇ ◇◇◇◇◇ ◇◇◇◇◇ ◇◇◇◇◇ ◇◇◇◇◇ ◇◇◇◇◇ ◇◇◇◇◇ ◇◇◇◇◇ ◇◇◇◇◇	148,750
◇◇◇◇◇ ◇◇◇◇◇ ◇◇◇◇◇ ◇◇◇◇◇ ◇◇◇◇◇ ◇◇◇◇◇ ◇◇◇◇◇ ◇◇◇◇◇ ◇◇◇◇◇ ◇◇◇◇◇	148,800
◇◇◇◇◇ ◇◇◇◇◇ ◇◇◇◇◇ ◇◇◇◇◇ ◇◇◇◇◇ ◇◇◇◇◇ ◇◇◇◇◇ ◇◇◇◇◇ ◇◇◇◇◇ ◇◇◇◇◇	148,850
◇◇◇◇◇ ◇◇◇◇◇ ◇◇◇◇◇ ◇◇◇◇◇ ◇◇◇◇◇ ◇◇◇◇◇ ◇◇◇◇◇ ◇◇◇◇◇ ◇◇◇◇◇ ◇◇◇◇◇	148,900
◇◇◇◇◇ ◇◇◇◇◇ ◇◇◇◇◇ ◇◇◇◇◇ ◇◇◇◇◇ ◇◇◇◇◇ ◇◇◇◇◇ ◇◇◇◇◇ ◇◇◇◇◇ ◇◇◇◇◇	148,950
◇◇◇◇◇ ◇◇◇◇◇ ◇◇◇◇◇ ◇◇◇◇◇ ◇◇◇◇◇ ◇◇◇◇◇ ◇◇◇◇◇ ◇◇◇◇◇ ◇◇◇◇◇ ◇◇◇◇◇	**149,000**
◇◇◇◇◇ ◇◇◇◇◇ ◇◇◇◇◇ ◇◇◇◇◇ ◇◇◇◇◇ ◇◇◇◇◇ ◇◇◇◇◇ ◇◇◇◇◇ ◇◇◇◇◇ ◇◇◇◇◇	149,050
◇◇◇◇◇ ◇◇◇◇◇ ◇◇◇◇◇ ◇◇◇◇◇ ◇◇◇◇◇ ◇◇◇◇◇ ◇◇◇◇◇ ◇◇◇◇◇ ◇◇◇◇◇ ◇◇◇◇◇	149,100
◇◇◇◇◇ ◇◇◇◇◇ ◇◇◇◇◇ ◇◇◇◇◇ ◇◇◇◇◇ ◇◇◇◇◇ ◇◇◇◇◇ ◇◇◇◇◇ ◇◇◇◇◇ ◇◇◇◇◇	149,150
◇◇◇◇◇ ◇◇◇◇◇ ◇◇◇◇◇ ◇◇◇◇◇ ◇◇◇◇◇ ◇◇◇◇◇ ◇◇◇◇◇ ◇◇◇◇◇ ◇◇◇◇◇ ◇◇◇◇◇	149,200
◇◇◇◇◇ ◇◇◇◇◇ ◇◇◇◇◇ ◇◇◇◇◇ ◇◇◇◇◇ ◇◇◇◇◇ ◇◇◇◇◇ ◇◇◇◇◇ ◇◇◇◇◇ ◇◇◇◇◇	149,250
◇◇◇◇◇ ◇◇◇◇◇ ◇◇◇◇◇ ◇◇◇◇◇ ◇◇◇◇◇ ◇◇◇◇◇ ◇◇◇◇◇ ◇◇◇◇◇ ◇◇◇◇◇ ◇◇◇◇◇	149,300
◇◇◇◇◇ ◇◇◇◇◇ ◇◇◇◇◇ ◇◇◇◇◇ ◇◇◇◇◇ ◇◇◇◇◇ ◇◇◇◇◇ ◇◇◇◇◇ ◇◇◇◇◇ ◇◇◇◇◇	149,350
◇◇◇◇◇ ◇◇◇◇◇ ◇◇◇◇◇ ◇◇◇◇◇ ◇◇◇◇◇ ◇◇◇◇◇ ◇◇◇◇◇ ◇◇◇◇◇ ◇◇◇◇◇ ◇◇◇◇◇	149,400
◇◇◇◇◇ ◇◇◇◇◇ ◇◇◇◇◇ ◇◇◇◇◇ ◇◇◇◇◇ ◇◇◇◇◇ ◇◇◇◇◇ ◇◇◇◇◇ ◇◇◇◇◇ ◇◇◇◇◇	149,450
◇◇◇◇◇ ◇◇◇◇◇ ◇◇◇◇◇ ◇◇◇◇◇ ◇◇◇◇◇ ◇◇◇◇◇ ◇◇◇◇◇ ◇◇◇◇◇ ◇◇◇◇◇ ◇◇◇◇◇	149,500
◇◇◇◇◇ ◇◇◇◇◇ ◇◇◇◇◇ ◇◇◇◇◇ ◇◇◇◇◇ ◇◇◇◇◇ ◇◇◇◇◇ ◇◇◇◇◇ ◇◇◇◇◇ ◇◇◇◇◇	149,550
◇◇◇◇◇ ◇◇◇◇◇ ◇◇◇◇◇ ◇◇◇◇◇ ◇◇◇◇◇ ◇◇◇◇◇ ◇◇◇◇◇ ◇◇◇◇◇ ◇◇◇◇◇ ◇◇◇◇◇	149,600
◇◇◇◇◇ ◇◇◇◇◇ ◇◇◇◇◇ ◇◇◇◇◇ ◇◇◇◇◇ ◇◇◇◇◇ ◇◇◇◇◇ ◇◇◇◇◇ ◇◇◇◇◇ ◇◇◇◇◇	149,650
◇◇◇◇◇ ◇◇◇◇◇ ◇◇◇◇◇ ◇◇◇◇◇ ◇◇◇◇◇ ◇◇◇◇◇ ◇◇◇◇◇ ◇◇◇◇◇ ◇◇◇◇◇ ◇◇◇◇◇	149,700
◇◇◇◇◇ ◇◇◇◇◇ ◇◇◇◇◇ ◇◇◇◇◇ ◇◇◇◇◇ ◇◇◇◇◇ ◇◇◇◇◇ ◇◇◇◇◇ ◇◇◇◇◇ ◇◇◇◇◇	149,750
◇◇◇◇◇ ◇◇◇◇◇ ◇◇◇◇◇ ◇◇◇◇◇ ◇◇◇◇◇ ◇◇◇◇◇ ◇◇◇◇◇ ◇◇◇◇◇ ◇◇◇◇◇ ◇◇◇◇◇	149,800
◇◇◇◇◇ ◇◇◇◇◇ ◇◇◇◇◇ ◇◇◇◇◇ ◇◇◇◇◇ ◇◇◇◇◇ ◇◇◇◇◇ ◇◇◇◇◇ ◇◇◇◇◇ ◇◇◇◇◇	149,850
◇◇◇◇◇ ◇◇◇◇◇ ◇◇◇◇◇ ◇◇◇◇◇ ◇◇◇◇◇ ◇◇◇◇◇ ◇◇◇◇◇ ◇◇◇◇◇ ◇◇◇◇◇ ◇◇◇◇◇	149,900
◇◇◇◇◇ ◇◇◇◇◇ ◇◇◇◇◇ ◇◇◇◇◇ ◇◇◇◇◇ ◇◇◇◇◇ ◇◇◇◇◇ ◇◇◇◇◇ ◇◇◇◇◇ ◇◇◇◇◇	149,950
◇◇◇◇◇ ◇◇◇◇◇ ◇◇◇◇◇ ◇◇◇◇◇ ◇◇◇◇◇ ◇◇◇◇◇ ◇◇◇◇◇ ◇◇◇◇◇ ◇◇◇◇◇ ◇◇◇◇◇	**150,000**

◇ The Number 150,001 to the Number 152,000

Start | | Notes

150,050
150,100
150,150
150,200
150,250
150,300
150,350
150,400
150,450
150,500
150,550
150,600
150,650
150,700
150,750
150,800
150,850
150,900
150,950
151,000
151,050
151,100
151,150
151,200
151,250
151,300
151,350
151,400
151,450
151,500
151,550
151,600
151,650
151,700
151,750
151,800
151,850
151,900
151,950
152,000

◇ The Number 152,001 to the Number 154,000

Start		Notes
◇◇◇◇◇ ◇◇◇◇◇ ◇◇◇◇◇ ◇◇◇◇◇ ◇◇◇◇◇ ◇◇◇◇◇ ◇◇◇◇◇ ◇◇◇◇◇ ◇◇◇◇◇ ◇◇◇◇◇	152,050
◇◇◇◇◇ ◇◇◇◇◇ ◇◇◇◇◇ ◇◇◇◇◇ ◇◇◇◇◇ ◇◇◇◇◇ ◇◇◇◇◇ ◇◇◇◇◇ ◇◇◇◇◇ ◇◇◇◇◇	152,100
◇◇◇◇◇ ◇◇◇◇◇ ◇◇◇◇◇ ◇◇◇◇◇ ◇◇◇◇◇ ◇◇◇◇◇ ◇◇◇◇◇ ◇◇◇◇◇ ◇◇◇◇◇ ◇◇◇◇◇	152,150
◇◇◇◇◇ ◇◇◇◇◇ ◇◇◇◇◇ ◇◇◇◇◇ ◇◇◇◇◇ ◇◇◇◇◇ ◇◇◇◇◇ ◇◇◇◇◇ ◇◇◇◇◇ ◇◇◇◇◇	152,200
◇◇◇◇◇ ◇◇◇◇◇ ◇◇◇◇◇ ◇◇◇◇◇ ◇◇◇◇◇ ◇◇◇◇◇ ◇◇◇◇◇ ◇◇◇◇◇ ◇◇◇◇◇ ◇◇◇◇◇	152,250
◇◇◇◇◇ ◇◇◇◇◇ ◇◇◇◇◇ ◇◇◇◇◇ ◇◇◇◇◇ ◇◇◇◇◇ ◇◇◇◇◇ ◇◇◇◇◇ ◇◇◇◇◇ ◇◇◇◇◇	152,300
◇◇◇◇◇ ◇◇◇◇◇ ◇◇◇◇◇ ◇◇◇◇◇ ◇◇◇◇◇ ◇◇◇◇◇ ◇◇◇◇◇ ◇◇◇◇◇ ◇◇◇◇◇ ◇◇◇◇◇	152,350
◇◇◇◇◇ ◇◇◇◇◇ ◇◇◇◇◇ ◇◇◇◇◇ ◇◇◇◇◇ ◇◇◇◇◇ ◇◇◇◇◇ ◇◇◇◇◇ ◇◇◇◇◇ ◇◇◇◇◇	152,400
◇◇◇◇◇ ◇◇◇◇◇ ◇◇◇◇◇ ◇◇◇◇◇ ◇◇◇◇◇ ◇◇◇◇◇ ◇◇◇◇◇ ◇◇◇◇◇ ◇◇◇◇◇ ◇◇◇◇◇	152,450
◇◇◇◇◇ ◇◇◇◇◇ ◇◇◇◇◇ ◇◇◇◇◇ ◇◇◇◇◇ ◇◇◇◇◇ ◇◇◇◇◇ ◇◇◇◇◇ ◇◇◇◇◇ ◇◇◇◇◇	152,500
◇◇◇◇◇ ◇◇◇◇◇ ◇◇◇◇◇ ◇◇◇◇◇ ◇◇◇◇◇ ◇◇◇◇◇ ◇◇◇◇◇ ◇◇◇◇◇ ◇◇◇◇◇ ◇◇◇◇◇	152,550
◇◇◇◇◇ ◇◇◇◇◇ ◇◇◇◇◇ ◇◇◇◇◇ ◇◇◇◇◇ ◇◇◇◇◇ ◇◇◇◇◇ ◇◇◇◇◇ ◇◇◇◇◇ ◇◇◇◇◇	152,600
◇◇◇◇◇ ◇◇◇◇◇ ◇◇◇◇◇ ◇◇◇◇◇ ◇◇◇◇◇ ◇◇◇◇◇ ◇◇◇◇◇ ◇◇◇◇◇ ◇◇◇◇◇ ◇◇◇◇◇	152,650
◇◇◇◇◇ ◇◇◇◇◇ ◇◇◇◇◇ ◇◇◇◇◇ ◇◇◇◇◇ ◇◇◇◇◇ ◇◇◇◇◇ ◇◇◇◇◇ ◇◇◇◇◇ ◇◇◇◇◇	152,700
◇◇◇◇◇ ◇◇◇◇◇ ◇◇◇◇◇ ◇◇◇◇◇ ◇◇◇◇◇ ◇◇◇◇◇ ◇◇◇◇◇ ◇◇◇◇◇ ◇◇◇◇◇ ◇◇◇◇◇	152,750
◇◇◇◇◇ ◇◇◇◇◇ ◇◇◇◇◇ ◇◇◇◇◇ ◇◇◇◇◇ ◇◇◇◇◇ ◇◇◇◇◇ ◇◇◇◇◇ ◇◇◇◇◇ ◇◇◇◇◇	152,800
◇◇◇◇◇ ◇◇◇◇◇ ◇◇◇◇◇ ◇◇◇◇◇ ◇◇◇◇◇ ◇◇◇◇◇ ◇◇◇◇◇ ◇◇◇◇◇ ◇◇◇◇◇ ◇◇◇◇◇	152,850
◇◇◇◇◇ ◇◇◇◇◇ ◇◇◇◇◇ ◇◇◇◇◇ ◇◇◇◇◇ ◇◇◇◇◇ ◇◇◇◇◇ ◇◇◇◇◇ ◇◇◇◇◇ ◇◇◇◇◇	152,900
◇◇◇◇◇ ◇◇◇◇◇ ◇◇◇◇◇ ◇◇◇◇◇ ◇◇◇◇◇ ◇◇◇◇◇ ◇◇◇◇◇ ◇◇◇◇◇ ◇◇◇◇◇ ◇◇◇◇◇	152,950
◇◇◇◇◇ ◇◇◇◇◇ ◇◇◇◇◇ ◇◇◇◇◇ ◇◇◇◇◇ ◇◇◇◇◇ ◇◇◇◇◇ ◇◇◇◇◇ ◇◇◇◇◇ ◇◇◇◇◇	**153,000**
◇◇◇◇◇ ◇◇◇◇◇ ◇◇◇◇◇ ◇◇◇◇◇ ◇◇◇◇◇ ◇◇◇◇◇ ◇◇◇◇◇ ◇◇◇◇◇ ◇◇◇◇◇ ◇◇◇◇◇	153,050
◇◇◇◇◇ ◇◇◇◇◇ ◇◇◇◇◇ ◇◇◇◇◇ ◇◇◇◇◇ ◇◇◇◇◇ ◇◇◇◇◇ ◇◇◇◇◇ ◇◇◇◇◇ ◇◇◇◇◇	153,100
◇◇◇◇◇ ◇◇◇◇◇ ◇◇◇◇◇ ◇◇◇◇◇ ◇◇◇◇◇ ◇◇◇◇◇ ◇◇◇◇◇ ◇◇◇◇◇ ◇◇◇◇◇ ◇◇◇◇◇	153,150
◇◇◇◇◇ ◇◇◇◇◇ ◇◇◇◇◇ ◇◇◇◇◇ ◇◇◇◇◇ ◇◇◇◇◇ ◇◇◇◇◇ ◇◇◇◇◇ ◇◇◇◇◇ ◇◇◇◇◇	153,200
◇◇◇◇◇ ◇◇◇◇◇ ◇◇◇◇◇ ◇◇◇◇◇ ◇◇◇◇◇ ◇◇◇◇◇ ◇◇◇◇◇ ◇◇◇◇◇ ◇◇◇◇◇ ◇◇◇◇◇	153,250
◇◇◇◇◇ ◇◇◇◇◇ ◇◇◇◇◇ ◇◇◇◇◇ ◇◇◇◇◇ ◇◇◇◇◇ ◇◇◇◇◇ ◇◇◇◇◇ ◇◇◇◇◇ ◇◇◇◇◇	153,300
◇◇◇◇◇ ◇◇◇◇◇ ◇◇◇◇◇ ◇◇◇◇◇ ◇◇◇◇◇ ◇◇◇◇◇ ◇◇◇◇◇ ◇◇◇◇◇ ◇◇◇◇◇ ◇◇◇◇◇	153,350
◇◇◇◇◇ ◇◇◇◇◇ ◇◇◇◇◇ ◇◇◇◇◇ ◇◇◇◇◇ ◇◇◇◇◇ ◇◇◇◇◇ ◇◇◇◇◇ ◇◇◇◇◇ ◇◇◇◇◇	153,400
◇◇◇◇◇ ◇◇◇◇◇ ◇◇◇◇◇ ◇◇◇◇◇ ◇◇◇◇◇ ◇◇◇◇◇ ◇◇◇◇◇ ◇◇◇◇◇ ◇◇◇◇◇ ◇◇◇◇◇	153,450
◇◇◇◇◇ ◇◇◇◇◇ ◇◇◇◇◇ ◇◇◇◇◇ ◇◇◇◇◇ ◇◇◇◇◇ ◇◇◇◇◇ ◇◇◇◇◇ ◇◇◇◇◇ ◇◇◇◇◇	153,500
◇◇◇◇◇ ◇◇◇◇◇ ◇◇◇◇◇ ◇◇◇◇◇ ◇◇◇◇◇ ◇◇◇◇◇ ◇◇◇◇◇ ◇◇◇◇◇ ◇◇◇◇◇ ◇◇◇◇◇	153,550
◇◇◇◇◇ ◇◇◇◇◇ ◇◇◇◇◇ ◇◇◇◇◇ ◇◇◇◇◇ ◇◇◇◇◇ ◇◇◇◇◇ ◇◇◇◇◇ ◇◇◇◇◇ ◇◇◇◇◇	153,600
◇◇◇◇◇ ◇◇◇◇◇ ◇◇◇◇◇ ◇◇◇◇◇ ◇◇◇◇◇ ◇◇◇◇◇ ◇◇◇◇◇ ◇◇◇◇◇ ◇◇◇◇◇ ◇◇◇◇◇	153,650
◇◇◇◇◇ ◇◇◇◇◇ ◇◇◇◇◇ ◇◇◇◇◇ ◇◇◇◇◇ ◇◇◇◇◇ ◇◇◇◇◇ ◇◇◇◇◇ ◇◇◇◇◇ ◇◇◇◇◇	153,700
◇◇◇◇◇ ◇◇◇◇◇ ◇◇◇◇◇ ◇◇◇◇◇ ◇◇◇◇◇ ◇◇◇◇◇ ◇◇◇◇◇ ◇◇◇◇◇ ◇◇◇◇◇ ◇◇◇◇◇	153,750
◇◇◇◇◇ ◇◇◇◇◇ ◇◇◇◇◇ ◇◇◇◇◇ ◇◇◇◇◇ ◇◇◇◇◇ ◇◇◇◇◇ ◇◇◇◇◇ ◇◇◇◇◇ ◇◇◇◇◇	153,800
◇◇◇◇◇ ◇◇◇◇◇ ◇◇◇◇◇ ◇◇◇◇◇ ◇◇◇◇◇ ◇◇◇◇◇ ◇◇◇◇◇ ◇◇◇◇◇ ◇◇◇◇◇ ◇◇◇◇◇	153,850
◇◇◇◇◇ ◇◇◇◇◇ ◇◇◇◇◇ ◇◇◇◇◇ ◇◇◇◇◇ ◇◇◇◇◇ ◇◇◇◇◇ ◇◇◇◇◇ ◇◇◇◇◇ ◇◇◇◇◇	153,900
◇◇◇◇◇ ◇◇◇◇◇ ◇◇◇◇◇ ◇◇◇◇◇ ◇◇◇◇◇ ◇◇◇◇◇ ◇◇◇◇◇ ◇◇◇◇◇ ◇◇◇◇◇ ◇◇◇◇◇	153,950
◇◇◇◇◇ ◇◇◇◇◇ ◇◇◇◇◇ ◇◇◇◇◇ ◇◇◇◇◇ ◇◇◇◇◇ ◇◇◇◇◇ ◇◇◇◇◇ ◇◇◇◇◇ ◇◇◇◇◇	**154,000**

The Number 154,001 to the Number 156,000

Start		Notes
	154,050	
	154,100	
	154,150	
	154,200	
	154,250	
	154,300	
	154,350	
	154,400	
	154,450	
	154,500	
	154,550	
	154,600	
	154,650	
	154,700	
	154,750	
	154,800	
	154,850	
	154,900	
	154,950	
	155,000	
	155,050	
	155,100	
	155,150	
	155,200	
	155,250	
	155,300	
	155,350	
	155,400	
	155,450	
	155,500	
	155,550	
	155,600	
	155,650	
	155,700	
	155,750	
	155,800	
	155,850	
	155,900	
	155,950	
	156,000	

The Number 156,001 to the Number 158,000

Start		Notes
	156,050
	156,100
	156,150
	156,200
	156,250
	156,300
	156,350
	156,400
	156,450
	156,500
	156,550
	156,600
	156,650
	156,700
	156,750
	156,800
	156,850
	156,900
	156,950
	157,000
	157,050
	157,100
	157,150
	157,200
	157,250
	157,300
	157,350
	157,400
	157,450
	157,500
	157,550
	157,600
	157,650
	157,700
	157,750
	157,800
	157,850
	157,900
	157,950
	158,000

The Number 158,001 to the Number 160,000

	Start		Notes
		158,050	
		158,100	
		158,150	
		158,200	
		158,250	
		158,300	
		158,350	
		158,400	
		158,450	
		158,500	
		158,550	
		158,600	
		158,650	
		158,700	
		158,750	
		158,800	
		158,850	
		158,900	
		158,950	
		159,000	
		159,050	
		159,100	
		159,150	
		159,200	
		159,250	
		159,300	
		159,350	
		159,400	
		159,450	
		159,500	
		159,550	
		159,600	
		159,650	
		159,700	
		159,750	
		159,800	
		159,850	
		159,900	
		159,950	
		160,000	

◊ The Number 160,001 to the Number 162,000

Start   Notes

	160,050	
	160,100	
	160,150	
	160,200	
	160,250	
	160,300	
	160,350	
	160,400	
	160,450	
	160,500	
	160,550	
	160,600	
	160,650	
	160,700	
	160,750	
	160,800	
	160,850	
	160,900	
	160,950	
	161,000	
	161,050	
	161,100	
	161,150	
	161,200	
	161,250	
	161,300	
	161,350	
	161,400	
	161,450	
	161,500	
	161,550	
	161,600	
	161,650	
	161,700	
	161,750	
	161,800	
	161,850	
	161,900	
	161,950	
	162,000	

The Number 162,001 to the Number 164,000

Start		Notes
	162,050	
	162,100	
	162,150	
	162,200	
	162,250	
	162,300	
	162,350	
	162,400	
	162,450	
	162,500	
	162,550	
	162,600	
	162,650	
	162,700	
	162,750	
	162,800	
	162,850	
	162,900	
	162,950	
	163,000	
	163,050	
	163,100	
	163,150	
	163,200	
	163,250	
	163,300	
	163,350	
	163,400	
	163,450	
	163,500	
	163,550	
	163,600	
	163,650	
	163,700	
	163,750	
	163,800	
	163,850	
	163,900	
	163,950	
	164,000	

The Number 164,001 to the Number 166,000

Start | | | | | | | | | | | Notes
| 164,050
| 164,100
| 164,150
| 164,200
| 164,250
| 164,300
| 164,350
| 164,400
| 164,450
| 164,500
| 164,550
| 164,600
| 164,650
| 164,700
| 164,750
| 164,800
| 164,850
| 164,900
| 164,950
| **165,000**
| 165,050
| 165,100
| 165,150
| 165,200
| 165,250
| 165,300
| 165,350
| 165,400
| 165,450
| 165,500
| 165,550
| 165,600
| 165,650
| 165,700
| 165,750
| 165,800
| 165,850
| 165,900
| 165,950
| **166,000**

The Number 166,001 to the Number 168,000

Start

	Notes
166,050	
166,100	
166,150	
166,200	
166,250	
166,300	
166,350	
166,400	
166,450	
166,500	
166,550	
166,600	
166,650	
166,700	
166,750	
166,800	
166,850	
166,900	
166,950	
167,000	
167,050	
167,100	
167,150	
167,200	
167,250	
167,300	
167,350	
167,400	
167,450	
167,500	
167,550	
167,600	
167,650	
167,700	
167,750	
167,800	
167,850	
167,900	
167,950	
168,000	

The Number 168,001 to the Number 170,000

Start   Notes

	168,050
	168,100
	168,150
	168,200
	168,250
	168,300
	168,350
	168,400
	168,450
	168,500
	168,550
	168,600
	168,650
	168,700
	168,750
	168,800
	168,850
	168,900
	168,950
	169,000
	169,050
	169,100
	169,150
	169,200
	169,250
	169,300
	169,350
	169,400
	169,450
	169,500
	169,550
	169,600
	169,650
	169,700
	169,750
	169,800
	169,850
	169,900
	169,950
	170,000

The Number 170,001 to the Number 172,000

Start	Notes
	170,050
	170,100
	170,150
	170,200
	170,250
	170,300
	170,350
	170,400
	170,450
	170,500
	170,550
	170,600
	170,650
	170,700
	170,750
	170,800
	170,850
	170,900
	170,950
	171,000
	171,050
	171,100
	171,150
	171,200
	171,250
	171,300
	171,350
	171,400
	171,450
	171,500
	171,550
	171,600
	171,650
	171,700
	171,750
	171,800
	171,850
	171,900
	171,950
	172,000

The Number 172,001 to the Number 174,000

Start

	Notes
172,050	
172,100	
172,150	
172,200	
172,250	
172,300	
172,350	
172,400	
172,450	
172,500	
172,550	
172,600	
172,650	
172,700	
172,750	
172,800	
172,850	
172,900	
172,950	
173,000	
173,050	
173,100	
173,150	
173,200	
173,250	
173,300	
173,350	
173,400	
173,450	
173,500	
173,550	
173,600	
173,650	
173,700	
173,750	
173,800	
173,850	
173,900	
173,950	
174,000	

The Number 174,001 to the Number 176,000

Start | | Notes

Number
174,050
174,100
174,150
174,200
174,250
174,300
174,350
174,400
174,450
174,500
174,550
174,600
174,650
174,700
174,750
174,800
174,850
174,900
174,950
175,000
175,050
175,100
175,150
175,200
175,250
175,300
175,350
175,400
175,450
175,500
175,550
175,600
175,650
175,700
175,750
175,800
175,850
175,900
175,950
176,000

The Number 176,001 to the Number 178,000

Start

Notes

	176,050
	176,100
	176,150
	176,200
	176,250
	176,300
	176,350
	176,400
	176,450
	176,500
	176,550
	176,600
	176,650
	176,700
	176,750
	176,800
	176,850
	176,900
	176,950
	177,000
	177,050
	177,100
	177,150
	177,200
	177,250
	177,300
	177,350
	177,400
	177,450
	177,500
	177,550
	177,600
	177,650
	177,700
	177,750
	177,800
	177,850
	177,900
	177,950
	178,000

The Number 178,001 to the Number 180,000

Start   Notes

	178,050
	178,100
	178,150
	178,200
	178,250
	178,300
	178,350
	178,400
	178,450
	178,500
	178,550
	178,600
	178,650
	178,700
	178,750
	178,800
	178,850
	178,900
	178,950
	179,000
	179,050
	179,100
	179,150
	179,200
	179,250
	179,300
	179,350
	179,400
	179,450
	179,500
	179,550
	179,600
	179,650
	179,700
	179,750
	179,800
	179,850
	179,900
	179,950
	180,000

The Number 180,001 to the Number 182,000

Start   Notes

	180,050
	180,100
	180,150
	180,200
	180,250
	180,300
	180,350
	180,400
	180,450
	180,500
	180,550
	180,600
	180,650
	180,700
	180,750
	180,800
	180,850
	180,900
	180,950
	181,000
	181,050
	181,100
	181,150
	181,200
	181,250
	181,300
	181,350
	181,400
	181,450
	181,500
	181,550
	181,600
	181,650
	181,700
	181,750
	181,800
	181,850
	181,900
	181,950
	182,000

The Number 182,001 to the Number 184,000

Start | Notes

	182,050
	182,100
	182,150
	182,200
	182,250
	182,300
	182,350
	182,400
	182,450
	182,500
	182,550
	182,600
	182,650
	182,700
	182,750
	182,800
	182,850
	182,900
	182,950
	183,000
	183,050
	183,100
	183,150
	183,200
	183,250
	183,300
	183,350
	183,400
	183,450
	183,500
	183,550
	183,600
	183,650
	183,700
	183,750
	183,800
	183,850
	183,900
	183,950
	184,000

The Number 184,001 to the Number 186,000

Start

Notes

	184,050
	184,100
	184,150
	184,200
	184,250
	184,300
	184,350
	184,400
	184,450
	184,500
	184,550
	184,600
	184,650
	184,700
	184,750
	184,800
	184,850
	184,900
	184,950
	185,000
	185,050
	185,100
	185,150
	185,200
	185,250
	185,300
	185,350
	185,400
	185,450
	185,500
	185,550
	185,600
	185,650
	185,700
	185,750
	185,800
	185,850
	185,900
	185,950
	186,000

The Number 186,001 to the Number 188,000

Start	Notes
186,050	
186,100	
186,150	
186,200	
186,250	
186,300	
186,350	
186,400	
186,450	
186,500	
186,550	
186,600	
186,650	
186,700	
186,750	
186,800	
186,850	
186,900	
186,950	
187,000	
187,050	
187,100	
187,150	
187,200	
187,250	
187,300	
187,350	
187,400	
187,450	
187,500	
187,550	
187,600	
187,650	
187,700	
187,750	
187,800	
187,850	
187,900	
187,950	
188,000	

The Number 188,001 to the Number 190,000

Start		Notes
	188,050	
	188,100	
	188,150	
	188,200	
	188,250	
	188,300	
	188,350	
	188,400	
	188,450	
	188,500	
	188,550	
	188,600	
	188,650	
	188,700	
	188,750	
	188,800	
	188,850	
	188,900	
	188,950	
	189,000	
	189,050	
	189,100	
	189,150	
	189,200	
	189,250	
	189,300	
	189,350	
	189,400	
	189,450	
	189,500	
	189,550	
	189,600	
	189,650	
	189,700	
	189,750	
	189,800	
	189,850	
	189,900	
	189,950	
	190,000	

The Number 190,001 to the Number 192,000

Start

Notes

	190,050
	190,100
	190,150
	190,200
	190,250
	190,300
	190,350
	190,400
	190,450
	190,500
	190,550
	190,600
	190,650
	190,700
	190,750
	190,800
	190,850
	190,900
	190,950
	191,000
	191,050
	191,100
	191,150
	191,200
	191,250
	191,300
	191,350
	191,400
	191,450
	191,500
	191,550
	191,600
	191,650
	191,700
	191,750
	191,800
	191,850
	191,900
	191,950
	192,000

The Number 192,001 to the Number 194,000

Start   Notes

	192,050
	192,100
	192,150
	192,200
	192,250
	192,300
	192,350
	192,400
	192,450
	192,500
	192,550
	192,600
	192,650
	192,700
	192,750
	192,800
	192,850
	192,900
	192,950
	193,000
	193,050
	193,100
	193,150
	193,200
	193,250
	193,300
	193,350
	193,400
	193,450
	193,500
	193,550
	193,600
	193,650
	193,700
	193,750
	193,800
	193,850
	193,900
	193,950
	194,000

The Number 194,001 to the Number 196,000

Start	Notes
	194,050
	194,100
	194,150
	194,200
	194,250
	194,300
	194,350
	194,400
	194,450
	194,500
	194,550
	194,600
	194,650
	194,700
	194,750
	194,800
	194,850
	194,900
	194,950
	195,000
	195,050
	195,100
	195,150
	195,200
	195,250
	195,300
	195,350
	195,400
	195,450
	195,500
	195,550
	195,600
	195,650
	195,700
	195,750
	195,800
	195,850
	195,900
	195,950
	196,000

The Number 196,001 to the Number 198,000

Start

	Notes
196,050	
196,100	
196,150	
196,200	
196,250	
196,300	
196,350	
196,400	
196,450	
196,500	
196,550	
196,600	
196,650	
196,700	
196,750	
196,800	
196,850	
196,900	
196,950	
197,000	
197,050	
197,100	
197,150	
197,200	
197,250	
197,300	
197,350	
197,400	
197,450	
197,500	
197,550	
197,600	
197,650	
197,700	
197,750	
197,800	
197,850	
197,900	
197,950	
198,000	

The Number 198,001 to the Number 200,000

Start

Notes

	198,050
	198,100
	198,150
	198,200
	198,250
	198,300
	198,350
	198,400
	198,450
	198,500
	198,550
	198,600
	198,650
	198,700
	198,750
	198,800
	198,850
	198,900
	198,950
	199,000
	199,050
	199,100
	199,150
	199,200
	199,250
	199,300
	199,350
	199,400
	199,450
	199,500
	199,550
	199,600
	199,650
	199,700
	199,750
	199,800
	199,850
	199,900
	199,950
	200,000

The Number 200,001 to the Number 202,000

Start   Notes

◇◇◇◇◇ ◇◇◇◇◇ ◇◇◇◇◇ ◇◇◇◇◇ ◇◇◇◇◇ ◇◇◇◇◇ ◇◇◇◇◇ ◇◇◇◇◇ ◇◇◇◇◇ ◇◇◇◇◇	200,050
◇◇◇◇◇ ◇◇◇◇◇ ◇◇◇◇◇ ◇◇◇◇◇ ◇◇◇◇◇ ◇◇◇◇◇ ◇◇◇◇◇ ◇◇◇◇◇ ◇◇◇◇◇ ◇◇◇◇◇	200,100
◇◇◇◇◇ ◇◇◇◇◇ ◇◇◇◇◇ ◇◇◇◇◇ ◇◇◇◇◇ ◇◇◇◇◇ ◇◇◇◇◇ ◇◇◇◇◇ ◇◇◇◇◇ ◇◇◇◇◇	200,150
◇◇◇◇◇ ◇◇◇◇◇ ◇◇◇◇◇ ◇◇◇◇◇ ◇◇◇◇◇ ◇◇◇◇◇ ◇◇◇◇◇ ◇◇◇◇◇ ◇◇◇◇◇ ◇◇◇◇◇	200,200
◇◇◇◇◇ ◇◇◇◇◇ ◇◇◇◇◇ ◇◇◇◇◇ ◇◇◇◇◇ ◇◇◇◇◇ ◇◇◇◇◇ ◇◇◇◇◇ ◇◇◇◇◇ ◇◇◇◇◇	200,250
◇◇◇◇◇ ◇◇◇◇◇ ◇◇◇◇◇ ◇◇◇◇◇ ◇◇◇◇◇ ◇◇◇◇◇ ◇◇◇◇◇ ◇◇◇◇◇ ◇◇◇◇◇ ◇◇◇◇◇	200,300
◇◇◇◇◇ ◇◇◇◇◇ ◇◇◇◇◇ ◇◇◇◇◇ ◇◇◇◇◇ ◇◇◇◇◇ ◇◇◇◇◇ ◇◇◇◇◇ ◇◇◇◇◇ ◇◇◇◇◇	200,350
◇◇◇◇◇ ◇◇◇◇◇ ◇◇◇◇◇ ◇◇◇◇◇ ◇◇◇◇◇ ◇◇◇◇◇ ◇◇◇◇◇ ◇◇◇◇◇ ◇◇◇◇◇ ◇◇◇◇◇	200,400
◇◇◇◇◇ ◇◇◇◇◇ ◇◇◇◇◇ ◇◇◇◇◇ ◇◇◇◇◇ ◇◇◇◇◇ ◇◇◇◇◇ ◇◇◇◇◇ ◇◇◇◇◇ ◇◇◇◇◇	200,450
◇◇◇◇◇ ◇◇◇◇◇ ◇◇◇◇◇ ◇◇◇◇◇ ◇◇◇◇◇ ◇◇◇◇◇ ◇◇◇◇◇ ◇◇◇◇◇ ◇◇◇◇◇ ◇◇◇◇◇	200,500
◇◇◇◇◇ ◇◇◇◇◇ ◇◇◇◇◇ ◇◇◇◇◇ ◇◇◇◇◇ ◇◇◇◇◇ ◇◇◇◇◇ ◇◇◇◇◇ ◇◇◇◇◇ ◇◇◇◇◇	200,550
◇◇◇◇◇ ◇◇◇◇◇ ◇◇◇◇◇ ◇◇◇◇◇ ◇◇◇◇◇ ◇◇◇◇◇ ◇◇◇◇◇ ◇◇◇◇◇ ◇◇◇◇◇ ◇◇◇◇◇	200,600
◇◇◇◇◇ ◇◇◇◇◇ ◇◇◇◇◇ ◇◇◇◇◇ ◇◇◇◇◇ ◇◇◇◇◇ ◇◇◇◇◇ ◇◇◇◇◇ ◇◇◇◇◇ ◇◇◇◇◇	200,650
◇◇◇◇◇ ◇◇◇◇◇ ◇◇◇◇◇ ◇◇◇◇◇ ◇◇◇◇◇ ◇◇◇◇◇ ◇◇◇◇◇ ◇◇◇◇◇ ◇◇◇◇◇ ◇◇◇◇◇	200,700
◇◇◇◇◇ ◇◇◇◇◇ ◇◇◇◇◇ ◇◇◇◇◇ ◇◇◇◇◇ ◇◇◇◇◇ ◇◇◇◇◇ ◇◇◇◇◇ ◇◇◇◇◇ ◇◇◇◇◇	200,750
◇◇◇◇◇ ◇◇◇◇◇ ◇◇◇◇◇ ◇◇◇◇◇ ◇◇◇◇◇ ◇◇◇◇◇ ◇◇◇◇◇ ◇◇◇◇◇ ◇◇◇◇◇ ◇◇◇◇◇	200,800
◇◇◇◇◇ ◇◇◇◇◇ ◇◇◇◇◇ ◇◇◇◇◇ ◇◇◇◇◇ ◇◇◇◇◇ ◇◇◇◇◇ ◇◇◇◇◇ ◇◇◇◇◇ ◇◇◇◇◇	200,850
◇◇◇◇◇ ◇◇◇◇◇ ◇◇◇◇◇ ◇◇◇◇◇ ◇◇◇◇◇ ◇◇◇◇◇ ◇◇◇◇◇ ◇◇◇◇◇ ◇◇◇◇◇ ◇◇◇◇◇	200,900
◇◇◇◇◇ ◇◇◇◇◇ ◇◇◇◇◇ ◇◇◇◇◇ ◇◇◇◇◇ ◇◇◇◇◇ ◇◇◇◇◇ ◇◇◇◇◇ ◇◇◇◇◇ ◇◇◇◇◇	200,950
◇◇◇◇◇ ◇◇◇◇◇ ◇◇◇◇◇ ◇◇◇◇◇ ◇◇◇◇◇ ◇◇◇◇◇ ◇◇◇◇◇ ◇◇◇◇◇ ◇◇◇◇◇ ◇◇◇◇◇	**201,000**
◇◇◇◇◇ ◇◇◇◇◇ ◇◇◇◇◇ ◇◇◇◇◇ ◇◇◇◇◇ ◇◇◇◇◇ ◇◇◇◇◇ ◇◇◇◇◇ ◇◇◇◇◇ ◇◇◇◇◇	201,050
◇◇◇◇◇ ◇◇◇◇◇ ◇◇◇◇◇ ◇◇◇◇◇ ◇◇◇◇◇ ◇◇◇◇◇ ◇◇◇◇◇ ◇◇◇◇◇ ◇◇◇◇◇ ◇◇◇◇◇	201,100
◇◇◇◇◇ ◇◇◇◇◇ ◇◇◇◇◇ ◇◇◇◇◇ ◇◇◇◇◇ ◇◇◇◇◇ ◇◇◇◇◇ ◇◇◇◇◇ ◇◇◇◇◇ ◇◇◇◇◇	201,150
◇◇◇◇◇ ◇◇◇◇◇ ◇◇◇◇◇ ◇◇◇◇◇ ◇◇◇◇◇ ◇◇◇◇◇ ◇◇◇◇◇ ◇◇◇◇◇ ◇◇◇◇◇ ◇◇◇◇◇	201,200
◇◇◇◇◇ ◇◇◇◇◇ ◇◇◇◇◇ ◇◇◇◇◇ ◇◇◇◇◇ ◇◇◇◇◇ ◇◇◇◇◇ ◇◇◇◇◇ ◇◇◇◇◇ ◇◇◇◇◇	201,250
◇◇◇◇◇ ◇◇◇◇◇ ◇◇◇◇◇ ◇◇◇◇◇ ◇◇◇◇◇ ◇◇◇◇◇ ◇◇◇◇◇ ◇◇◇◇◇ ◇◇◇◇◇ ◇◇◇◇◇	201,300
◇◇◇◇◇ ◇◇◇◇◇ ◇◇◇◇◇ ◇◇◇◇◇ ◇◇◇◇◇ ◇◇◇◇◇ ◇◇◇◇◇ ◇◇◇◇◇ ◇◇◇◇◇ ◇◇◇◇◇	201,350
◇◇◇◇◇ ◇◇◇◇◇ ◇◇◇◇◇ ◇◇◇◇◇ ◇◇◇◇◇ ◇◇◇◇◇ ◇◇◇◇◇ ◇◇◇◇◇ ◇◇◇◇◇ ◇◇◇◇◇	201,400
◇◇◇◇◇ ◇◇◇◇◇ ◇◇◇◇◇ ◇◇◇◇◇ ◇◇◇◇◇ ◇◇◇◇◇ ◇◇◇◇◇ ◇◇◇◇◇ ◇◇◇◇◇ ◇◇◇◇◇	201,450
◇◇◇◇◇ ◇◇◇◇◇ ◇◇◇◇◇ ◇◇◇◇◇ ◇◇◇◇◇ ◇◇◇◇◇ ◇◇◇◇◇ ◇◇◇◇◇ ◇◇◇◇◇ ◇◇◇◇◇	201,500
◇◇◇◇◇ ◇◇◇◇◇ ◇◇◇◇◇ ◇◇◇◇◇ ◇◇◇◇◇ ◇◇◇◇◇ ◇◇◇◇◇ ◇◇◇◇◇ ◇◇◇◇◇ ◇◇◇◇◇	201,550
◇◇◇◇◇ ◇◇◇◇◇ ◇◇◇◇◇ ◇◇◇◇◇ ◇◇◇◇◇ ◇◇◇◇◇ ◇◇◇◇◇ ◇◇◇◇◇ ◇◇◇◇◇ ◇◇◇◇◇	201,600
◇◇◇◇◇ ◇◇◇◇◇ ◇◇◇◇◇ ◇◇◇◇◇ ◇◇◇◇◇ ◇◇◇◇◇ ◇◇◇◇◇ ◇◇◇◇◇ ◇◇◇◇◇ ◇◇◇◇◇	201,650
◇◇◇◇◇ ◇◇◇◇◇ ◇◇◇◇◇ ◇◇◇◇◇ ◇◇◇◇◇ ◇◇◇◇◇ ◇◇◇◇◇ ◇◇◇◇◇ ◇◇◇◇◇ ◇◇◇◇◇	201,700
◇◇◇◇◇ ◇◇◇◇◇ ◇◇◇◇◇ ◇◇◇◇◇ ◇◇◇◇◇ ◇◇◇◇◇ ◇◇◇◇◇ ◇◇◇◇◇ ◇◇◇◇◇ ◇◇◇◇◇	201,750
◇◇◇◇◇ ◇◇◇◇◇ ◇◇◇◇◇ ◇◇◇◇◇ ◇◇◇◇◇ ◇◇◇◇◇ ◇◇◇◇◇ ◇◇◇◇◇ ◇◇◇◇◇ ◇◇◇◇◇	201,800
◇◇◇◇◇ ◇◇◇◇◇ ◇◇◇◇◇ ◇◇◇◇◇ ◇◇◇◇◇ ◇◇◇◇◇ ◇◇◇◇◇ ◇◇◇◇◇ ◇◇◇◇◇ ◇◇◇◇◇	201,850
◇◇◇◇◇ ◇◇◇◇◇ ◇◇◇◇◇ ◇◇◇◇◇ ◇◇◇◇◇ ◇◇◇◇◇ ◇◇◇◇◇ ◇◇◇◇◇ ◇◇◇◇◇ ◇◇◇◇◇	201,900
◇◇◇◇◇ ◇◇◇◇◇ ◇◇◇◇◇ ◇◇◇◇◇ ◇◇◇◇◇ ◇◇◇◇◇ ◇◇◇◇◇ ◇◇◇◇◇ ◇◇◇◇◇ ◇◇◇◇◇	201,950
◇◇◇◇◇ ◇◇◇◇◇ ◇◇◇◇◇ ◇◇◇◇◇ ◇◇◇◇◇ ◇◇◇◇◇ ◇◇◇◇◇ ◇◇◇◇◇ ◇◇◇◇◇ ◇◇◇◇◇	**202,000**

The Number 202,001 to the Number 204,000

Start		Notes
	202,050	
	202,100	
	202,150	
	202,200	
	202,250	
	202,300	
	202,350	
	202,400	
	202,450	
	202,500	
	202,550	
	202,600	
	202,650	
	202,700	
	202,750	
	202,800	
	202,850	
	202,900	
	202,950	
	203,000	
	203,050	
	203,100	
	203,150	
	203,200	
	203,250	
	203,300	
	203,350	
	203,400	
	203,450	
	203,500	
	203,550	
	203,600	
	203,650	
	203,700	
	203,750	
	203,800	
	203,850	
	203,900	
	203,950	
	204,000	

The Number 204,001 to the Number 206,000

Start

Notes

	204,050
	204,100
	204,150
	204,200
	204,250
	204,300
	204,350
	204,400
	204,450
	204,500
	204,550
	204,600
	204,650
	204,700
	204,750
	204,800
	204,850
	204,900
	204,950
	205,000
	205,050
	205,100
	205,150
	205,200
	205,250
	205,300
	205,350
	205,400
	205,450
	205,500
	205,550
	205,600
	205,650
	205,700
	205,750
	205,800
	205,850
	205,900
	205,950
	206,000

◇ The Number 206,001 to the Number 208,000

Start   **Notes**

	206,050
	206,100
	206,150
	206,200
	206,250
	206,300
	206,350
	206,400
	206,450
	206,500
	206,550
	206,600
	206,650
	206,700
	206,750
	206,800
	206,850
	206,900
	206,950
	207,000
	207,050
	207,100
	207,150
	207,200
	207,250
	207,300
	207,350
	207,400
	207,450
	207,500
	207,550
	207,600
	207,650
	207,700
	207,750
	207,800
	207,850
	207,900
	207,950
	208,000

The Number 208,001 to the Number 210,000

Start		Notes

		Notes
	208,050	
	208,100	
	208,150	
	208,200	
	208,250	
	208,300	
	208,350	
	208,400	
	208,450	
	208,500	
	208,550	
	208,600	
	208,650	
	208,700	
	208,750	
	208,800	
	208,850	
	208,900	
	208,950	
	209,000	
	209,050	
	209,100	
	209,150	
	209,200	
	209,250	
	209,300	
	209,350	
	209,400	
	209,450	
	209,500	
	209,550	
	209,600	
	209,650	
	209,700	
	209,750	
	209,800	
	209,850	
	209,900	
	209,950	
	210,000	

◇ The Number 210,001 to the Number 212,000

Start		Notes
	210,050	
	210,100	
	210,150	
	210,200	
	210,250	
	210,300	
	210,350	
	210,400	
	210,450	
	210,500	
	210,550	
	210,600	
	210,650	
	210,700	
	210,750	
	210,800	
	210,850	
	210,900	
	210,950	
	211,000	
	211,050	
	211,100	
	211,150	
	211,200	
	211,250	
	211,300	
	211,350	
	211,400	
	211,450	
	211,500	
	211,550	
	211,600	
	211,650	
	211,700	
	211,750	
	211,800	
	211,850	
	211,900	
	211,950	
	212,000	

The Number 212,001 to the Number 214,000

Start
Notes

212,050	
212,100	
212,150	
212,200	
212,250	
212,300	
212,350	
212,400	
212,450	
212,500	
212,550	
212,600	
212,650	
212,700	
212,750	
212,800	
212,850	
212,900	
212,950	
213,000	
213,050	
213,100	
213,150	
213,200	
213,250	
213,300	
213,350	
213,400	
213,450	
213,500	
213,550	
213,600	
213,650	
213,700	
213,750	
213,800	
213,850	
213,900	
213,950	
214,000	

The Number 214,001 to the Number 216,000

Start		Notes
	214,050	
	214,100	
	214,150	
	214,200	
	214,250	
	214,300	
	214,350	
	214,400	
	214,450	
	214,500	
	214,550	
	214,600	
	214,650	
	214,700	
	214,750	
	214,800	
	214,850	
	214,900	
	214,950	
	215,000	
	215,050	
	215,100	
	215,150	
	215,200	
	215,250	
	215,300	
	215,350	
	215,400	
	215,450	
	215,500	
	215,550	
	215,600	
	215,650	
	215,700	
	215,750	
	215,800	
	215,850	
	215,900	
	215,950	
	216,000	

The Number 216,001 to the Number 218,000

Start   Notes

	216,050
	216,100
	216,150
	216,200
	216,250
	216,300
	216,350
	216,400
	216,450
	216,500
	216,550
	216,600
	216,650
	216,700
	216,750
	216,800
	216,850
	216,900
	216,950
	217,000
	217,050
	217,100
	217,150
	217,200
	217,250
	217,300
	217,350
	217,400
	217,450
	217,500
	217,550
	217,600
	217,650
	217,700
	217,750
	217,800
	217,850
	217,900
	217,950
	218,000

The Number 218,001 to the Number 220,000

Start		Notes
	218,050	
	218,100	
	218,150	
	218,200	
	218,250	
	218,300	
	218,350	
	218,400	
	218,450	
	218,500	
	218,550	
	218,600	
	218,650	
	218,700	
	218,750	
	218,800	
	218,850	
	218,900	
	218,950	
	219,000	
	219,050	
	219,100	
	219,150	
	219,200	
	219,250	
	219,300	
	219,350	
	219,400	
	219,450	
	219,500	
	219,550	
	219,600	
	219,650	
	219,700	
	219,750	
	219,800	
	219,850	
	219,900	
	219,950	
	220,000	

The Number 220,001 to the Number 222,000

Start   Notes

	220,050
	220,100
	220,150
	220,200
	220,250
	220,300
	220,350
	220,400
	220,450
	220,500
	220,550
	220,600
	220,650
	220,700
	220,750
	220,800
	220,850
	220,900
	220,950
	221,000
	221,050
	221,100
	221,150
	221,200
	221,250
	221,300
	221,350
	221,400
	221,450
	221,500
	221,550
	221,600
	221,650
	221,700
	221,750
	221,800
	221,850
	221,900
	221,950
	222,000

The Number 222,001 to the Number 224,000

Start		Notes
	222,050	
	222,100	
	222,150	
	222,200	
	222,250	
	222,300	
	222,350	
	222,400	
	222,450	
	222,500	
	222,550	
	222,600	
	222,650	
	222,700	
	222,750	
	222,800	
	222,850	
	222,900	
	222,950	
	223,000	
	223,050	
	223,100	
	223,150	
	223,200	
	223,250	
	223,300	
	223,350	
	223,400	
	223,450	
	223,500	
	223,550	
	223,600	
	223,650	
	223,700	
	223,750	
	223,800	
	223,850	
	223,900	
	223,950	
	224,000	

The Number 224,001 to the Number 226,000

Start

Notes

	224,050
	224,100
	224,150
	224,200
	224,250
	224,300
	224,350
	224,400
	224,450
	224,500
	224,550
	224,600
	224,650
	224,700
	224,750
	224,800
	224,850
	224,900
	224,950
	225,000
	225,050
	225,100
	225,150
	225,200
	225,250
	225,300
	225,350
	225,400
	225,450
	225,500
	225,550
	225,600
	225,650
	225,700
	225,750
	225,800
	225,850
	225,900
	225,950
	226,000

The Number 226,001 to the Number 228,000

Start	Notes
	226,050
	226,100
	226,150
	226,200
	226,250
	226,300
	226,350
	226,400
	226,450
	226,500
	226,550
	226,600
	226,650
	226,700
	226,750
	226,800
	226,850
	226,900
	226,950
	227,000
	227,050
	227,100
	227,150
	227,200
	227,250
	227,300
	227,350
	227,400
	227,450
	227,500
	227,550
	227,600
	227,650
	227,700
	227,750
	227,800
	227,850
	227,900
	227,950
	228,000

The Number 228,001 to the Number 230,000

Start

	Number
	228,050
	228,100
	228,150
	228,200
	228,250
	228,300
	228,350
	228,400
	228,450
	228,500
	228,550
	228,600
	228,650
	228,700
	228,750
	228,800
	228,850
	228,900
	228,950
	229,000
	229,050
	229,100
	229,150
	229,200
	229,250
	229,300
	229,350
	229,400
	229,450
	229,500
	229,550
	229,600
	229,650
	229,700
	229,750
	229,800
	229,850
	229,900
	229,950
	230,000

The Number 230,001 to the Number 232,000

Start   Notes

	230,050
	230,100
	230,150
	230,200
	230,250
	230,300
	230,350
	230,400
	230,450
	230,500
	230,550
	230,600
	230,650
	230,700
	230,750
	230,800
	230,850
	230,900
	230,950
	231,000
	231,050
	231,100
	231,150
	231,200
	231,250
	231,300
	231,350
	231,400
	231,450
	231,500
	231,550
	231,600
	231,650
	231,700
	231,750
	231,800
	231,850
	231,900
	231,950
	232,000

The Number 232,001 to the Number 234,000

Start

Notes

	Number
	232,050
	232,100
	232,150
	232,200
	232,250
	232,300
	232,350
	232,400
	232,450
	232,500
	232,550
	232,600
	232,650
	232,700
	232,750
	232,800
	232,850
	232,900
	232,950
	233,000
	233,050
	233,100
	233,150
	233,200
	233,250
	233,300
	233,350
	233,400
	233,450
	233,500
	233,550
	233,600
	233,650
	233,700
	233,750
	233,800
	233,850
	233,900
	233,950
	234,000

The Number 234,001 to the Number 236,000

Start   Notes

◊◊◊◊◊ ◊◊◊◊◊ ◊◊◊◊◊ ◊◊◊◊◊ ◊◊◊◊◊ ◊◊◊◊◊ ◊◊◊◊◊ ◊◊◊◊◊ ◊◊◊◊◊ ◊◊◊◊◊	234,050
◊◊◊◊◊ ◊◊◊◊◊ ◊◊◊◊◊ ◊◊◊◊◊ ◊◊◊◊◊ ◊◊◊◊◊ ◊◊◊◊◊ ◊◊◊◊◊ ◊◊◊◊◊ ◊◊◊◊◊	234,100
◊◊◊◊◊ ◊◊◊◊◊ ◊◊◊◊◊ ◊◊◊◊◊ ◊◊◊◊◊ ◊◊◊◊◊ ◊◊◊◊◊ ◊◊◊◊◊ ◊◊◊◊◊ ◊◊◊◊◊	234,150
◊◊◊◊◊ ◊◊◊◊◊ ◊◊◊◊◊ ◊◊◊◊◊ ◊◊◊◊◊ ◊◊◊◊◊ ◊◊◊◊◊ ◊◊◊◊◊ ◊◊◊◊◊ ◊◊◊◊◊	234,200
◊◊◊◊◊ ◊◊◊◊◊ ◊◊◊◊◊ ◊◊◊◊◊ ◊◊◊◊◊ ◊◊◊◊◊ ◊◊◊◊◊ ◊◊◊◊◊ ◊◊◊◊◊ ◊◊◊◊◊	234,250
◊◊◊◊◊ ◊◊◊◊◊ ◊◊◊◊◊ ◊◊◊◊◊ ◊◊◊◊◊ ◊◊◊◊◊ ◊◊◊◊◊ ◊◊◊◊◊ ◊◊◊◊◊ ◊◊◊◊◊	234,300
◊◊◊◊◊ ◊◊◊◊◊ ◊◊◊◊◊ ◊◊◊◊◊ ◊◊◊◊◊ ◊◊◊◊◊ ◊◊◊◊◊ ◊◊◊◊◊ ◊◊◊◊◊ ◊◊◊◊◊	234,350
◊◊◊◊◊ ◊◊◊◊◊ ◊◊◊◊◊ ◊◊◊◊◊ ◊◊◊◊◊ ◊◊◊◊◊ ◊◊◊◊◊ ◊◊◊◊◊ ◊◊◊◊◊ ◊◊◊◊◊	234,400
◊◊◊◊◊ ◊◊◊◊◊ ◊◊◊◊◊ ◊◊◊◊◊ ◊◊◊◊◊ ◊◊◊◊◊ ◊◊◊◊◊ ◊◊◊◊◊ ◊◊◊◊◊ ◊◊◊◊◊	234,450
◊◊◊◊◊ ◊◊◊◊◊ ◊◊◊◊◊ ◊◊◊◊◊ ◊◊◊◊◊ ◊◊◊◊◊ ◊◊◊◊◊ ◊◊◊◊◊ ◊◊◊◊◊ ◊◊◊◊◊	234,500
◊◊◊◊◊ ◊◊◊◊◊ ◊◊◊◊◊ ◊◊◊◊◊ ◊◊◊◊◊ ◊◊◊◊◊ ◊◊◊◊◊ ◊◊◊◊◊ ◊◊◊◊◊ ◊◊◊◊◊	234,550
◊◊◊◊◊ ◊◊◊◊◊ ◊◊◊◊◊ ◊◊◊◊◊ ◊◊◊◊◊ ◊◊◊◊◊ ◊◊◊◊◊ ◊◊◊◊◊ ◊◊◊◊◊ ◊◊◊◊◊	234,600
◊◊◊◊◊ ◊◊◊◊◊ ◊◊◊◊◊ ◊◊◊◊◊ ◊◊◊◊◊ ◊◊◊◊◊ ◊◊◊◊◊ ◊◊◊◊◊ ◊◊◊◊◊ ◊◊◊◊◊	234,650
◊◊◊◊◊ ◊◊◊◊◊ ◊◊◊◊◊ ◊◊◊◊◊ ◊◊◊◊◊ ◊◊◊◊◊ ◊◊◊◊◊ ◊◊◊◊◊ ◊◊◊◊◊ ◊◊◊◊◊	234,700
◊◊◊◊◊ ◊◊◊◊◊ ◊◊◊◊◊ ◊◊◊◊◊ ◊◊◊◊◊ ◊◊◊◊◊ ◊◊◊◊◊ ◊◊◊◊◊ ◊◊◊◊◊ ◊◊◊◊◊	234,750
◊◊◊◊◊ ◊◊◊◊◊ ◊◊◊◊◊ ◊◊◊◊◊ ◊◊◊◊◊ ◊◊◊◊◊ ◊◊◊◊◊ ◊◊◊◊◊ ◊◊◊◊◊ ◊◊◊◊◊	234,800
◊◊◊◊◊ ◊◊◊◊◊ ◊◊◊◊◊ ◊◊◊◊◊ ◊◊◊◊◊ ◊◊◊◊◊ ◊◊◊◊◊ ◊◊◊◊◊ ◊◊◊◊◊ ◊◊◊◊◊	234,850
◊◊◊◊◊ ◊◊◊◊◊ ◊◊◊◊◊ ◊◊◊◊◊ ◊◊◊◊◊ ◊◊◊◊◊ ◊◊◊◊◊ ◊◊◊◊◊ ◊◊◊◊◊ ◊◊◊◊◊	234,900
◊◊◊◊◊ ◊◊◊◊◊ ◊◊◊◊◊ ◊◊◊◊◊ ◊◊◊◊◊ ◊◊◊◊◊ ◊◊◊◊◊ ◊◊◊◊◊ ◊◊◊◊◊ ◊◊◊◊◊	234,950
◊◊◊◊◊ ◊◊◊◊◊ ◊◊◊◊◊ ◊◊◊◊◊ ◊◊◊◊◊ ◊◊◊◊◊ ◊◊◊◊◊ ◊◊◊◊◊ ◊◊◊◊◊ ◊◊◊◊◊	**235,000**
◊◊◊◊◊ ◊◊◊◊◊ ◊◊◊◊◊ ◊◊◊◊◊ ◊◊◊◊◊ ◊◊◊◊◊ ◊◊◊◊◊ ◊◊◊◊◊ ◊◊◊◊◊ ◊◊◊◊◊	235,050
◊◊◊◊◊ ◊◊◊◊◊ ◊◊◊◊◊ ◊◊◊◊◊ ◊◊◊◊◊ ◊◊◊◊◊ ◊◊◊◊◊ ◊◊◊◊◊ ◊◊◊◊◊ ◊◊◊◊◊	235,100
◊◊◊◊◊ ◊◊◊◊◊ ◊◊◊◊◊ ◊◊◊◊◊ ◊◊◊◊◊ ◊◊◊◊◊ ◊◊◊◊◊ ◊◊◊◊◊ ◊◊◊◊◊ ◊◊◊◊◊	235,150
◊◊◊◊◊ ◊◊◊◊◊ ◊◊◊◊◊ ◊◊◊◊◊ ◊◊◊◊◊ ◊◊◊◊◊ ◊◊◊◊◊ ◊◊◊◊◊ ◊◊◊◊◊ ◊◊◊◊◊	235,200
◊◊◊◊◊ ◊◊◊◊◊ ◊◊◊◊◊ ◊◊◊◊◊ ◊◊◊◊◊ ◊◊◊◊◊ ◊◊◊◊◊ ◊◊◊◊◊ ◊◊◊◊◊ ◊◊◊◊◊	235,250
◊◊◊◊◊ ◊◊◊◊◊ ◊◊◊◊◊ ◊◊◊◊◊ ◊◊◊◊◊ ◊◊◊◊◊ ◊◊◊◊◊ ◊◊◊◊◊ ◊◊◊◊◊ ◊◊◊◊◊	235,300
◊◊◊◊◊ ◊◊◊◊◊ ◊◊◊◊◊ ◊◊◊◊◊ ◊◊◊◊◊ ◊◊◊◊◊ ◊◊◊◊◊ ◊◊◊◊◊ ◊◊◊◊◊ ◊◊◊◊◊	235,350
◊◊◊◊◊ ◊◊◊◊◊ ◊◊◊◊◊ ◊◊◊◊◊ ◊◊◊◊◊ ◊◊◊◊◊ ◊◊◊◊◊ ◊◊◊◊◊ ◊◊◊◊◊ ◊◊◊◊◊	235,400
◊◊◊◊◊ ◊◊◊◊◊ ◊◊◊◊◊ ◊◊◊◊◊ ◊◊◊◊◊ ◊◊◊◊◊ ◊◊◊◊◊ ◊◊◊◊◊ ◊◊◊◊◊ ◊◊◊◊◊	235,450
◊◊◊◊◊ ◊◊◊◊◊ ◊◊◊◊◊ ◊◊◊◊◊ ◊◊◊◊◊ ◊◊◊◊◊ ◊◊◊◊◊ ◊◊◊◊◊ ◊◊◊◊◊ ◊◊◊◊◊	235,500
◊◊◊◊◊ ◊◊◊◊◊ ◊◊◊◊◊ ◊◊◊◊◊ ◊◊◊◊◊ ◊◊◊◊◊ ◊◊◊◊◊ ◊◊◊◊◊ ◊◊◊◊◊ ◊◊◊◊◊	235,550
◊◊◊◊◊ ◊◊◊◊◊ ◊◊◊◊◊ ◊◊◊◊◊ ◊◊◊◊◊ ◊◊◊◊◊ ◊◊◊◊◊ ◊◊◊◊◊ ◊◊◊◊◊ ◊◊◊◊◊	235,600
◊◊◊◊◊ ◊◊◊◊◊ ◊◊◊◊◊ ◊◊◊◊◊ ◊◊◊◊◊ ◊◊◊◊◊ ◊◊◊◊◊ ◊◊◊◊◊ ◊◊◊◊◊ ◊◊◊◊◊	235,650
◊◊◊◊◊ ◊◊◊◊◊ ◊◊◊◊◊ ◊◊◊◊◊ ◊◊◊◊◊ ◊◊◊◊◊ ◊◊◊◊◊ ◊◊◊◊◊ ◊◊◊◊◊ ◊◊◊◊◊	235,700
◊◊◊◊◊ ◊◊◊◊◊ ◊◊◊◊◊ ◊◊◊◊◊ ◊◊◊◊◊ ◊◊◊◊◊ ◊◊◊◊◊ ◊◊◊◊◊ ◊◊◊◊◊ ◊◊◊◊◊	235,750
◊◊◊◊◊ ◊◊◊◊◊ ◊◊◊◊◊ ◊◊◊◊◊ ◊◊◊◊◊ ◊◊◊◊◊ ◊◊◊◊◊ ◊◊◊◊◊ ◊◊◊◊◊ ◊◊◊◊◊	235,800
◊◊◊◊◊ ◊◊◊◊◊ ◊◊◊◊◊ ◊◊◊◊◊ ◊◊◊◊◊ ◊◊◊◊◊ ◊◊◊◊◊ ◊◊◊◊◊ ◊◊◊◊◊ ◊◊◊◊◊	235,850
◊◊◊◊◊ ◊◊◊◊◊ ◊◊◊◊◊ ◊◊◊◊◊ ◊◊◊◊◊ ◊◊◊◊◊ ◊◊◊◊◊ ◊◊◊◊◊ ◊◊◊◊◊ ◊◊◊◊◊	235,900
◊◊◊◊◊ ◊◊◊◊◊ ◊◊◊◊◊ ◊◊◊◊◊ ◊◊◊◊◊ ◊◊◊◊◊ ◊◊◊◊◊ ◊◊◊◊◊ ◊◊◊◊◊ ◊◊◊◊◊	235,950
◊◊◊◊◊ ◊◊◊◊◊ ◊◊◊◊◊ ◊◊◊◊◊ ◊◊◊◊◊ ◊◊◊◊◊ ◊◊◊◊◊ ◊◊◊◊◊ ◊◊◊◊◊ ◊◊◊◊◊	**236,000**

The Number 236,001 to the Number 238,000

Start		Notes

	Value	Notes
	236,050	
	236,100	
	236,150	
	236,200	
	236,250	
	236,300	
	236,350	
	236,400	
	236,450	
	236,500	
	236,550	
	236,600	
	236,650	
	236,700	
	236,750	
	236,800	
	236,850	
	236,900	
	236,950	
	237,000	
	237,050	
	237,100	
	237,150	
	237,200	
	237,250	
	237,300	
	237,350	
	237,400	
	237,450	
	237,500	
	237,550	
	237,600	
	237,650	
	237,700	
	237,750	
	237,800	
	237,850	
	237,900	
	237,950	
	238,000	

The Number 238,001 to the Number 240,000

Start   **Notes**

	238,050
	238,100
	238,150
	238,200
	238,250
	238,300
	238,350
	238,400
	238,450
	238,500
	238,550
	238,600
	238,650
	238,700
	238,750
	238,800
	238,850
	238,900
	238,950
	239,000
	239,050
	239,100
	239,150
	239,200
	239,250
	239,300
	239,350
	239,400
	239,450
	239,500
	239,550
	239,600
	239,650
	239,700
	239,750
	239,800
	239,850
	239,900
	239,950
	240,000

The Number 240,001 to the Number 242,000

Start   Notes

	240,050
	240,100
	240,150
	240,200
	240,250
	240,300
	240,350
	240,400
	240,450
	240,500
	240,550
	240,600
	240,650
	240,700
	240,750
	240,800
	240,850
	240,900
	240,950
	241,000
	241,050
	241,100
	241,150
	241,200
	241,250
	241,300
	241,350
	241,400
	241,450
	241,500
	241,550
	241,600
	241,650
	241,700
	241,750
	241,800
	241,850
	241,900
	241,950
	242,000

The Number 242,001 to the Number 244,000

Start	Notes
	242,050
	242,100
	242,150
	242,200
	242,250
	242,300
	242,350
	242,400
	242,450
	242,500
	242,550
	242,600
	242,650
	242,700
	242,750
	242,800
	242,850
	242,900
	242,950
	243,000
	243,050
	243,100
	243,150
	243,200
	243,250
	243,300
	243,350
	243,400
	243,450
	243,500
	243,550
	243,600
	243,650
	243,700
	243,750
	243,800
	243,850
	243,900
	243,950
	244,000

The Number 244,001 to the Number 246,000

Start | | Notes

	244,050
	244,100
	244,150
	244,200
	244,250
	244,300
	244,350
	244,400
	244,450
	244,500
	244,550
	244,600
	244,650
	244,700
	244,750
	244,800
	244,850
	244,900
	244,950
	245,000
	245,050
	245,100
	245,150
	245,200
	245,250
	245,300
	245,350
	245,400
	245,450
	245,500
	245,550
	245,600
	245,650
	245,700
	245,750
	245,800
	245,850
	245,900
	245,950
	246,000

◊ The Number 246,001 to the Number 248,000

Start | | Notes

	Number
	246,050
	246,100
	246,150
	246,200
	246,250
	246,300
	246,350
	246,400
	246,450
	246,500
	246,550
	246,600
	246,650
	246,700
	246,750
	246,800
	246,850
	246,900
	246,950
	247,000
	247,050
	247,100
	247,150
	247,200
	247,250
	247,300
	247,350
	247,400
	247,450
	247,500
	247,550
	247,600
	247,650
	247,700
	247,750
	247,800
	247,850
	247,900
	247,950
	248,000

The Number 248,001 to the Number 250,000

Start   Notes

	248,050
	248,100
	248,150
	248,200
	248,250
	248,300
	248,350
	248,400
	248,450
	248,500
	248,550
	248,600
	248,650
	248,700
	248,750
	248,800
	248,850
	248,900
	248,950
	249,000
	249,050
	249,100
	249,150
	249,200
	249,250
	249,300
	249,350
	249,400
	249,450
	249,500
	249,550
	249,600
	249,650
	249,700
	249,750
	249,800
	249,850
	249,900
	249,950
	250,000

The Number 250,001 to the Number 252,000

Start	Notes
	250,050
	250,100
	250,150
	250,200
	250,250
	250,300
	250,350
	250,400
	250,450
	250,500
	250,550
	250,600
	250,650
	250,700
	250,750
	250,800
	250,850
	250,900
	250,950
	251,000
	251,050
	251,100
	251,150
	251,200
	251,250
	251,300
	251,350
	251,400
	251,450
	251,500
	251,550
	251,600
	251,650
	251,700
	251,750
	251,800
	251,850
	251,900
	251,950
	252,000

The Number 252,001 to the Number 254,000

Start		Notes

	252,050	
	252,100	
	252,150	
	252,200	
	252,250	
	252,300	
	252,350	
	252,400	
	252,450	
	252,500	
	252,550	
	252,600	
	252,650	
	252,700	
	252,750	
	252,800	
	252,850	
	252,900	
	252,950	
	253,000	
	253,050	
	253,100	
	253,150	
	253,200	
	253,250	
	253,300	
	253,350	
	253,400	
	253,450	
	253,500	
	253,550	
	253,600	
	253,650	
	253,700	
	253,750	
	253,800	
	253,850	
	253,900	
	253,950	
	254,000	

The Number 254,001 to the Number 256,000

Start		Notes
	254,050	
	254,100	
	254,150	
	254,200	
	254,250	
	254,300	
	254,350	
	254,400	
	254,450	
	254,500	
	254,550	
	254,600	
	254,650	
	254,700	
	254,750	
	254,800	
	254,850	
	254,900	
	254,950	
	255,000	
	255,050	
	255,100	
	255,150	
	255,200	
	255,250	
	255,300	
	255,350	
	255,400	
	255,450	
	255,500	
	255,550	
	255,600	
	255,650	
	255,700	
	255,750	
	255,800	
	255,850	
	255,900	
	255,950	
	256,000	

The Number 256,001 to the Number 258,000

Start   **Notes**

	256,050
	256,100
	256,150
	256,200
	256,250
	256,300
	256,350
	256,400
	256,450
	256,500
	256,550
	256,600
	256,650
	256,700
	256,750
	256,800
	256,850
	256,900
	256,950
	257,000
	257,050
	257,100
	257,150
	257,200
	257,250
	257,300
	257,350
	257,400
	257,450
	257,500
	257,550
	257,600
	257,650
	257,700
	257,750
	257,800
	257,850
	257,900
	257,950
	258,000

The Number 258,001 to the Number 260,000

Start	Notes
	258,050
	258,100
	258,150
	258,200
	258,250
	258,300
	258,350
	258,400
	258,450
	258,500
	258,550
	258,600
	258,650
	258,700
	258,750
	258,800
	258,850
	258,900
	258,950
	259,000
	259,050
	259,100
	259,150
	259,200
	259,250
	259,300
	259,350
	259,400
	259,450
	259,500
	259,550
	259,600
	259,650
	259,700
	259,750
	259,800
	259,850
	259,900
	259,950
	260,000

The Number 260,001 to the Number 262,000

Start

Notes

	260,050
	260,100
	260,150
	260,200
	260,250
	260,300
	260,350
	260,400
	260,450
	260,500
	260,550
	260,600
	260,650
	260,700
	260,750
	260,800
	260,850
	260,900
	260,950
	261,000
	261,050
	261,100
	261,150
	261,200
	261,250
	261,300
	261,350
	261,400
	261,450
	261,500
	261,550
	261,600
	261,650
	261,700
	261,750
	261,800
	261,850
	261,900
	261,950
	262,000

The Number 262,001 to the Number 264,000

Start		Notes
	262,050	
	262,100	
	262,150	
	262,200	
	262,250	
	262,300	
	262,350	
	262,400	
	262,450	
	262,500	
	262,550	
	262,600	
	262,650	
	262,700	
	262,750	
	262,800	
	262,850	
	262,900	
	262,950	
	263,000	
	263,050	
	263,100	
	263,150	
	263,200	
	263,250	
	263,300	
	263,350	
	263,400	
	263,450	
	263,500	
	263,550	
	263,600	
	263,650	
	263,700	
	263,750	
	263,800	
	263,850	
	263,900	
	263,950	
	264,000	

The Number 264,001 to the Number 266,000

Start

Notes

	264,050
	264,100
	264,150
	264,200
	264,250
	264,300
	264,350
	264,400
	264,450
	264,500
	264,550
	264,600
	264,650
	264,700
	264,750
	264,800
	264,850
	264,900
	264,950
	265,000
	265,050
	265,100
	265,150
	265,200
	265,250
	265,300
	265,350
	265,400
	265,450
	265,500
	265,550
	265,600
	265,650
	265,700
	265,750
	265,800
	265,850
	265,900
	265,950
	266,000

The Number 266,001 to the Number 268,000

Start		Notes
	266,050	
	266,100	
	266,150	
	266,200	
	266,250	
	266,300	
	266,350	
	266,400	
	266,450	
	266,500	
	266,550	
	266,600	
	266,650	
	266,700	
	266,750	
	266,800	
	266,850	
	266,900	
	266,950	
	267,000	
	267,050	
	267,100	
	267,150	
	267,200	
	267,250	
	267,300	
	267,350	
	267,400	
	267,450	
	267,500	
	267,550	
	267,600	
	267,650	
	267,700	
	267,750	
	267,800	
	267,850	
	267,900	
	267,950	
	268,000	

◇ The Number 268,001 to the Number 270,000

Start   **Notes**

	268,050
	268,100
	268,150
	268,200
	268,250
	268,300
	268,350
	268,400
	268,450
	268,500
	268,550
	268,600
	268,650
	268,700
	268,750
	268,800
	268,850
	268,900
	268,950
	269,000
	269,050
	269,100
	269,150
	269,200
	269,250
	269,300
	269,350
	269,400
	269,450
	269,500
	269,550
	269,600
	269,650
	269,700
	269,750
	269,800
	269,850
	269,900
	269,950
	270,000

The Number 270,001 to the Number 272,000

Start	Notes
	270,050
	270,100
	270,150
	270,200
	270,250
	270,300
	270,350
	270,400
	270,450
	270,500
	270,550
	270,600
	270,650
	270,700
	270,750
	270,800
	270,850
	270,900
	270,950
	271,000
	271,050
	271,100
	271,150
	271,200
	271,250
	271,300
	271,350
	271,400
	271,450
	271,500
	271,550
	271,600
	271,650
	271,700
	271,750
	271,800
	271,850
	271,900
	271,950
	272,000

◇ The Number 272,001 to the Number 274,000

Start   **Notes**

	272,050
	272,100
	272,150
	272,200
	272,250
	272,300
	272,350
	272,400
	272,450
	272,500
	272,550
	272,600
	272,650
	272,700
	272,750
	272,800
	272,850
	272,900
	272,950
	273,000
	273,050
	273,100
	273,150
	273,200
	273,250
	273,300
	273,350
	273,400
	273,450
	273,500
	273,550
	273,600
	273,650
	273,700
	273,750
	273,800
	273,850
	273,900
	273,950
	274,000

The Number 274,001 to the Number 276,000

Start	Notes
	274,050
	274,100
	274,150
	274,200
	274,250
	274,300
	274,350
	274,400
	274,450
	274,500
	274,550
	274,600
	274,650
	274,700
	274,750
	274,800
	274,850
	274,900
	274,950
	275,000
	275,050
	275,100
	275,150
	275,200
	275,250
	275,300
	275,350
	275,400
	275,450
	275,500
	275,550
	275,600
	275,650
	275,700
	275,750
	275,800
	275,850
	275,900
	275,950
	276,000

The Number 276,001 to the Number 278,000

Start

Notes

	276,050
	276,100
	276,150
	276,200
	276,250
	276,300
	276,350
	276,400
	276,450
	276,500
	276,550
	276,600
	276,650
	276,700
	276,750
	276,800
	276,850
	276,900
	276,950
	277,000
	277,050
	277,100
	277,150
	277,200
	277,250
	277,300
	277,350
	277,400
	277,450
	277,500
	277,550
	277,600
	277,650
	277,700
	277,750
	277,800
	277,850
	277,900
	277,950
	278,000

The Number 278,001 to the Number 280,000

Start   Notes

Number	Notes
278,050	
278,100	
278,150	
278,200	
278,250	
278,300	
278,350	
278,400	
278,450	
278,500	
278,550	
278,600	
278,650	
278,700	
278,750	
278,800	
278,850	
278,900	
278,950	
279,000	
279,050	
279,100	
279,150	
279,200	
279,250	
279,300	
279,350	
279,400	
279,450	
279,500	
279,550	
279,600	
279,650	
279,700	
279,750	
279,800	
279,850	
279,900	
279,950	
280,000	

The Number 280,001 to the Number 282,000

Start | | Notes

	Number
◊◊◊◊◊ ◊◊◊◊◊ ◊◊◊◊◊ ◊◊◊◊◊ ◊◊◊◊◊ ◊◊◊◊◊ ◊◊◊◊◊ ◊◊◊◊◊ ◊◊◊◊◊ ◊◊◊◊◊	280,050
◊◊◊◊◊ ◊◊◊◊◊ ◊◊◊◊◊ ◊◊◊◊◊ ◊◊◊◊◊ ◊◊◊◊◊ ◊◊◊◊◊ ◊◊◊◊◊ ◊◊◊◊◊ ◊◊◊◊◊	280,100
◊◊◊◊◊ ◊◊◊◊◊ ◊◊◊◊◊ ◊◊◊◊◊ ◊◊◊◊◊ ◊◊◊◊◊ ◊◊◊◊◊ ◊◊◊◊◊ ◊◊◊◊◊ ◊◊◊◊◊	280,150
◊◊◊◊◊ ◊◊◊◊◊ ◊◊◊◊◊ ◊◊◊◊◊ ◊◊◊◊◊ ◊◊◊◊◊ ◊◊◊◊◊ ◊◊◊◊◊ ◊◊◊◊◊ ◊◊◊◊◊	280,200
◊◊◊◊◊ ◊◊◊◊◊ ◊◊◊◊◊ ◊◊◊◊◊ ◊◊◊◊◊ ◊◊◊◊◊ ◊◊◊◊◊ ◊◊◊◊◊ ◊◊◊◊◊ ◊◊◊◊◊	280,250
◊◊◊◊◊ ◊◊◊◊◊ ◊◊◊◊◊ ◊◊◊◊◊ ◊◊◊◊◊ ◊◊◊◊◊ ◊◊◊◊◊ ◊◊◊◊◊ ◊◊◊◊◊ ◊◊◊◊◊	280,300
◊◊◊◊◊ ◊◊◊◊◊ ◊◊◊◊◊ ◊◊◊◊◊ ◊◊◊◊◊ ◊◊◊◊◊ ◊◊◊◊◊ ◊◊◊◊◊ ◊◊◊◊◊ ◊◊◊◊◊	280,350
◊◊◊◊◊ ◊◊◊◊◊ ◊◊◊◊◊ ◊◊◊◊◊ ◊◊◊◊◊ ◊◊◊◊◊ ◊◊◊◊◊ ◊◊◊◊◊ ◊◊◊◊◊ ◊◊◊◊◊	280,400
◊◊◊◊◊ ◊◊◊◊◊ ◊◊◊◊◊ ◊◊◊◊◊ ◊◊◊◊◊ ◊◊◊◊◊ ◊◊◊◊◊ ◊◊◊◊◊ ◊◊◊◊◊ ◊◊◊◊◊	280,450
◊◊◊◊◊ ◊◊◊◊◊ ◊◊◊◊◊ ◊◊◊◊◊ ◊◊◊◊◊ ◊◊◊◊◊ ◊◊◊◊◊ ◊◊◊◊◊ ◊◊◊◊◊ ◊◊◊◊◊	280,500
◊◊◊◊◊ ◊◊◊◊◊ ◊◊◊◊◊ ◊◊◊◊◊ ◊◊◊◊◊ ◊◊◊◊◊ ◊◊◊◊◊ ◊◊◊◊◊ ◊◊◊◊◊ ◊◊◊◊◊	280,550
◊◊◊◊◊ ◊◊◊◊◊ ◊◊◊◊◊ ◊◊◊◊◊ ◊◊◊◊◊ ◊◊◊◊◊ ◊◊◊◊◊ ◊◊◊◊◊ ◊◊◊◊◊ ◊◊◊◊◊	280,600
◊◊◊◊◊ ◊◊◊◊◊ ◊◊◊◊◊ ◊◊◊◊◊ ◊◊◊◊◊ ◊◊◊◊◊ ◊◊◊◊◊ ◊◊◊◊◊ ◊◊◊◊◊ ◊◊◊◊◊	280,650
◊◊◊◊◊ ◊◊◊◊◊ ◊◊◊◊◊ ◊◊◊◊◊ ◊◊◊◊◊ ◊◊◊◊◊ ◊◊◊◊◊ ◊◊◊◊◊ ◊◊◊◊◊ ◊◊◊◊◊	280,700
◊◊◊◊◊ ◊◊◊◊◊ ◊◊◊◊◊ ◊◊◊◊◊ ◊◊◊◊◊ ◊◊◊◊◊ ◊◊◊◊◊ ◊◊◊◊◊ ◊◊◊◊◊ ◊◊◊◊◊	280,750
◊◊◊◊◊ ◊◊◊◊◊ ◊◊◊◊◊ ◊◊◊◊◊ ◊◊◊◊◊ ◊◊◊◊◊ ◊◊◊◊◊ ◊◊◊◊◊ ◊◊◊◊◊ ◊◊◊◊◊	280,800
◊◊◊◊◊ ◊◊◊◊◊ ◊◊◊◊◊ ◊◊◊◊◊ ◊◊◊◊◊ ◊◊◊◊◊ ◊◊◊◊◊ ◊◊◊◊◊ ◊◊◊◊◊ ◊◊◊◊◊	280,850
◊◊◊◊◊ ◊◊◊◊◊ ◊◊◊◊◊ ◊◊◊◊◊ ◊◊◊◊◊ ◊◊◊◊◊ ◊◊◊◊◊ ◊◊◊◊◊ ◊◊◊◊◊ ◊◊◊◊◊	280,900
◊◊◊◊◊ ◊◊◊◊◊ ◊◊◊◊◊ ◊◊◊◊◊ ◊◊◊◊◊ ◊◊◊◊◊ ◊◊◊◊◊ ◊◊◊◊◊ ◊◊◊◊◊ ◊◊◊◊◊	280,950
◊◊◊◊◊ ◊◊◊◊◊ ◊◊◊◊◊ ◊◊◊◊◊ ◊◊◊◊◊ ◊◊◊◊◊ ◊◊◊◊◊ ◊◊◊◊◊ ◊◊◊◊◊ ◊◊◊◊◊	**281,000**
◊◊◊◊◊ ◊◊◊◊◊ ◊◊◊◊◊ ◊◊◊◊◊ ◊◊◊◊◊ ◊◊◊◊◊ ◊◊◊◊◊ ◊◊◊◊◊ ◊◊◊◊◊ ◊◊◊◊◊	281,050
◊◊◊◊◊ ◊◊◊◊◊ ◊◊◊◊◊ ◊◊◊◊◊ ◊◊◊◊◊ ◊◊◊◊◊ ◊◊◊◊◊ ◊◊◊◊◊ ◊◊◊◊◊ ◊◊◊◊◊	281,100
◊◊◊◊◊ ◊◊◊◊◊ ◊◊◊◊◊ ◊◊◊◊◊ ◊◊◊◊◊ ◊◊◊◊◊ ◊◊◊◊◊ ◊◊◊◊◊ ◊◊◊◊◊ ◊◊◊◊◊	281,150
◊◊◊◊◊ ◊◊◊◊◊ ◊◊◊◊◊ ◊◊◊◊◊ ◊◊◊◊◊ ◊◊◊◊◊ ◊◊◊◊◊ ◊◊◊◊◊ ◊◊◊◊◊ ◊◊◊◊◊	281,200
◊◊◊◊◊ ◊◊◊◊◊ ◊◊◊◊◊ ◊◊◊◊◊ ◊◊◊◊◊ ◊◊◊◊◊ ◊◊◊◊◊ ◊◊◊◊◊ ◊◊◊◊◊ ◊◊◊◊◊	281,250
◊◊◊◊◊ ◊◊◊◊◊ ◊◊◊◊◊ ◊◊◊◊◊ ◊◊◊◊◊ ◊◊◊◊◊ ◊◊◊◊◊ ◊◊◊◊◊ ◊◊◊◊◊ ◊◊◊◊◊	281,300
◊◊◊◊◊ ◊◊◊◊◊ ◊◊◊◊◊ ◊◊◊◊◊ ◊◊◊◊◊ ◊◊◊◊◊ ◊◊◊◊◊ ◊◊◊◊◊ ◊◊◊◊◊ ◊◊◊◊◊	281,350
◊◊◊◊◊ ◊◊◊◊◊ ◊◊◊◊◊ ◊◊◊◊◊ ◊◊◊◊◊ ◊◊◊◊◊ ◊◊◊◊◊ ◊◊◊◊◊ ◊◊◊◊◊ ◊◊◊◊◊	281,400
◊◊◊◊◊ ◊◊◊◊◊ ◊◊◊◊◊ ◊◊◊◊◊ ◊◊◊◊◊ ◊◊◊◊◊ ◊◊◊◊◊ ◊◊◊◊◊ ◊◊◊◊◊ ◊◊◊◊◊	281,450
◊◊◊◊◊ ◊◊◊◊◊ ◊◊◊◊◊ ◊◊◊◊◊ ◊◊◊◊◊ ◊◊◊◊◊ ◊◊◊◊◊ ◊◊◊◊◊ ◊◊◊◊◊ ◊◊◊◊◊	281,500
◊◊◊◊◊ ◊◊◊◊◊ ◊◊◊◊◊ ◊◊◊◊◊ ◊◊◊◊◊ ◊◊◊◊◊ ◊◊◊◊◊ ◊◊◊◊◊ ◊◊◊◊◊ ◊◊◊◊◊	281,550
◊◊◊◊◊ ◊◊◊◊◊ ◊◊◊◊◊ ◊◊◊◊◊ ◊◊◊◊◊ ◊◊◊◊◊ ◊◊◊◊◊ ◊◊◊◊◊ ◊◊◊◊◊ ◊◊◊◊◊	281,600
◊◊◊◊◊ ◊◊◊◊◊ ◊◊◊◊◊ ◊◊◊◊◊ ◊◊◊◊◊ ◊◊◊◊◊ ◊◊◊◊◊ ◊◊◊◊◊ ◊◊◊◊◊ ◊◊◊◊◊	281,650
◊◊◊◊◊ ◊◊◊◊◊ ◊◊◊◊◊ ◊◊◊◊◊ ◊◊◊◊◊ ◊◊◊◊◊ ◊◊◊◊◊ ◊◊◊◊◊ ◊◊◊◊◊ ◊◊◊◊◊	281,700
◊◊◊◊◊ ◊◊◊◊◊ ◊◊◊◊◊ ◊◊◊◊◊ ◊◊◊◊◊ ◊◊◊◊◊ ◊◊◊◊◊ ◊◊◊◊◊ ◊◊◊◊◊ ◊◊◊◊◊	281,750
◊◊◊◊◊ ◊◊◊◊◊ ◊◊◊◊◊ ◊◊◊◊◊ ◊◊◊◊◊ ◊◊◊◊◊ ◊◊◊◊◊ ◊◊◊◊◊ ◊◊◊◊◊ ◊◊◊◊◊	281,800
◊◊◊◊◊ ◊◊◊◊◊ ◊◊◊◊◊ ◊◊◊◊◊ ◊◊◊◊◊ ◊◊◊◊◊ ◊◊◊◊◊ ◊◊◊◊◊ ◊◊◊◊◊ ◊◊◊◊◊	281,850
◊◊◊◊◊ ◊◊◊◊◊ ◊◊◊◊◊ ◊◊◊◊◊ ◊◊◊◊◊ ◊◊◊◊◊ ◊◊◊◊◊ ◊◊◊◊◊ ◊◊◊◊◊ ◊◊◊◊◊	281,900
◊◊◊◊◊ ◊◊◊◊◊ ◊◊◊◊◊ ◊◊◊◊◊ ◊◊◊◊◊ ◊◊◊◊◊ ◊◊◊◊◊ ◊◊◊◊◊ ◊◊◊◊◊ ◊◊◊◊◊	281,950
◊◊◊◊◊ ◊◊◊◊◊ ◊◊◊◊◊ ◊◊◊◊◊ ◊◊◊◊◊ ◊◊◊◊◊ ◊◊◊◊◊ ◊◊◊◊◊ ◊◊◊◊◊ ◊◊◊◊◊	**282,000**

The Number 282,001 to the Number 284,000

Start	Notes
282,050	
282,100	
282,150	
282,200	
282,250	
282,300	
282,350	
282,400	
282,450	
282,500	
282,550	
282,600	
282,650	
282,700	
282,750	
282,800	
282,850	
282,900	
282,950	
283,000	
283,050	
283,100	
283,150	
283,200	
283,250	
283,300	
283,350	
283,400	
283,450	
283,500	
283,550	
283,600	
283,650	
283,700	
283,750	
283,800	
283,850	
283,900	
283,950	
284,000	

The Number 284,001 to the Number 286,000

Start

	Number
	284,050
	284,100
	284,150
	284,200
	284,250
	284,300
	284,350
	284,400
	284,450
	284,500
	284,550
	284,600
	284,650
	284,700
	284,750
	284,800
	284,850
	284,900
	284,950
	285,000
	285,050
	285,100
	285,150
	285,200
	285,250
	285,300
	285,350
	285,400
	285,450
	285,500
	285,550
	285,600
	285,650
	285,700
	285,750
	285,800
	285,850
	285,900
	285,950
	286,000

The Number 286,001 to the Number 288,000

Start   Notes

	286,050
	286,100
	286,150
	286,200
	286,250
	286,300
	286,350
	286,400
	286,450
	286,500
	286,550
	286,600
	286,650
	286,700
	286,750
	286,800
	286,850
	286,900
	286,950
	287,000
	287,050
	287,100
	287,150
	287,200
	287,250
	287,300
	287,350
	287,400
	287,450
	287,500
	287,550
	287,600
	287,650
	287,700
	287,750
	287,800
	287,850
	287,900
	287,950
	288,000

The Number 288,001 to the Number 290,000

Start		Notes
288,050
288,100
288,150
288,200
288,250
288,300
288,350
288,400
288,450
288,500
288,550
288,600
288,650
288,700
288,750
288,800
288,850
288,900
288,950
289,000
289,050
289,100
289,150
289,200
289,250
289,300
289,350
289,400
289,450
289,500
289,550
289,600
289,650
289,700
289,750
289,800
289,850
289,900
289,950
290,000

The Number 290,001 to the Number 292,000

Start	Notes
	290,050
	290,100
	290,150
	290,200
	290,250
	290,300
	290,350
	290,400
	290,450
	290,500
	290,550
	290,600
	290,650
	290,700
	290,750
	290,800
	290,850
	290,900
	290,950
	291,000
	291,050
	291,100
	291,150
	291,200
	291,250
	291,300
	291,350
	291,400
	291,450
	291,500
	291,550
	291,600
	291,650
	291,700
	291,750
	291,800
	291,850
	291,900
	291,950
	292,000

The Number 292,001 to the Number 294,000

Start		Notes

	292,050	
	292,100	
	292,150	
	292,200	
	292,250	
	292,300	
	292,350	
	292,400	
	292,450	
	292,500	
	292,550	
	292,600	
	292,650	
	292,700	
	292,750	
	292,800	
	292,850	
	292,900	
	292,950	
	293,000	
	293,050	
	293,100	
	293,150	
	293,200	
	293,250	
	293,300	
	293,350	
	293,400	
	293,450	
	293,500	
	293,550	
	293,600	
	293,650	
	293,700	
	293,750	
	293,800	
	293,850	
	293,900	
	293,950	
	294,000	

The Number 294,001 to the Number 296,000

Start

Notes

294,050	
294,100	
294,150	
294,200	
294,250	
294,300	
294,350	
294,400	
294,450	
294,500	
294,550	
294,600	
294,650	
294,700	
294,750	
294,800	
294,850	
294,900	
294,950	
295,000	
295,050	
295,100	
295,150	
295,200	
295,250	
295,300	
295,350	
295,400	
295,450	
295,500	
295,550	
295,600	
295,650	
295,700	
295,750	
295,800	
295,850	
295,900	
295,950	
296,000	

The Number 296,001 to the Number 298,000

Start		Notes

	296,050
	296,100
	296,150
	296,200
	296,250
	296,300
	296,350
	296,400
	296,450
	296,500
	296,550
	296,600
	296,650
	296,700
	296,750
	296,800
	296,850
	296,900
	296,950
	297,000
	297,050
	297,100
	297,150
	297,200
	297,250
	297,300
	297,350
	297,400
	297,450
	297,500
	297,550
	297,600
	297,650
	297,700
	297,750
	297,800
	297,850
	297,900
	297,950
	298,000

The Number 298,001 to the Number 300,000

Start	Notes
	298,050
	298,100
	298,150
	298,200
	298,250
	298,300
	298,350
	298,400
	298,450
	298,500
	298,550
	298,600
	298,650
	298,700
	298,750
	298,800
	298,850
	298,900
	298,950
	299,000
	299,050
	299,100
	299,150
	299,200
	299,250
	299,300
	299,350
	299,400
	299,450
	299,500
	299,550
	299,600
	299,650
	299,700
	299,750
	299,800
	299,850
	299,900
	299,950
	300,000

The Number 300,001 to the Number 302,000

Start   **Notes**

	300,050
	300,100
	300,150
	300,200
	300,250
	300,300
	300,350
	300,400
	300,450
	300,500
	300,550
	300,600
	300,650
	300,700
	300,750
	300,800
	300,850
	300,900
	300,950
	301,000
	301,050
	301,100
	301,150
	301,200
	301,250
	301,300
	301,350
	301,400
	301,450
	301,500
	301,550
	301,600
	301,650
	301,700
	301,750
	301,800
	301,850
	301,900
	301,950
	302,000

The Number 302,001 to the Number 304,000

Start

	Notes
	302,050
	302,100
	302,150
	302,200
	302,250
	302,300
	302,350
	302,400
	302,450
	302,500
	302,550
	302,600
	302,650
	302,700
	302,750
	302,800
	302,850
	302,900
	302,950
	303,000
	303,050
	303,100
	303,150
	303,200
	303,250
	303,300
	303,350
	303,400
	303,450
	303,500
	303,550
	303,600
	303,650
	303,700
	303,750
	303,800
	303,850
	303,900
	303,950
	304,000

The Number 304,001 to the Number 306,000

Start

	Notes
304,050	
304,100	
304,150	
304,200	
304,250	
304,300	
304,350	
304,400	
304,450	
304,500	
304,550	
304,600	
304,650	
304,700	
304,750	
304,800	
304,850	
304,900	
304,950	
305,000	
305,050	
305,100	
305,150	
305,200	
305,250	
305,300	
305,350	
305,400	
305,450	
305,500	
305,550	
305,600	
305,650	
305,700	
305,750	
305,800	
305,850	
305,900	
305,950	
306,000	

The Number 306,001 to the Number 308,000

Start

Notes

◇◇◇◇◇ ◇◇◇◇◇ ◇◇◇◇◇ ◇◇◇◇◇ ◇◇◇◇◇ ◇◇◇◇◇ ◇◇◇◇◇ ◇◇◇◇◇ ◇◇◇◇◇ ◇◇◇◇◇	306,050
◇◇◇◇◇ ◇◇◇◇◇ ◇◇◇◇◇ ◇◇◇◇◇ ◇◇◇◇◇ ◇◇◇◇◇ ◇◇◇◇◇ ◇◇◇◇◇ ◇◇◇◇◇ ◇◇◇◇◇	306,100
◇◇◇◇◇ ◇◇◇◇◇ ◇◇◇◇◇ ◇◇◇◇◇ ◇◇◇◇◇ ◇◇◇◇◇ ◇◇◇◇◇ ◇◇◇◇◇ ◇◇◇◇◇ ◇◇◇◇◇	306,150
◇◇◇◇◇ ◇◇◇◇◇ ◇◇◇◇◇ ◇◇◇◇◇ ◇◇◇◇◇ ◇◇◇◇◇ ◇◇◇◇◇ ◇◇◇◇◇ ◇◇◇◇◇ ◇◇◇◇◇	306,200
◇◇◇◇◇ ◇◇◇◇◇ ◇◇◇◇◇ ◇◇◇◇◇ ◇◇◇◇◇ ◇◇◇◇◇ ◇◇◇◇◇ ◇◇◇◇◇ ◇◇◇◇◇ ◇◇◇◇◇	306,250
◇◇◇◇◇ ◇◇◇◇◇ ◇◇◇◇◇ ◇◇◇◇◇ ◇◇◇◇◇ ◇◇◇◇◇ ◇◇◇◇◇ ◇◇◇◇◇ ◇◇◇◇◇ ◇◇◇◇◇	306,300
◇◇◇◇◇ ◇◇◇◇◇ ◇◇◇◇◇ ◇◇◇◇◇ ◇◇◇◇◇ ◇◇◇◇◇ ◇◇◇◇◇ ◇◇◇◇◇ ◇◇◇◇◇ ◇◇◇◇◇	306,350
◇◇◇◇◇ ◇◇◇◇◇ ◇◇◇◇◇ ◇◇◇◇◇ ◇◇◇◇◇ ◇◇◇◇◇ ◇◇◇◇◇ ◇◇◇◇◇ ◇◇◇◇◇ ◇◇◇◇◇	306,400
◇◇◇◇◇ ◇◇◇◇◇ ◇◇◇◇◇ ◇◇◇◇◇ ◇◇◇◇◇ ◇◇◇◇◇ ◇◇◇◇◇ ◇◇◇◇◇ ◇◇◇◇◇ ◇◇◇◇◇	306,450
◇◇◇◇◇ ◇◇◇◇◇ ◇◇◇◇◇ ◇◇◇◇◇ ◇◇◇◇◇ ◇◇◇◇◇ ◇◇◇◇◇ ◇◇◇◇◇ ◇◇◇◇◇ ◇◇◇◇◇	306,500
◇◇◇◇◇ ◇◇◇◇◇ ◇◇◇◇◇ ◇◇◇◇◇ ◇◇◇◇◇ ◇◇◇◇◇ ◇◇◇◇◇ ◇◇◇◇◇ ◇◇◇◇◇ ◇◇◇◇◇	306,550
◇◇◇◇◇ ◇◇◇◇◇ ◇◇◇◇◇ ◇◇◇◇◇ ◇◇◇◇◇ ◇◇◇◇◇ ◇◇◇◇◇ ◇◇◇◇◇ ◇◇◇◇◇ ◇◇◇◇◇	306,600
◇◇◇◇◇ ◇◇◇◇◇ ◇◇◇◇◇ ◇◇◇◇◇ ◇◇◇◇◇ ◇◇◇◇◇ ◇◇◇◇◇ ◇◇◇◇◇ ◇◇◇◇◇ ◇◇◇◇◇	306,650
◇◇◇◇◇ ◇◇◇◇◇ ◇◇◇◇◇ ◇◇◇◇◇ ◇◇◇◇◇ ◇◇◇◇◇ ◇◇◇◇◇ ◇◇◇◇◇ ◇◇◇◇◇ ◇◇◇◇◇	306,700
◇◇◇◇◇ ◇◇◇◇◇ ◇◇◇◇◇ ◇◇◇◇◇ ◇◇◇◇◇ ◇◇◇◇◇ ◇◇◇◇◇ ◇◇◇◇◇ ◇◇◇◇◇ ◇◇◇◇◇	306,750
◇◇◇◇◇ ◇◇◇◇◇ ◇◇◇◇◇ ◇◇◇◇◇ ◇◇◇◇◇ ◇◇◇◇◇ ◇◇◇◇◇ ◇◇◇◇◇ ◇◇◇◇◇ ◇◇◇◇◇	306,800
◇◇◇◇◇ ◇◇◇◇◇ ◇◇◇◇◇ ◇◇◇◇◇ ◇◇◇◇◇ ◇◇◇◇◇ ◇◇◇◇◇ ◇◇◇◇◇ ◇◇◇◇◇ ◇◇◇◇◇	306,850
◇◇◇◇◇ ◇◇◇◇◇ ◇◇◇◇◇ ◇◇◇◇◇ ◇◇◇◇◇ ◇◇◇◇◇ ◇◇◇◇◇ ◇◇◇◇◇ ◇◇◇◇◇ ◇◇◇◇◇	306,900
◇◇◇◇◇ ◇◇◇◇◇ ◇◇◇◇◇ ◇◇◇◇◇ ◇◇◇◇◇ ◇◇◇◇◇ ◇◇◇◇◇ ◇◇◇◇◇ ◇◇◇◇◇ ◇◇◇◇◇	306,950
◇◇◇◇◇ ◇◇◇◇◇ ◇◇◇◇◇ ◇◇◇◇◇ ◇◇◇◇◇ ◇◇◇◇◇ ◇◇◇◇◇ ◇◇◇◇◇ ◇◇◇◇◇ ◇◇◇◇◇	**307,000**
◇◇◇◇◇ ◇◇◇◇◇ ◇◇◇◇◇ ◇◇◇◇◇ ◇◇◇◇◇ ◇◇◇◇◇ ◇◇◇◇◇ ◇◇◇◇◇ ◇◇◇◇◇ ◇◇◇◇◇	307,050
◇◇◇◇◇ ◇◇◇◇◇ ◇◇◇◇◇ ◇◇◇◇◇ ◇◇◇◇◇ ◇◇◇◇◇ ◇◇◇◇◇ ◇◇◇◇◇ ◇◇◇◇◇ ◇◇◇◇◇	307,100
◇◇◇◇◇ ◇◇◇◇◇ ◇◇◇◇◇ ◇◇◇◇◇ ◇◇◇◇◇ ◇◇◇◇◇ ◇◇◇◇◇ ◇◇◇◇◇ ◇◇◇◇◇ ◇◇◇◇◇	307,150
◇◇◇◇◇ ◇◇◇◇◇ ◇◇◇◇◇ ◇◇◇◇◇ ◇◇◇◇◇ ◇◇◇◇◇ ◇◇◇◇◇ ◇◇◇◇◇ ◇◇◇◇◇ ◇◇◇◇◇	307,200
◇◇◇◇◇ ◇◇◇◇◇ ◇◇◇◇◇ ◇◇◇◇◇ ◇◇◇◇◇ ◇◇◇◇◇ ◇◇◇◇◇ ◇◇◇◇◇ ◇◇◇◇◇ ◇◇◇◇◇	307,250
◇◇◇◇◇ ◇◇◇◇◇ ◇◇◇◇◇ ◇◇◇◇◇ ◇◇◇◇◇ ◇◇◇◇◇ ◇◇◇◇◇ ◇◇◇◇◇ ◇◇◇◇◇ ◇◇◇◇◇	307,300
◇◇◇◇◇ ◇◇◇◇◇ ◇◇◇◇◇ ◇◇◇◇◇ ◇◇◇◇◇ ◇◇◇◇◇ ◇◇◇◇◇ ◇◇◇◇◇ ◇◇◇◇◇ ◇◇◇◇◇	307,350
◇◇◇◇◇ ◇◇◇◇◇ ◇◇◇◇◇ ◇◇◇◇◇ ◇◇◇◇◇ ◇◇◇◇◇ ◇◇◇◇◇ ◇◇◇◇◇ ◇◇◇◇◇ ◇◇◇◇◇	307,400
◇◇◇◇◇ ◇◇◇◇◇ ◇◇◇◇◇ ◇◇◇◇◇ ◇◇◇◇◇ ◇◇◇◇◇ ◇◇◇◇◇ ◇◇◇◇◇ ◇◇◇◇◇ ◇◇◇◇◇	307,450
◇◇◇◇◇ ◇◇◇◇◇ ◇◇◇◇◇ ◇◇◇◇◇ ◇◇◇◇◇ ◇◇◇◇◇ ◇◇◇◇◇ ◇◇◇◇◇ ◇◇◇◇◇ ◇◇◇◇◇	307,500
◇◇◇◇◇ ◇◇◇◇◇ ◇◇◇◇◇ ◇◇◇◇◇ ◇◇◇◇◇ ◇◇◇◇◇ ◇◇◇◇◇ ◇◇◇◇◇ ◇◇◇◇◇ ◇◇◇◇◇	307,550
◇◇◇◇◇ ◇◇◇◇◇ ◇◇◇◇◇ ◇◇◇◇◇ ◇◇◇◇◇ ◇◇◇◇◇ ◇◇◇◇◇ ◇◇◇◇◇ ◇◇◇◇◇ ◇◇◇◇◇	307,600
◇◇◇◇◇ ◇◇◇◇◇ ◇◇◇◇◇ ◇◇◇◇◇ ◇◇◇◇◇ ◇◇◇◇◇ ◇◇◇◇◇ ◇◇◇◇◇ ◇◇◇◇◇ ◇◇◇◇◇	307,650
◇◇◇◇◇ ◇◇◇◇◇ ◇◇◇◇◇ ◇◇◇◇◇ ◇◇◇◇◇ ◇◇◇◇◇ ◇◇◇◇◇ ◇◇◇◇◇ ◇◇◇◇◇ ◇◇◇◇◇	307,700
◇◇◇◇◇ ◇◇◇◇◇ ◇◇◇◇◇ ◇◇◇◇◇ ◇◇◇◇◇ ◇◇◇◇◇ ◇◇◇◇◇ ◇◇◇◇◇ ◇◇◇◇◇ ◇◇◇◇◇	307,750
◇◇◇◇◇ ◇◇◇◇◇ ◇◇◇◇◇ ◇◇◇◇◇ ◇◇◇◇◇ ◇◇◇◇◇ ◇◇◇◇◇ ◇◇◇◇◇ ◇◇◇◇◇ ◇◇◇◇◇	307,800
◇◇◇◇◇ ◇◇◇◇◇ ◇◇◇◇◇ ◇◇◇◇◇ ◇◇◇◇◇ ◇◇◇◇◇ ◇◇◇◇◇ ◇◇◇◇◇ ◇◇◇◇◇ ◇◇◇◇◇	307,850
◇◇◇◇◇ ◇◇◇◇◇ ◇◇◇◇◇ ◇◇◇◇◇ ◇◇◇◇◇ ◇◇◇◇◇ ◇◇◇◇◇ ◇◇◇◇◇ ◇◇◇◇◇ ◇◇◇◇◇	307,900
◇◇◇◇◇ ◇◇◇◇◇ ◇◇◇◇◇ ◇◇◇◇◇ ◇◇◇◇◇ ◇◇◇◇◇ ◇◇◇◇◇ ◇◇◇◇◇ ◇◇◇◇◇ ◇◇◇◇◇	307,950
◇◇◇◇◇ ◇◇◇◇◇ ◇◇◇◇◇ ◇◇◇◇◇ ◇◇◇◇◇ ◇◇◇◇◇ ◇◇◇◇◇ ◇◇◇◇◇ ◇◇◇◇◇ ◇◇◇◇◇	**308,000**

The Number 308,001 to the Number 310,000

Start   Notes

Diamonds	Number
◊◊◊◊◊ ◊◊◊◊◊ ◊◊◊◊◊ ◊◊◊◊◊ ◊◊◊◊◊ ◊◊◊◊◊ ◊◊◊◊◊ ◊◊◊◊◊ ◊◊◊◊◊ ◊◊◊◊◊	308,050
◊◊◊◊◊ ◊◊◊◊◊ ◊◊◊◊◊ ◊◊◊◊◊ ◊◊◊◊◊ ◊◊◊◊◊ ◊◊◊◊◊ ◊◊◊◊◊ ◊◊◊◊◊ ◊◊◊◊◊	308,100
◊◊◊◊◊ ◊◊◊◊◊ ◊◊◊◊◊ ◊◊◊◊◊ ◊◊◊◊◊ ◊◊◊◊◊ ◊◊◊◊◊ ◊◊◊◊◊ ◊◊◊◊◊ ◊◊◊◊◊	308,150
◊◊◊◊◊ ◊◊◊◊◊ ◊◊◊◊◊ ◊◊◊◊◊ ◊◊◊◊◊ ◊◊◊◊◊ ◊◊◊◊◊ ◊◊◊◊◊ ◊◊◊◊◊ ◊◊◊◊◊	308,200
◊◊◊◊◊ ◊◊◊◊◊ ◊◊◊◊◊ ◊◊◊◊◊ ◊◊◊◊◊ ◊◊◊◊◊ ◊◊◊◊◊ ◊◊◊◊◊ ◊◊◊◊◊ ◊◊◊◊◊	308,250
◊◊◊◊◊ ◊◊◊◊◊ ◊◊◊◊◊ ◊◊◊◊◊ ◊◊◊◊◊ ◊◊◊◊◊ ◊◊◊◊◊ ◊◊◊◊◊ ◊◊◊◊◊ ◊◊◊◊◊	308,300
◊◊◊◊◊ ◊◊◊◊◊ ◊◊◊◊◊ ◊◊◊◊◊ ◊◊◊◊◊ ◊◊◊◊◊ ◊◊◊◊◊ ◊◊◊◊◊ ◊◊◊◊◊ ◊◊◊◊◊	308,350
◊◊◊◊◊ ◊◊◊◊◊ ◊◊◊◊◊ ◊◊◊◊◊ ◊◊◊◊◊ ◊◊◊◊◊ ◊◊◊◊◊ ◊◊◊◊◊ ◊◊◊◊◊ ◊◊◊◊◊	308,400
◊◊◊◊◊ ◊◊◊◊◊ ◊◊◊◊◊ ◊◊◊◊◊ ◊◊◊◊◊ ◊◊◊◊◊ ◊◊◊◊◊ ◊◊◊◊◊ ◊◊◊◊◊ ◊◊◊◊◊	308,450
◊◊◊◊◊ ◊◊◊◊◊ ◊◊◊◊◊ ◊◊◊◊◊ ◊◊◊◊◊ ◊◊◊◊◊ ◊◊◊◊◊ ◊◊◊◊◊ ◊◊◊◊◊ ◊◊◊◊◊	308,500
◊◊◊◊◊ ◊◊◊◊◊ ◊◊◊◊◊ ◊◊◊◊◊ ◊◊◊◊◊ ◊◊◊◊◊ ◊◊◊◊◊ ◊◊◊◊◊ ◊◊◊◊◊ ◊◊◊◊◊	308,550
◊◊◊◊◊ ◊◊◊◊◊ ◊◊◊◊◊ ◊◊◊◊◊ ◊◊◊◊◊ ◊◊◊◊◊ ◊◊◊◊◊ ◊◊◊◊◊ ◊◊◊◊◊ ◊◊◊◊◊	308,600
◊◊◊◊◊ ◊◊◊◊◊ ◊◊◊◊◊ ◊◊◊◊◊ ◊◊◊◊◊ ◊◊◊◊◊ ◊◊◊◊◊ ◊◊◊◊◊ ◊◊◊◊◊ ◊◊◊◊◊	308,650
◊◊◊◊◊ ◊◊◊◊◊ ◊◊◊◊◊ ◊◊◊◊◊ ◊◊◊◊◊ ◊◊◊◊◊ ◊◊◊◊◊ ◊◊◊◊◊ ◊◊◊◊◊ ◊◊◊◊◊	308,700
◊◊◊◊◊ ◊◊◊◊◊ ◊◊◊◊◊ ◊◊◊◊◊ ◊◊◊◊◊ ◊◊◊◊◊ ◊◊◊◊◊ ◊◊◊◊◊ ◊◊◊◊◊ ◊◊◊◊◊	308,750
◊◊◊◊◊ ◊◊◊◊◊ ◊◊◊◊◊ ◊◊◊◊◊ ◊◊◊◊◊ ◊◊◊◊◊ ◊◊◊◊◊ ◊◊◊◊◊ ◊◊◊◊◊ ◊◊◊◊◊	308,800
◊◊◊◊◊ ◊◊◊◊◊ ◊◊◊◊◊ ◊◊◊◊◊ ◊◊◊◊◊ ◊◊◊◊◊ ◊◊◊◊◊ ◊◊◊◊◊ ◊◊◊◊◊ ◊◊◊◊◊	308,850
◊◊◊◊◊ ◊◊◊◊◊ ◊◊◊◊◊ ◊◊◊◊◊ ◊◊◊◊◊ ◊◊◊◊◊ ◊◊◊◊◊ ◊◊◊◊◊ ◊◊◊◊◊ ◊◊◊◊◊	308,900
◊◊◊◊◊ ◊◊◊◊◊ ◊◊◊◊◊ ◊◊◊◊◊ ◊◊◊◊◊ ◊◊◊◊◊ ◊◊◊◊◊ ◊◊◊◊◊ ◊◊◊◊◊ ◊◊◊◊◊	308,950
◊◊◊◊◊ ◊◊◊◊◊ ◊◊◊◊◊ ◊◊◊◊◊ ◊◊◊◊◊ ◊◊◊◊◊ ◊◊◊◊◊ ◊◊◊◊◊ ◊◊◊◊◊ ◊◊◊◊◊	**309,000**
◊◊◊◊◊ ◊◊◊◊◊ ◊◊◊◊◊ ◊◊◊◊◊ ◊◊◊◊◊ ◊◊◊◊◊ ◊◊◊◊◊ ◊◊◊◊◊ ◊◊◊◊◊ ◊◊◊◊◊	309,050
◊◊◊◊◊ ◊◊◊◊◊ ◊◊◊◊◊ ◊◊◊◊◊ ◊◊◊◊◊ ◊◊◊◊◊ ◊◊◊◊◊ ◊◊◊◊◊ ◊◊◊◊◊ ◊◊◊◊◊	309,100
◊◊◊◊◊ ◊◊◊◊◊ ◊◊◊◊◊ ◊◊◊◊◊ ◊◊◊◊◊ ◊◊◊◊◊ ◊◊◊◊◊ ◊◊◊◊◊ ◊◊◊◊◊ ◊◊◊◊◊	309,150
◊◊◊◊◊ ◊◊◊◊◊ ◊◊◊◊◊ ◊◊◊◊◊ ◊◊◊◊◊ ◊◊◊◊◊ ◊◊◊◊◊ ◊◊◊◊◊ ◊◊◊◊◊ ◊◊◊◊◊	309,200
◊◊◊◊◊ ◊◊◊◊◊ ◊◊◊◊◊ ◊◊◊◊◊ ◊◊◊◊◊ ◊◊◊◊◊ ◊◊◊◊◊ ◊◊◊◊◊ ◊◊◊◊◊ ◊◊◊◊◊	309,250
◊◊◊◊◊ ◊◊◊◊◊ ◊◊◊◊◊ ◊◊◊◊◊ ◊◊◊◊◊ ◊◊◊◊◊ ◊◊◊◊◊ ◊◊◊◊◊ ◊◊◊◊◊ ◊◊◊◊◊	309,300
◊◊◊◊◊ ◊◊◊◊◊ ◊◊◊◊◊ ◊◊◊◊◊ ◊◊◊◊◊ ◊◊◊◊◊ ◊◊◊◊◊ ◊◊◊◊◊ ◊◊◊◊◊ ◊◊◊◊◊	309,350
◊◊◊◊◊ ◊◊◊◊◊ ◊◊◊◊◊ ◊◊◊◊◊ ◊◊◊◊◊ ◊◊◊◊◊ ◊◊◊◊◊ ◊◊◊◊◊ ◊◊◊◊◊ ◊◊◊◊◊	309,400
◊◊◊◊◊ ◊◊◊◊◊ ◊◊◊◊◊ ◊◊◊◊◊ ◊◊◊◊◊ ◊◊◊◊◊ ◊◊◊◊◊ ◊◊◊◊◊ ◊◊◊◊◊ ◊◊◊◊◊	309,450
◊◊◊◊◊ ◊◊◊◊◊ ◊◊◊◊◊ ◊◊◊◊◊ ◊◊◊◊◊ ◊◊◊◊◊ ◊◊◊◊◊ ◊◊◊◊◊ ◊◊◊◊◊ ◊◊◊◊◊	309,500
◊◊◊◊◊ ◊◊◊◊◊ ◊◊◊◊◊ ◊◊◊◊◊ ◊◊◊◊◊ ◊◊◊◊◊ ◊◊◊◊◊ ◊◊◊◊◊ ◊◊◊◊◊ ◊◊◊◊◊	309,550
◊◊◊◊◊ ◊◊◊◊◊ ◊◊◊◊◊ ◊◊◊◊◊ ◊◊◊◊◊ ◊◊◊◊◊ ◊◊◊◊◊ ◊◊◊◊◊ ◊◊◊◊◊ ◊◊◊◊◊	309,600
◊◊◊◊◊ ◊◊◊◊◊ ◊◊◊◊◊ ◊◊◊◊◊ ◊◊◊◊◊ ◊◊◊◊◊ ◊◊◊◊◊ ◊◊◊◊◊ ◊◊◊◊◊ ◊◊◊◊◊	309,650
◊◊◊◊◊ ◊◊◊◊◊ ◊◊◊◊◊ ◊◊◊◊◊ ◊◊◊◊◊ ◊◊◊◊◊ ◊◊◊◊◊ ◊◊◊◊◊ ◊◊◊◊◊ ◊◊◊◊◊	309,700
◊◊◊◊◊ ◊◊◊◊◊ ◊◊◊◊◊ ◊◊◊◊◊ ◊◊◊◊◊ ◊◊◊◊◊ ◊◊◊◊◊ ◊◊◊◊◊ ◊◊◊◊◊ ◊◊◊◊◊	309,750
◊◊◊◊◊ ◊◊◊◊◊ ◊◊◊◊◊ ◊◊◊◊◊ ◊◊◊◊◊ ◊◊◊◊◊ ◊◊◊◊◊ ◊◊◊◊◊ ◊◊◊◊◊ ◊◊◊◊◊	309,800
◊◊◊◊◊ ◊◊◊◊◊ ◊◊◊◊◊ ◊◊◊◊◊ ◊◊◊◊◊ ◊◊◊◊◊ ◊◊◊◊◊ ◊◊◊◊◊ ◊◊◊◊◊ ◊◊◊◊◊	309,850
◊◊◊◊◊ ◊◊◊◊◊ ◊◊◊◊◊ ◊◊◊◊◊ ◊◊◊◊◊ ◊◊◊◊◊ ◊◊◊◊◊ ◊◊◊◊◊ ◊◊◊◊◊ ◊◊◊◊◊	309,900
◊◊◊◊◊ ◊◊◊◊◊ ◊◊◊◊◊ ◊◊◊◊◊ ◊◊◊◊◊ ◊◊◊◊◊ ◊◊◊◊◊ ◊◊◊◊◊ ◊◊◊◊◊ ◊◊◊◊◊	309,950
◊◊◊◊◊ ◊◊◊◊◊ ◊◊◊◊◊ ◊◊◊◊◊ ◊◊◊◊◊ ◊◊◊◊◊ ◊◊◊◊◊ ◊◊◊◊◊ ◊◊◊◊◊ ◊◊◊◊◊	**310,000**

The Number 310,001 to the Number 312,000

Start		Notes
◊◊◊◊◊ ◊◊◊◊◊ ◊◊◊◊◊ ◊◊◊◊◊ ◊◊◊◊◊ ◊◊◊◊◊ ◊◊◊◊◊ ◊◊◊◊◊ ◊◊◊◊◊ ◊◊◊◊◊	310,050
◊◊◊◊◊ ◊◊◊◊◊ ◊◊◊◊◊ ◊◊◊◊◊ ◊◊◊◊◊ ◊◊◊◊◊ ◊◊◊◊◊ ◊◊◊◊◊ ◊◊◊◊◊ ◊◊◊◊◊	310,100
◊◊◊◊◊ ◊◊◊◊◊ ◊◊◊◊◊ ◊◊◊◊◊ ◊◊◊◊◊ ◊◊◊◊◊ ◊◊◊◊◊ ◊◊◊◊◊ ◊◊◊◊◊ ◊◊◊◊◊	310,150
◊◊◊◊◊ ◊◊◊◊◊ ◊◊◊◊◊ ◊◊◊◊◊ ◊◊◊◊◊ ◊◊◊◊◊ ◊◊◊◊◊ ◊◊◊◊◊ ◊◊◊◊◊ ◊◊◊◊◊	310,200
◊◊◊◊◊ ◊◊◊◊◊ ◊◊◊◊◊ ◊◊◊◊◊ ◊◊◊◊◊ ◊◊◊◊◊ ◊◊◊◊◊ ◊◊◊◊◊ ◊◊◊◊◊ ◊◊◊◊◊	310,250
◊◊◊◊◊ ◊◊◊◊◊ ◊◊◊◊◊ ◊◊◊◊◊ ◊◊◊◊◊ ◊◊◊◊◊ ◊◊◊◊◊ ◊◊◊◊◊ ◊◊◊◊◊ ◊◊◊◊◊	310,300
◊◊◊◊◊ ◊◊◊◊◊ ◊◊◊◊◊ ◊◊◊◊◊ ◊◊◊◊◊ ◊◊◊◊◊ ◊◊◊◊◊ ◊◊◊◊◊ ◊◊◊◊◊ ◊◊◊◊◊	310,350
◊◊◊◊◊ ◊◊◊◊◊ ◊◊◊◊◊ ◊◊◊◊◊ ◊◊◊◊◊ ◊◊◊◊◊ ◊◊◊◊◊ ◊◊◊◊◊ ◊◊◊◊◊ ◊◊◊◊◊	310,400
◊◊◊◊◊ ◊◊◊◊◊ ◊◊◊◊◊ ◊◊◊◊◊ ◊◊◊◊◊ ◊◊◊◊◊ ◊◊◊◊◊ ◊◊◊◊◊ ◊◊◊◊◊ ◊◊◊◊◊	310,450
◊◊◊◊◊ ◊◊◊◊◊ ◊◊◊◊◊ ◊◊◊◊◊ ◊◊◊◊◊ ◊◊◊◊◊ ◊◊◊◊◊ ◊◊◊◊◊ ◊◊◊◊◊ ◊◊◊◊◊	310,500
◊◊◊◊◊ ◊◊◊◊◊ ◊◊◊◊◊ ◊◊◊◊◊ ◊◊◊◊◊ ◊◊◊◊◊ ◊◊◊◊◊ ◊◊◊◊◊ ◊◊◊◊◊ ◊◊◊◊◊	310,550
◊◊◊◊◊ ◊◊◊◊◊ ◊◊◊◊◊ ◊◊◊◊◊ ◊◊◊◊◊ ◊◊◊◊◊ ◊◊◊◊◊ ◊◊◊◊◊ ◊◊◊◊◊ ◊◊◊◊◊	310,600
◊◊◊◊◊ ◊◊◊◊◊ ◊◊◊◊◊ ◊◊◊◊◊ ◊◊◊◊◊ ◊◊◊◊◊ ◊◊◊◊◊ ◊◊◊◊◊ ◊◊◊◊◊ ◊◊◊◊◊	310,650
◊◊◊◊◊ ◊◊◊◊◊ ◊◊◊◊◊ ◊◊◊◊◊ ◊◊◊◊◊ ◊◊◊◊◊ ◊◊◊◊◊ ◊◊◊◊◊ ◊◊◊◊◊ ◊◊◊◊◊	310,700
◊◊◊◊◊ ◊◊◊◊◊ ◊◊◊◊◊ ◊◊◊◊◊ ◊◊◊◊◊ ◊◊◊◊◊ ◊◊◊◊◊ ◊◊◊◊◊ ◊◊◊◊◊ ◊◊◊◊◊	310,750
◊◊◊◊◊ ◊◊◊◊◊ ◊◊◊◊◊ ◊◊◊◊◊ ◊◊◊◊◊ ◊◊◊◊◊ ◊◊◊◊◊ ◊◊◊◊◊ ◊◊◊◊◊ ◊◊◊◊◊	310,800
◊◊◊◊◊ ◊◊◊◊◊ ◊◊◊◊◊ ◊◊◊◊◊ ◊◊◊◊◊ ◊◊◊◊◊ ◊◊◊◊◊ ◊◊◊◊◊ ◊◊◊◊◊ ◊◊◊◊◊	310,850
◊◊◊◊◊ ◊◊◊◊◊ ◊◊◊◊◊ ◊◊◊◊◊ ◊◊◊◊◊ ◊◊◊◊◊ ◊◊◊◊◊ ◊◊◊◊◊ ◊◊◊◊◊ ◊◊◊◊◊	310,900
◊◊◊◊◊ ◊◊◊◊◊ ◊◊◊◊◊ ◊◊◊◊◊ ◊◊◊◊◊ ◊◊◊◊◊ ◊◊◊◊◊ ◊◊◊◊◊ ◊◊◊◊◊ ◊◊◊◊◊	310,950
◊◊◊◊◊ ◊◊◊◊◊ ◊◊◊◊◊ ◊◊◊◊◊ ◊◊◊◊◊ ◊◊◊◊◊ ◊◊◊◊◊ ◊◊◊◊◊ ◊◊◊◊◊ ◊◊◊◊◊	**311,000**
◊◊◊◊◊ ◊◊◊◊◊ ◊◊◊◊◊ ◊◊◊◊◊ ◊◊◊◊◊ ◊◊◊◊◊ ◊◊◊◊◊ ◊◊◊◊◊ ◊◊◊◊◊ ◊◊◊◊◊	311,050
◊◊◊◊◊ ◊◊◊◊◊ ◊◊◊◊◊ ◊◊◊◊◊ ◊◊◊◊◊ ◊◊◊◊◊ ◊◊◊◊◊ ◊◊◊◊◊ ◊◊◊◊◊ ◊◊◊◊◊	311,100
◊◊◊◊◊ ◊◊◊◊◊ ◊◊◊◊◊ ◊◊◊◊◊ ◊◊◊◊◊ ◊◊◊◊◊ ◊◊◊◊◊ ◊◊◊◊◊ ◊◊◊◊◊ ◊◊◊◊◊	311,150
◊◊◊◊◊ ◊◊◊◊◊ ◊◊◊◊◊ ◊◊◊◊◊ ◊◊◊◊◊ ◊◊◊◊◊ ◊◊◊◊◊ ◊◊◊◊◊ ◊◊◊◊◊ ◊◊◊◊◊	311,200
◊◊◊◊◊ ◊◊◊◊◊ ◊◊◊◊◊ ◊◊◊◊◊ ◊◊◊◊◊ ◊◊◊◊◊ ◊◊◊◊◊ ◊◊◊◊◊ ◊◊◊◊◊ ◊◊◊◊◊	311,250
◊◊◊◊◊ ◊◊◊◊◊ ◊◊◊◊◊ ◊◊◊◊◊ ◊◊◊◊◊ ◊◊◊◊◊ ◊◊◊◊◊ ◊◊◊◊◊ ◊◊◊◊◊ ◊◊◊◊◊	311,300
◊◊◊◊◊ ◊◊◊◊◊ ◊◊◊◊◊ ◊◊◊◊◊ ◊◊◊◊◊ ◊◊◊◊◊ ◊◊◊◊◊ ◊◊◊◊◊ ◊◊◊◊◊ ◊◊◊◊◊	311,350
◊◊◊◊◊ ◊◊◊◊◊ ◊◊◊◊◊ ◊◊◊◊◊ ◊◊◊◊◊ ◊◊◊◊◊ ◊◊◊◊◊ ◊◊◊◊◊ ◊◊◊◊◊ ◊◊◊◊◊	311,400
◊◊◊◊◊ ◊◊◊◊◊ ◊◊◊◊◊ ◊◊◊◊◊ ◊◊◊◊◊ ◊◊◊◊◊ ◊◊◊◊◊ ◊◊◊◊◊ ◊◊◊◊◊ ◊◊◊◊◊	311,450
◊◊◊◊◊ ◊◊◊◊◊ ◊◊◊◊◊ ◊◊◊◊◊ ◊◊◊◊◊ ◊◊◊◊◊ ◊◊◊◊◊ ◊◊◊◊◊ ◊◊◊◊◊ ◊◊◊◊◊	311,500
◊◊◊◊◊ ◊◊◊◊◊ ◊◊◊◊◊ ◊◊◊◊◊ ◊◊◊◊◊ ◊◊◊◊◊ ◊◊◊◊◊ ◊◊◊◊◊ ◊◊◊◊◊ ◊◊◊◊◊	311,550
◊◊◊◊◊ ◊◊◊◊◊ ◊◊◊◊◊ ◊◊◊◊◊ ◊◊◊◊◊ ◊◊◊◊◊ ◊◊◊◊◊ ◊◊◊◊◊ ◊◊◊◊◊ ◊◊◊◊◊	311,600
◊◊◊◊◊ ◊◊◊◊◊ ◊◊◊◊◊ ◊◊◊◊◊ ◊◊◊◊◊ ◊◊◊◊◊ ◊◊◊◊◊ ◊◊◊◊◊ ◊◊◊◊◊ ◊◊◊◊◊	311,650
◊◊◊◊◊ ◊◊◊◊◊ ◊◊◊◊◊ ◊◊◊◊◊ ◊◊◊◊◊ ◊◊◊◊◊ ◊◊◊◊◊ ◊◊◊◊◊ ◊◊◊◊◊ ◊◊◊◊◊	311,700
◊◊◊◊◊ ◊◊◊◊◊ ◊◊◊◊◊ ◊◊◊◊◊ ◊◊◊◊◊ ◊◊◊◊◊ ◊◊◊◊◊ ◊◊◊◊◊ ◊◊◊◊◊ ◊◊◊◊◊	311,750
◊◊◊◊◊ ◊◊◊◊◊ ◊◊◊◊◊ ◊◊◊◊◊ ◊◊◊◊◊ ◊◊◊◊◊ ◊◊◊◊◊ ◊◊◊◊◊ ◊◊◊◊◊ ◊◊◊◊◊	311,800
◊◊◊◊◊ ◊◊◊◊◊ ◊◊◊◊◊ ◊◊◊◊◊ ◊◊◊◊◊ ◊◊◊◊◊ ◊◊◊◊◊ ◊◊◊◊◊ ◊◊◊◊◊ ◊◊◊◊◊	311,850
◊◊◊◊◊ ◊◊◊◊◊ ◊◊◊◊◊ ◊◊◊◊◊ ◊◊◊◊◊ ◊◊◊◊◊ ◊◊◊◊◊ ◊◊◊◊◊ ◊◊◊◊◊ ◊◊◊◊◊	311,900
◊◊◊◊◊ ◊◊◊◊◊ ◊◊◊◊◊ ◊◊◊◊◊ ◊◊◊◊◊ ◊◊◊◊◊ ◊◊◊◊◊ ◊◊◊◊◊ ◊◊◊◊◊ ◊◊◊◊◊	311,950
◊◊◊◊◊ ◊◊◊◊◊ ◊◊◊◊◊ ◊◊◊◊◊ ◊◊◊◊◊ ◊◊◊◊◊ ◊◊◊◊◊ ◊◊◊◊◊ ◊◊◊◊◊ ◊◊◊◊◊	**312,000**

The Number 312,001 to the Number 314,000

Start   Notes

	312,050
	312,100
	312,150
	312,200
	312,250
	312,300
	312,350
	312,400
	312,450
	312,500
	312,550
	312,600
	312,650
	312,700
	312,750
	312,800
	312,850
	312,900
	312,950
	313,000
	313,050
	313,100
	313,150
	313,200
	313,250
	313,300
	313,350
	313,400
	313,450
	313,500
	313,550
	313,600
	313,650
	313,700
	313,750
	313,800
	313,850
	313,900
	313,950
	314,000

The Number 314,001 to the Number 316,000

Start		Notes
	314,050	
	314,100	
	314,150	
	314,200	
	314,250	
	314,300	
	314,350	
	314,400	
	314,450	
	314,500	
	314,550	
	314,600	
	314,650	
	314,700	
	314,750	
	314,800	
	314,850	
	314,900	
	314,950	
	315,000	
	315,050	
	315,100	
	315,150	
	315,200	
	315,250	
	315,300	
	315,350	
	315,400	
	315,450	
	315,500	
	315,550	
	315,600	
	315,650	
	315,700	
	315,750	
	315,800	
	315,850	
	315,900	
	315,950	
	316,000	

The Number 316,001 to the Number 318,000

Start   Notes

	316,050
	316,100
	316,150
	316,200
	316,250
	316,300
	316,350
	316,400
	316,450
	316,500
	316,550
	316,600
	316,650
	316,700
	316,750
	316,800
	316,850
	316,900
	316,950
	317,000
	317,050
	317,100
	317,150
	317,200
	317,250
	317,300
	317,350
	317,400
	317,450
	317,500
	317,550
	317,600
	317,650
	317,700
	317,750
	317,800
	317,850
	317,900
	317,950
	318,000

The Number 318,001 to the Number 320,000

Start	Notes
	318,050
	318,100
	318,150
	318,200
	318,250
	318,300
	318,350
	318,400
	318,450
	318,500
	318,550
	318,600
	318,650
	318,700
	318,750
	318,800
	318,850
	318,900
	318,950
	319,000
	319,050
	319,100
	319,150
	319,200
	319,250
	319,300
	319,350
	319,400
	319,450
	319,500
	319,550
	319,600
	319,650
	319,700
	319,750
	319,800
	319,850
	319,900
	319,950
	320,000

The Number 320,001 to the Number 322,000

Start   **Notes**

	320,050
	320,100
	320,150
	320,200
	320,250
	320,300
	320,350
	320,400
	320,450
	320,500
	320,550
	320,600
	320,650
	320,700
	320,750
	320,800
	320,850
	320,900
	320,950
	321,000
	321,050
	321,100
	321,150
	321,200
	321,250
	321,300
	321,350
	321,400
	321,450
	321,500
	321,550
	321,600
	321,650
	321,700
	321,750
	321,800
	321,850
	321,900
	321,950
	322,000

The Number 322,001 to the Number 324,000

Start	Notes
322,050	
322,100	
322,150	
322,200	
322,250	
322,300	
322,350	
322,400	
322,450	
322,500	
322,550	
322,600	
322,650	
322,700	
322,750	
322,800	
322,850	
322,900	
322,950	
323,000	
323,050	
323,100	
323,150	
323,200	
323,250	
323,300	
323,350	
323,400	
323,450	
323,500	
323,550	
323,600	
323,650	
323,700	
323,750	
323,800	
323,850	
323,900	
323,950	
324,000	

The Number 324,001 to the Number 326,000

Start | Notes

	324,050
	324,100
	324,150
	324,200
	324,250
	324,300
	324,350
	324,400
	324,450
	324,500
	324,550
	324,600
	324,650
	324,700
	324,750
	324,800
	324,850
	324,900
	324,950
	325,000
	325,050
	325,100
	325,150
	325,200
	325,250
	325,300
	325,350
	325,400
	325,450
	325,500
	325,550
	325,600
	325,650
	325,700
	325,750
	325,800
	325,850
	325,900
	325,950
	326,000

The Number 326,001 to the Number 328,000

Start		Notes
	326,050	
	326,100	
	326,150	
	326,200	
	326,250	
	326,300	
	326,350	
	326,400	
	326,450	
	326,500	
	326,550	
	326,600	
	326,650	
	326,700	
	326,750	
	326,800	
	326,850	
	326,900	
	326,950	
	327,000	
	327,050	
	327,100	
	327,150	
	327,200	
	327,250	
	327,300	
	327,350	
	327,400	
	327,450	
	327,500	
	327,550	
	327,600	
	327,650	
	327,700	
	327,750	
	327,800	
	327,850	
	327,900	
	327,950	
	328,000	

The Number 328,001 to the Number 330,000

Start

Notes

328,050	
328,100	
328,150	
328,200	
328,250	
328,300	
328,350	
328,400	
328,450	
328,500	
328,550	
328,600	
328,650	
328,700	
328,750	
328,800	
328,850	
328,900	
328,950	
329,000	
329,050	
329,100	
329,150	
329,200	
329,250	
329,300	
329,350	
329,400	
329,450	
329,500	
329,550	
329,600	
329,650	
329,700	
329,750	
329,800	
329,850	
329,900	
329,950	
330,000	

The Number 330,001 to the Number 332,000

Start		Notes
330,050
330,100
330,150
330,200
330,250
330,300
330,350
330,400
330,450
330,500
330,550
330,600
330,650
330,700
330,750
330,800
330,850
330,900
330,950
331,000
331,050
331,100
331,150
331,200
331,250
331,300
331,350
331,400
331,450
331,500
331,550
331,600
331,650
331,700
331,750
331,800
331,850
331,900
331,950
332,000

The Number 332,001 to the Number 334,000

Start   Notes

	332,050
	332,100
	332,150
	332,200
	332,250
	332,300
	332,350
	332,400
	332,450
	332,500
	332,550
	332,600
	332,650
	332,700
	332,750
	332,800
	332,850
	332,900
	332,950
	333,000
	333,050
	333,100
	333,150
	333,200
	333,250
	333,300
	333,350
	333,400
	333,450
	333,500
	333,550
	333,600
	333,650
	333,700
	333,750
	333,800
	333,850
	333,900
	333,950
	334,000

The Number 334,001 to the Number 336,000

Start | | Notes

334,050
334,100
334,150
334,200
334,250
334,300
334,350
334,400
334,450
334,500
334,550
334,600
334,650
334,700
334,750
334,800
334,850
334,900
334,950
335,000
335,050
335,100
335,150
335,200
335,250
335,300
335,350
335,400
335,450
335,500
335,550
335,600
335,650
335,700
335,750
335,800
335,850
335,900
335,950
336,000

The Number 336,001 to the Number 338,000

Start   Notes

	336,050
	336,100
	336,150
	336,200
	336,250
	336,300
	336,350
	336,400
	336,450
	336,500
	336,550
	336,600
	336,650
	336,700
	336,750
	336,800
	336,850
	336,900
	336,950
	337,000
	337,050
	337,100
	337,150
	337,200
	337,250
	337,300
	337,350
	337,400
	337,450
	337,500
	337,550
	337,600
	337,650
	337,700
	337,750
	337,800
	337,850
	337,900
	337,950
	338,000

The Number 338,001 to the Number 340,000

Start

Notes

◊◊◊◊◊ ◊◊◊◊◊ ◊◊◊◊◊ ◊◊◊◊◊ ◊◊◊◊◊ ◊◊◊◊◊ ◊◊◊◊◊ ◊◊◊◊◊ ◊◊◊◊◊ ◊◊◊◊◊	338,050
◊◊◊◊◊ ◊◊◊◊◊ ◊◊◊◊◊ ◊◊◊◊◊ ◊◊◊◊◊ ◊◊◊◊◊ ◊◊◊◊◊ ◊◊◊◊◊ ◊◊◊◊◊ ◊◊◊◊◊	338,100
◊◊◊◊◊ ◊◊◊◊◊ ◊◊◊◊◊ ◊◊◊◊◊ ◊◊◊◊◊ ◊◊◊◊◊ ◊◊◊◊◊ ◊◊◊◊◊ ◊◊◊◊◊ ◊◊◊◊◊	338,150
◊◊◊◊◊ ◊◊◊◊◊ ◊◊◊◊◊ ◊◊◊◊◊ ◊◊◊◊◊ ◊◊◊◊◊ ◊◊◊◊◊ ◊◊◊◊◊ ◊◊◊◊◊ ◊◊◊◊◊	338,200
◊◊◊◊◊ ◊◊◊◊◊ ◊◊◊◊◊ ◊◊◊◊◊ ◊◊◊◊◊ ◊◊◊◊◊ ◊◊◊◊◊ ◊◊◊◊◊ ◊◊◊◊◊ ◊◊◊◊◊	338,250
◊◊◊◊◊ ◊◊◊◊◊ ◊◊◊◊◊ ◊◊◊◊◊ ◊◊◊◊◊ ◊◊◊◊◊ ◊◊◊◊◊ ◊◊◊◊◊ ◊◊◊◊◊ ◊◊◊◊◊	338,300
◊◊◊◊◊ ◊◊◊◊◊ ◊◊◊◊◊ ◊◊◊◊◊ ◊◊◊◊◊ ◊◊◊◊◊ ◊◊◊◊◊ ◊◊◊◊◊ ◊◊◊◊◊ ◊◊◊◊◊	338,350
◊◊◊◊◊ ◊◊◊◊◊ ◊◊◊◊◊ ◊◊◊◊◊ ◊◊◊◊◊ ◊◊◊◊◊ ◊◊◊◊◊ ◊◊◊◊◊ ◊◊◊◊◊ ◊◊◊◊◊	338,400
◊◊◊◊◊ ◊◊◊◊◊ ◊◊◊◊◊ ◊◊◊◊◊ ◊◊◊◊◊ ◊◊◊◊◊ ◊◊◊◊◊ ◊◊◊◊◊ ◊◊◊◊◊ ◊◊◊◊◊	338,450
◊◊◊◊◊ ◊◊◊◊◊ ◊◊◊◊◊ ◊◊◊◊◊ ◊◊◊◊◊ ◊◊◊◊◊ ◊◊◊◊◊ ◊◊◊◊◊ ◊◊◊◊◊ ◊◊◊◊◊	338,500
◊◊◊◊◊ ◊◊◊◊◊ ◊◊◊◊◊ ◊◊◊◊◊ ◊◊◊◊◊ ◊◊◊◊◊ ◊◊◊◊◊ ◊◊◊◊◊ ◊◊◊◊◊ ◊◊◊◊◊	338,550
◊◊◊◊◊ ◊◊◊◊◊ ◊◊◊◊◊ ◊◊◊◊◊ ◊◊◊◊◊ ◊◊◊◊◊ ◊◊◊◊◊ ◊◊◊◊◊ ◊◊◊◊◊ ◊◊◊◊◊	338,600
◊◊◊◊◊ ◊◊◊◊◊ ◊◊◊◊◊ ◊◊◊◊◊ ◊◊◊◊◊ ◊◊◊◊◊ ◊◊◊◊◊ ◊◊◊◊◊ ◊◊◊◊◊ ◊◊◊◊◊	338,650
◊◊◊◊◊ ◊◊◊◊◊ ◊◊◊◊◊ ◊◊◊◊◊ ◊◊◊◊◊ ◊◊◊◊◊ ◊◊◊◊◊ ◊◊◊◊◊ ◊◊◊◊◊ ◊◊◊◊◊	338,700
◊◊◊◊◊ ◊◊◊◊◊ ◊◊◊◊◊ ◊◊◊◊◊ ◊◊◊◊◊ ◊◊◊◊◊ ◊◊◊◊◊ ◊◊◊◊◊ ◊◊◊◊◊ ◊◊◊◊◊	338,750
◊◊◊◊◊ ◊◊◊◊◊ ◊◊◊◊◊ ◊◊◊◊◊ ◊◊◊◊◊ ◊◊◊◊◊ ◊◊◊◊◊ ◊◊◊◊◊ ◊◊◊◊◊ ◊◊◊◊◊	338,800
◊◊◊◊◊ ◊◊◊◊◊ ◊◊◊◊◊ ◊◊◊◊◊ ◊◊◊◊◊ ◊◊◊◊◊ ◊◊◊◊◊ ◊◊◊◊◊ ◊◊◊◊◊ ◊◊◊◊◊	338,850
◊◊◊◊◊ ◊◊◊◊◊ ◊◊◊◊◊ ◊◊◊◊◊ ◊◊◊◊◊ ◊◊◊◊◊ ◊◊◊◊◊ ◊◊◊◊◊ ◊◊◊◊◊ ◊◊◊◊◊	338,900
◊◊◊◊◊ ◊◊◊◊◊ ◊◊◊◊◊ ◊◊◊◊◊ ◊◊◊◊◊ ◊◊◊◊◊ ◊◊◊◊◊ ◊◊◊◊◊ ◊◊◊◊◊ ◊◊◊◊◊	338,950
◊◊◊◊◊ ◊◊◊◊◊ ◊◊◊◊◊ ◊◊◊◊◊ ◊◊◊◊◊ ◊◊◊◊◊ ◊◊◊◊◊ ◊◊◊◊◊ ◊◊◊◊◊ ◊◊◊◊◊	**339,000**
◊◊◊◊◊ ◊◊◊◊◊ ◊◊◊◊◊ ◊◊◊◊◊ ◊◊◊◊◊ ◊◊◊◊◊ ◊◊◊◊◊ ◊◊◊◊◊ ◊◊◊◊◊ ◊◊◊◊◊	339,050
◊◊◊◊◊ ◊◊◊◊◊ ◊◊◊◊◊ ◊◊◊◊◊ ◊◊◊◊◊ ◊◊◊◊◊ ◊◊◊◊◊ ◊◊◊◊◊ ◊◊◊◊◊ ◊◊◊◊◊	339,100
◊◊◊◊◊ ◊◊◊◊◊ ◊◊◊◊◊ ◊◊◊◊◊ ◊◊◊◊◊ ◊◊◊◊◊ ◊◊◊◊◊ ◊◊◊◊◊ ◊◊◊◊◊ ◊◊◊◊◊	339,150
◊◊◊◊◊ ◊◊◊◊◊ ◊◊◊◊◊ ◊◊◊◊◊ ◊◊◊◊◊ ◊◊◊◊◊ ◊◊◊◊◊ ◊◊◊◊◊ ◊◊◊◊◊ ◊◊◊◊◊	339,200
◊◊◊◊◊ ◊◊◊◊◊ ◊◊◊◊◊ ◊◊◊◊◊ ◊◊◊◊◊ ◊◊◊◊◊ ◊◊◊◊◊ ◊◊◊◊◊ ◊◊◊◊◊ ◊◊◊◊◊	339,250
◊◊◊◊◊ ◊◊◊◊◊ ◊◊◊◊◊ ◊◊◊◊◊ ◊◊◊◊◊ ◊◊◊◊◊ ◊◊◊◊◊ ◊◊◊◊◊ ◊◊◊◊◊ ◊◊◊◊◊	339,300
◊◊◊◊◊ ◊◊◊◊◊ ◊◊◊◊◊ ◊◊◊◊◊ ◊◊◊◊◊ ◊◊◊◊◊ ◊◊◊◊◊ ◊◊◊◊◊ ◊◊◊◊◊ ◊◊◊◊◊	339,350
◊◊◊◊◊ ◊◊◊◊◊ ◊◊◊◊◊ ◊◊◊◊◊ ◊◊◊◊◊ ◊◊◊◊◊ ◊◊◊◊◊ ◊◊◊◊◊ ◊◊◊◊◊ ◊◊◊◊◊	339,400
◊◊◊◊◊ ◊◊◊◊◊ ◊◊◊◊◊ ◊◊◊◊◊ ◊◊◊◊◊ ◊◊◊◊◊ ◊◊◊◊◊ ◊◊◊◊◊ ◊◊◊◊◊ ◊◊◊◊◊	339,450
◊◊◊◊◊ ◊◊◊◊◊ ◊◊◊◊◊ ◊◊◊◊◊ ◊◊◊◊◊ ◊◊◊◊◊ ◊◊◊◊◊ ◊◊◊◊◊ ◊◊◊◊◊ ◊◊◊◊◊	339,500
◊◊◊◊◊ ◊◊◊◊◊ ◊◊◊◊◊ ◊◊◊◊◊ ◊◊◊◊◊ ◊◊◊◊◊ ◊◊◊◊◊ ◊◊◊◊◊ ◊◊◊◊◊ ◊◊◊◊◊	339,550
◊◊◊◊◊ ◊◊◊◊◊ ◊◊◊◊◊ ◊◊◊◊◊ ◊◊◊◊◊ ◊◊◊◊◊ ◊◊◊◊◊ ◊◊◊◊◊ ◊◊◊◊◊ ◊◊◊◊◊	339,600
◊◊◊◊◊ ◊◊◊◊◊ ◊◊◊◊◊ ◊◊◊◊◊ ◊◊◊◊◊ ◊◊◊◊◊ ◊◊◊◊◊ ◊◊◊◊◊ ◊◊◊◊◊ ◊◊◊◊◊	339,650
◊◊◊◊◊ ◊◊◊◊◊ ◊◊◊◊◊ ◊◊◊◊◊ ◊◊◊◊◊ ◊◊◊◊◊ ◊◊◊◊◊ ◊◊◊◊◊ ◊◊◊◊◊ ◊◊◊◊◊	339,700
◊◊◊◊◊ ◊◊◊◊◊ ◊◊◊◊◊ ◊◊◊◊◊ ◊◊◊◊◊ ◊◊◊◊◊ ◊◊◊◊◊ ◊◊◊◊◊ ◊◊◊◊◊ ◊◊◊◊◊	339,750
◊◊◊◊◊ ◊◊◊◊◊ ◊◊◊◊◊ ◊◊◊◊◊ ◊◊◊◊◊ ◊◊◊◊◊ ◊◊◊◊◊ ◊◊◊◊◊ ◊◊◊◊◊ ◊◊◊◊◊	339,800
◊◊◊◊◊ ◊◊◊◊◊ ◊◊◊◊◊ ◊◊◊◊◊ ◊◊◊◊◊ ◊◊◊◊◊ ◊◊◊◊◊ ◊◊◊◊◊ ◊◊◊◊◊ ◊◊◊◊◊	339,850
◊◊◊◊◊ ◊◊◊◊◊ ◊◊◊◊◊ ◊◊◊◊◊ ◊◊◊◊◊ ◊◊◊◊◊ ◊◊◊◊◊ ◊◊◊◊◊ ◊◊◊◊◊ ◊◊◊◊◊	339,900
◊◊◊◊◊ ◊◊◊◊◊ ◊◊◊◊◊ ◊◊◊◊◊ ◊◊◊◊◊ ◊◊◊◊◊ ◊◊◊◊◊ ◊◊◊◊◊ ◊◊◊◊◊ ◊◊◊◊◊	339,950
◊◊◊◊◊ ◊◊◊◊◊ ◊◊◊◊◊ ◊◊◊◊◊ ◊◊◊◊◊ ◊◊◊◊◊ ◊◊◊◊◊ ◊◊◊◊◊ ◊◊◊◊◊ ◊◊◊◊◊	**340,000**

◊ The Number 340,001 to the Number 342,000

Start		Notes

	340,050
	340,100
	340,150
	340,200
	340,250
	340,300
	340,350
	340,400
	340,450
	340,500
	340,550
	340,600
	340,650
	340,700
	340,750
	340,800
	340,850
	340,900
	340,950
	341,000
	341,050
	341,100
	341,150
	341,200
	341,250
	341,300
	341,350
	341,400
	341,450
	341,500
	341,550
	341,600
	341,650
	341,700
	341,750
	341,800
	341,850
	341,900
	341,950
	342,000

The Number 342,001 to the Number 344,000

Start		Notes
◊◊◊◊◊ ◊◊◊◊◊ ◊◊◊◊◊ ◊◊◊◊◊ ◊◊◊◊◊ ◊◊◊◊◊ ◊◊◊◊◊ ◊◊◊◊◊ ◊◊◊◊◊ ◊◊◊◊◊	342,050
◊◊◊◊◊ ◊◊◊◊◊ ◊◊◊◊◊ ◊◊◊◊◊ ◊◊◊◊◊ ◊◊◊◊◊ ◊◊◊◊◊ ◊◊◊◊◊ ◊◊◊◊◊ ◊◊◊◊◊	342,100
◊◊◊◊◊ ◊◊◊◊◊ ◊◊◊◊◊ ◊◊◊◊◊ ◊◊◊◊◊ ◊◊◊◊◊ ◊◊◊◊◊ ◊◊◊◊◊ ◊◊◊◊◊ ◊◊◊◊◊	342,150
◊◊◊◊◊ ◊◊◊◊◊ ◊◊◊◊◊ ◊◊◊◊◊ ◊◊◊◊◊ ◊◊◊◊◊ ◊◊◊◊◊ ◊◊◊◊◊ ◊◊◊◊◊ ◊◊◊◊◊	342,200
◊◊◊◊◊ ◊◊◊◊◊ ◊◊◊◊◊ ◊◊◊◊◊ ◊◊◊◊◊ ◊◊◊◊◊ ◊◊◊◊◊ ◊◊◊◊◊ ◊◊◊◊◊ ◊◊◊◊◊	342,250
◊◊◊◊◊ ◊◊◊◊◊ ◊◊◊◊◊ ◊◊◊◊◊ ◊◊◊◊◊ ◊◊◊◊◊ ◊◊◊◊◊ ◊◊◊◊◊ ◊◊◊◊◊ ◊◊◊◊◊	342,300
◊◊◊◊◊ ◊◊◊◊◊ ◊◊◊◊◊ ◊◊◊◊◊ ◊◊◊◊◊ ◊◊◊◊◊ ◊◊◊◊◊ ◊◊◊◊◊ ◊◊◊◊◊ ◊◊◊◊◊	342,350
◊◊◊◊◊ ◊◊◊◊◊ ◊◊◊◊◊ ◊◊◊◊◊ ◊◊◊◊◊ ◊◊◊◊◊ ◊◊◊◊◊ ◊◊◊◊◊ ◊◊◊◊◊ ◊◊◊◊◊	342,400
◊◊◊◊◊ ◊◊◊◊◊ ◊◊◊◊◊ ◊◊◊◊◊ ◊◊◊◊◊ ◊◊◊◊◊ ◊◊◊◊◊ ◊◊◊◊◊ ◊◊◊◊◊ ◊◊◊◊◊	342,450
◊◊◊◊◊ ◊◊◊◊◊ ◊◊◊◊◊ ◊◊◊◊◊ ◊◊◊◊◊ ◊◊◊◊◊ ◊◊◊◊◊ ◊◊◊◊◊ ◊◊◊◊◊ ◊◊◊◊◊	342,500
◊◊◊◊◊ ◊◊◊◊◊ ◊◊◊◊◊ ◊◊◊◊◊ ◊◊◊◊◊ ◊◊◊◊◊ ◊◊◊◊◊ ◊◊◊◊◊ ◊◊◊◊◊ ◊◊◊◊◊	342,550
◊◊◊◊◊ ◊◊◊◊◊ ◊◊◊◊◊ ◊◊◊◊◊ ◊◊◊◊◊ ◊◊◊◊◊ ◊◊◊◊◊ ◊◊◊◊◊ ◊◊◊◊◊ ◊◊◊◊◊	342,600
◊◊◊◊◊ ◊◊◊◊◊ ◊◊◊◊◊ ◊◊◊◊◊ ◊◊◊◊◊ ◊◊◊◊◊ ◊◊◊◊◊ ◊◊◊◊◊ ◊◊◊◊◊ ◊◊◊◊◊	342,650
◊◊◊◊◊ ◊◊◊◊◊ ◊◊◊◊◊ ◊◊◊◊◊ ◊◊◊◊◊ ◊◊◊◊◊ ◊◊◊◊◊ ◊◊◊◊◊ ◊◊◊◊◊ ◊◊◊◊◊	342,700
◊◊◊◊◊ ◊◊◊◊◊ ◊◊◊◊◊ ◊◊◊◊◊ ◊◊◊◊◊ ◊◊◊◊◊ ◊◊◊◊◊ ◊◊◊◊◊ ◊◊◊◊◊ ◊◊◊◊◊	342,750
◊◊◊◊◊ ◊◊◊◊◊ ◊◊◊◊◊ ◊◊◊◊◊ ◊◊◊◊◊ ◊◊◊◊◊ ◊◊◊◊◊ ◊◊◊◊◊ ◊◊◊◊◊ ◊◊◊◊◊	342,800
◊◊◊◊◊ ◊◊◊◊◊ ◊◊◊◊◊ ◊◊◊◊◊ ◊◊◊◊◊ ◊◊◊◊◊ ◊◊◊◊◊ ◊◊◊◊◊ ◊◊◊◊◊ ◊◊◊◊◊	342,850
◊◊◊◊◊ ◊◊◊◊◊ ◊◊◊◊◊ ◊◊◊◊◊ ◊◊◊◊◊ ◊◊◊◊◊ ◊◊◊◊◊ ◊◊◊◊◊ ◊◊◊◊◊ ◊◊◊◊◊	342,900
◊◊◊◊◊ ◊◊◊◊◊ ◊◊◊◊◊ ◊◊◊◊◊ ◊◊◊◊◊ ◊◊◊◊◊ ◊◊◊◊◊ ◊◊◊◊◊ ◊◊◊◊◊ ◊◊◊◊◊	342,950
◊◊◊◊◊ ◊◊◊◊◊ ◊◊◊◊◊ ◊◊◊◊◊ ◊◊◊◊◊ ◊◊◊◊◊ ◊◊◊◊◊ ◊◊◊◊◊ ◊◊◊◊◊ ◊◊◊◊◊	**343,000**
◊◊◊◊◊ ◊◊◊◊◊ ◊◊◊◊◊ ◊◊◊◊◊ ◊◊◊◊◊ ◊◊◊◊◊ ◊◊◊◊◊ ◊◊◊◊◊ ◊◊◊◊◊ ◊◊◊◊◊	343,050
◊◊◊◊◊ ◊◊◊◊◊ ◊◊◊◊◊ ◊◊◊◊◊ ◊◊◊◊◊ ◊◊◊◊◊ ◊◊◊◊◊ ◊◊◊◊◊ ◊◊◊◊◊ ◊◊◊◊◊	343,100
◊◊◊◊◊ ◊◊◊◊◊ ◊◊◊◊◊ ◊◊◊◊◊ ◊◊◊◊◊ ◊◊◊◊◊ ◊◊◊◊◊ ◊◊◊◊◊ ◊◊◊◊◊ ◊◊◊◊◊	343,150
◊◊◊◊◊ ◊◊◊◊◊ ◊◊◊◊◊ ◊◊◊◊◊ ◊◊◊◊◊ ◊◊◊◊◊ ◊◊◊◊◊ ◊◊◊◊◊ ◊◊◊◊◊ ◊◊◊◊◊	343,200
◊◊◊◊◊ ◊◊◊◊◊ ◊◊◊◊◊ ◊◊◊◊◊ ◊◊◊◊◊ ◊◊◊◊◊ ◊◊◊◊◊ ◊◊◊◊◊ ◊◊◊◊◊ ◊◊◊◊◊	343,250
◊◊◊◊◊ ◊◊◊◊◊ ◊◊◊◊◊ ◊◊◊◊◊ ◊◊◊◊◊ ◊◊◊◊◊ ◊◊◊◊◊ ◊◊◊◊◊ ◊◊◊◊◊ ◊◊◊◊◊	343,300
◊◊◊◊◊ ◊◊◊◊◊ ◊◊◊◊◊ ◊◊◊◊◊ ◊◊◊◊◊ ◊◊◊◊◊ ◊◊◊◊◊ ◊◊◊◊◊ ◊◊◊◊◊ ◊◊◊◊◊	343,350
◊◊◊◊◊ ◊◊◊◊◊ ◊◊◊◊◊ ◊◊◊◊◊ ◊◊◊◊◊ ◊◊◊◊◊ ◊◊◊◊◊ ◊◊◊◊◊ ◊◊◊◊◊ ◊◊◊◊◊	343,400
◊◊◊◊◊ ◊◊◊◊◊ ◊◊◊◊◊ ◊◊◊◊◊ ◊◊◊◊◊ ◊◊◊◊◊ ◊◊◊◊◊ ◊◊◊◊◊ ◊◊◊◊◊ ◊◊◊◊◊	343,450
◊◊◊◊◊ ◊◊◊◊◊ ◊◊◊◊◊ ◊◊◊◊◊ ◊◊◊◊◊ ◊◊◊◊◊ ◊◊◊◊◊ ◊◊◊◊◊ ◊◊◊◊◊ ◊◊◊◊◊	343,500
◊◊◊◊◊ ◊◊◊◊◊ ◊◊◊◊◊ ◊◊◊◊◊ ◊◊◊◊◊ ◊◊◊◊◊ ◊◊◊◊◊ ◊◊◊◊◊ ◊◊◊◊◊ ◊◊◊◊◊	343,550
◊◊◊◊◊ ◊◊◊◊◊ ◊◊◊◊◊ ◊◊◊◊◊ ◊◊◊◊◊ ◊◊◊◊◊ ◊◊◊◊◊ ◊◊◊◊◊ ◊◊◊◊◊ ◊◊◊◊◊	343,600
◊◊◊◊◊ ◊◊◊◊◊ ◊◊◊◊◊ ◊◊◊◊◊ ◊◊◊◊◊ ◊◊◊◊◊ ◊◊◊◊◊ ◊◊◊◊◊ ◊◊◊◊◊ ◊◊◊◊◊	343,650
◊◊◊◊◊ ◊◊◊◊◊ ◊◊◊◊◊ ◊◊◊◊◊ ◊◊◊◊◊ ◊◊◊◊◊ ◊◊◊◊◊ ◊◊◊◊◊ ◊◊◊◊◊ ◊◊◊◊◊	343,700
◊◊◊◊◊ ◊◊◊◊◊ ◊◊◊◊◊ ◊◊◊◊◊ ◊◊◊◊◊ ◊◊◊◊◊ ◊◊◊◊◊ ◊◊◊◊◊ ◊◊◊◊◊ ◊◊◊◊◊	343,750
◊◊◊◊◊ ◊◊◊◊◊ ◊◊◊◊◊ ◊◊◊◊◊ ◊◊◊◊◊ ◊◊◊◊◊ ◊◊◊◊◊ ◊◊◊◊◊ ◊◊◊◊◊ ◊◊◊◊◊	343,800
◊◊◊◊◊ ◊◊◊◊◊ ◊◊◊◊◊ ◊◊◊◊◊ ◊◊◊◊◊ ◊◊◊◊◊ ◊◊◊◊◊ ◊◊◊◊◊ ◊◊◊◊◊ ◊◊◊◊◊	343,850
◊◊◊◊◊ ◊◊◊◊◊ ◊◊◊◊◊ ◊◊◊◊◊ ◊◊◊◊◊ ◊◊◊◊◊ ◊◊◊◊◊ ◊◊◊◊◊ ◊◊◊◊◊ ◊◊◊◊◊	343,900
◊◊◊◊◊ ◊◊◊◊◊ ◊◊◊◊◊ ◊◊◊◊◊ ◊◊◊◊◊ ◊◊◊◊◊ ◊◊◊◊◊ ◊◊◊◊◊ ◊◊◊◊◊ ◊◊◊◊◊	343,950
◊◊◊◊◊ ◊◊◊◊◊ ◊◊◊◊◊ ◊◊◊◊◊ ◊◊◊◊◊ ◊◊◊◊◊ ◊◊◊◊◊ ◊◊◊◊◊ ◊◊◊◊◊ ◊◊◊◊◊	**344,000**

The Number 344,001 to the Number 346,000

Start		Notes
344,050
344,100
344,150
344,200
344,250
344,300
344,350
344,400
344,450
344,500
344,550
344,600
344,650
344,700
344,750
344,800
344,850
344,900
344,950
345,000
345,050
345,100
345,150
345,200
345,250
345,300
345,350
345,400
345,450
345,500
345,550
345,600
345,650
345,700
345,750
345,800
345,850
345,900
345,950
346,000

The Number 346,001 to the Number 348,000

Start

Notes

	346,050
	346,100
	346,150
	346,200
	346,250
	346,300
	346,350
	346,400
	346,450
	346,500
	346,550
	346,600
	346,650
	346,700
	346,750
	346,800
	346,850
	346,900
	346,950
	347,000
	347,050
	347,100
	347,150
	347,200
	347,250
	347,300
	347,350
	347,400
	347,450
	347,500
	347,550
	347,600
	347,650
	347,700
	347,750
	347,800
	347,850
	347,900
	347,950
	348,000

◊ The Number 348,001 to the Number 350,000

Start   **Notes**

	348,050
	348,100
	348,150
	348,200
	348,250
	348,300
	348,350
	348,400
	348,450
	348,500
	348,550
	348,600
	348,650
	348,700
	348,750
	348,800
	348,850
	348,900
	348,950
	349,000
	349,050
	349,100
	349,150
	349,200
	349,250
	349,300
	349,350
	349,400
	349,450
	349,500
	349,550
	349,600
	349,650
	349,700
	349,750
	349,800
	349,850
	349,900
	349,950
	350,000

The Number 350,001 to the Number 352,000

	Notes
350,050	
350,100	
350,150	
350,200	
350,250	
350,300	
350,350	
350,400	
350,450	
350,500	
350,550	
350,600	
350,650	
350,700	
350,750	
350,800	
350,850	
350,900	
350,950	
351,000	
351,050	
351,100	
351,150	
351,200	
351,250	
351,300	
351,350	
351,400	
351,450	
351,500	
351,550	
351,600	
351,650	
351,700	
351,750	
351,800	
351,850	
351,900	
351,950	
352,000	

The Number 352,001 to the Number 354,000

Start   Notes

	352,050
	352,100
	352,150
	352,200
	352,250
	352,300
	352,350
	352,400
	352,450
	352,500
	352,550
	352,600
	352,650
	352,700
	352,750
	352,800
	352,850
	352,900
	352,950
	353,000
	353,050
	353,100
	353,150
	353,200
	353,250
	353,300
	353,350
	353,400
	353,450
	353,500
	353,550
	353,600
	353,650
	353,700
	353,750
	353,800
	353,850
	353,900
	353,950
	354,000

The Number 354,001 to the Number 356,000

Start		Notes
	354,050	
	354,100	
	354,150	
	354,200	
	354,250	
	354,300	
	354,350	
	354,400	
	354,450	
	354,500	
	354,550	
	354,600	
	354,650	
	354,700	
	354,750	
	354,800	
	354,850	
	354,900	
	354,950	
	355,000	
	355,050	
	355,100	
	355,150	
	355,200	
	355,250	
	355,300	
	355,350	
	355,400	
	355,450	
	355,500	
	355,550	
	355,600	
	355,650	
	355,700	
	355,750	
	355,800	
	355,850	
	355,900	
	355,950	
	356,000	

The Number 356,001 to the Number 358,000

Start   Notes

	356,050
	356,100
	356,150
	356,200
	356,250
	356,300
	356,350
	356,400
	356,450
	356,500
	356,550
	356,600
	356,650
	356,700
	356,750
	356,800
	356,850
	356,900
	356,950
	357,000
	357,050
	357,100
	357,150
	357,200
	357,250
	357,300
	357,350
	357,400
	357,450
	357,500
	357,550
	357,600
	357,650
	357,700
	357,750
	357,800
	357,850
	357,900
	357,950
	358,000

The Number 358,001 to the Number 360,000

	Notes
358,050	
358,100	
358,150	
358,200	
358,250	
358,300	
358,350	
358,400	
358,450	
358,500	
358,550	
358,600	
358,650	
358,700	
358,750	
358,800	
358,850	
358,900	
358,950	
359,000	
359,050	
359,100	
359,150	
359,200	
359,250	
359,300	
359,350	
359,400	
359,450	
359,500	
359,550	
359,600	
359,650	
359,700	
359,750	
359,800	
359,850	
359,900	
359,950	
360,000	

◇ The Number 360,001 to the Number 362,000

Start   **Notes**

	360,050
	360,100
	360,150
	360,200
	360,250
	360,300
	360,350
	360,400
	360,450
	360,500
	360,550
	360,600
	360,650
	360,700
	360,750
	360,800
	360,850
	360,900
	360,950
	361,000
	361,050
	361,100
	361,150
	361,200
	361,250
	361,300
	361,350
	361,400
	361,450
	361,500
	361,550
	361,600
	361,650
	361,700
	361,750
	361,800
	361,850
	361,900
	361,950
	362,000

The Number 362,001 to the Number 364,000

Start | | Notes

362,050	
362,100	
362,150	
362,200	
362,250	
362,300	
362,350	
362,400	
362,450	
362,500	
362,550	
362,600	
362,650	
362,700	
362,750	
362,800	
362,850	
362,900	
362,950	
363,000	
363,050	
363,100	
363,150	
363,200	
363,250	
363,300	
363,350	
363,400	
363,450	
363,500	
363,550	
363,600	
363,650	
363,700	
363,750	
363,800	
363,850	
363,900	
363,950	
364,000	

The Number 364,001 to the Number 366,000

Start   Notes

	364,050
	364,100
	364,150
	364,200
	364,250
	364,300
	364,350
	364,400
	364,450
	364,500
	364,550
	364,600
	364,650
	364,700
	364,750
	364,800
	364,850
	364,900
	364,950
	365,000
	365,050
	365,100
	365,150
	365,200
	365,250
	365,300
	365,350
	365,400
	365,450
	365,500
	365,550
	365,600
	365,650
	365,700
	365,750
	365,800
	365,850
	365,900
	365,950
	366,000

The Number 366,001 to the Number 368,000

Start   Notes

	366,050
	366,100
	366,150
	366,200
	366,250
	366,300
	366,350
	366,400
	366,450
	366,500
	366,550
	366,600
	366,650
	366,700
	366,750
	366,800
	366,850
	366,900
	366,950
	367,000
	367,050
	367,100
	367,150
	367,200
	367,250
	367,300
	367,350
	367,400
	367,450
	367,500
	367,550
	367,600
	367,650
	367,700
	367,750
	367,800
	367,850
	367,900
	367,950
	368,000

The Number 368,001 to the Number 370,000

Start   Notes

	368,050
	368,100
	368,150
	368,200
	368,250
	368,300
	368,350
	368,400
	368,450
	368,500
	368,550
	368,600
	368,650
	368,700
	368,750
	368,800
	368,850
	368,900
	368,950
	369,000
	369,050
	369,100
	369,150
	369,200
	369,250
	369,300
	369,350
	369,400
	369,450
	369,500
	369,550
	369,600
	369,650
	369,700
	369,750
	369,800
	369,850
	369,900
	369,950
	370,000

The Number 370,001 to the Number 372,000

Start		Notes
◊◊◊◊◊ ◊◊◊◊◊ ◊◊◊◊◊ ◊◊◊◊◊ ◊◊◊◊◊ ◊◊◊◊◊ ◊◊◊◊◊ ◊◊◊◊◊ ◊◊◊◊◊ ◊◊◊◊◊	370,050
◊◊◊◊◊ ◊◊◊◊◊ ◊◊◊◊◊ ◊◊◊◊◊ ◊◊◊◊◊ ◊◊◊◊◊ ◊◊◊◊◊ ◊◊◊◊◊ ◊◊◊◊◊ ◊◊◊◊◊	370,100
◊◊◊◊◊ ◊◊◊◊◊ ◊◊◊◊◊ ◊◊◊◊◊ ◊◊◊◊◊ ◊◊◊◊◊ ◊◊◊◊◊ ◊◊◊◊◊ ◊◊◊◊◊ ◊◊◊◊◊	370,150
◊◊◊◊◊ ◊◊◊◊◊ ◊◊◊◊◊ ◊◊◊◊◊ ◊◊◊◊◊ ◊◊◊◊◊ ◊◊◊◊◊ ◊◊◊◊◊ ◊◊◊◊◊ ◊◊◊◊◊	370,200
◊◊◊◊◊ ◊◊◊◊◊ ◊◊◊◊◊ ◊◊◊◊◊ ◊◊◊◊◊ ◊◊◊◊◊ ◊◊◊◊◊ ◊◊◊◊◊ ◊◊◊◊◊ ◊◊◊◊◊	370,250
◊◊◊◊◊ ◊◊◊◊◊ ◊◊◊◊◊ ◊◊◊◊◊ ◊◊◊◊◊ ◊◊◊◊◊ ◊◊◊◊◊ ◊◊◊◊◊ ◊◊◊◊◊ ◊◊◊◊◊	370,300
◊◊◊◊◊ ◊◊◊◊◊ ◊◊◊◊◊ ◊◊◊◊◊ ◊◊◊◊◊ ◊◊◊◊◊ ◊◊◊◊◊ ◊◊◊◊◊ ◊◊◊◊◊ ◊◊◊◊◊	370,350
◊◊◊◊◊ ◊◊◊◊◊ ◊◊◊◊◊ ◊◊◊◊◊ ◊◊◊◊◊ ◊◊◊◊◊ ◊◊◊◊◊ ◊◊◊◊◊ ◊◊◊◊◊ ◊◊◊◊◊	370,400
◊◊◊◊◊ ◊◊◊◊◊ ◊◊◊◊◊ ◊◊◊◊◊ ◊◊◊◊◊ ◊◊◊◊◊ ◊◊◊◊◊ ◊◊◊◊◊ ◊◊◊◊◊ ◊◊◊◊◊	370,450
◊◊◊◊◊ ◊◊◊◊◊ ◊◊◊◊◊ ◊◊◊◊◊ ◊◊◊◊◊ ◊◊◊◊◊ ◊◊◊◊◊ ◊◊◊◊◊ ◊◊◊◊◊ ◊◊◊◊◊	370,500
◊◊◊◊◊ ◊◊◊◊◊ ◊◊◊◊◊ ◊◊◊◊◊ ◊◊◊◊◊ ◊◊◊◊◊ ◊◊◊◊◊ ◊◊◊◊◊ ◊◊◊◊◊ ◊◊◊◊◊	370,550
◊◊◊◊◊ ◊◊◊◊◊ ◊◊◊◊◊ ◊◊◊◊◊ ◊◊◊◊◊ ◊◊◊◊◊ ◊◊◊◊◊ ◊◊◊◊◊ ◊◊◊◊◊ ◊◊◊◊◊	370,600
◊◊◊◊◊ ◊◊◊◊◊ ◊◊◊◊◊ ◊◊◊◊◊ ◊◊◊◊◊ ◊◊◊◊◊ ◊◊◊◊◊ ◊◊◊◊◊ ◊◊◊◊◊ ◊◊◊◊◊	370,650
◊◊◊◊◊ ◊◊◊◊◊ ◊◊◊◊◊ ◊◊◊◊◊ ◊◊◊◊◊ ◊◊◊◊◊ ◊◊◊◊◊ ◊◊◊◊◊ ◊◊◊◊◊ ◊◊◊◊◊	370,700
◊◊◊◊◊ ◊◊◊◊◊ ◊◊◊◊◊ ◊◊◊◊◊ ◊◊◊◊◊ ◊◊◊◊◊ ◊◊◊◊◊ ◊◊◊◊◊ ◊◊◊◊◊ ◊◊◊◊◊	370,750
◊◊◊◊◊ ◊◊◊◊◊ ◊◊◊◊◊ ◊◊◊◊◊ ◊◊◊◊◊ ◊◊◊◊◊ ◊◊◊◊◊ ◊◊◊◊◊ ◊◊◊◊◊ ◊◊◊◊◊	370,800
◊◊◊◊◊ ◊◊◊◊◊ ◊◊◊◊◊ ◊◊◊◊◊ ◊◊◊◊◊ ◊◊◊◊◊ ◊◊◊◊◊ ◊◊◊◊◊ ◊◊◊◊◊ ◊◊◊◊◊	370,850
◊◊◊◊◊ ◊◊◊◊◊ ◊◊◊◊◊ ◊◊◊◊◊ ◊◊◊◊◊ ◊◊◊◊◊ ◊◊◊◊◊ ◊◊◊◊◊ ◊◊◊◊◊ ◊◊◊◊◊	370,900
◊◊◊◊◊ ◊◊◊◊◊ ◊◊◊◊◊ ◊◊◊◊◊ ◊◊◊◊◊ ◊◊◊◊◊ ◊◊◊◊◊ ◊◊◊◊◊ ◊◊◊◊◊ ◊◊◊◊◊	370,950
◊◊◊◊◊ ◊◊◊◊◊ ◊◊◊◊◊ ◊◊◊◊◊ ◊◊◊◊◊ ◊◊◊◊◊ ◊◊◊◊◊ ◊◊◊◊◊ ◊◊◊◊◊ ◊◊◊◊◊	**371,000**
◊◊◊◊◊ ◊◊◊◊◊ ◊◊◊◊◊ ◊◊◊◊◊ ◊◊◊◊◊ ◊◊◊◊◊ ◊◊◊◊◊ ◊◊◊◊◊ ◊◊◊◊◊ ◊◊◊◊◊	371,050
◊◊◊◊◊ ◊◊◊◊◊ ◊◊◊◊◊ ◊◊◊◊◊ ◊◊◊◊◊ ◊◊◊◊◊ ◊◊◊◊◊ ◊◊◊◊◊ ◊◊◊◊◊ ◊◊◊◊◊	371,100
◊◊◊◊◊ ◊◊◊◊◊ ◊◊◊◊◊ ◊◊◊◊◊ ◊◊◊◊◊ ◊◊◊◊◊ ◊◊◊◊◊ ◊◊◊◊◊ ◊◊◊◊◊ ◊◊◊◊◊	371,150
◊◊◊◊◊ ◊◊◊◊◊ ◊◊◊◊◊ ◊◊◊◊◊ ◊◊◊◊◊ ◊◊◊◊◊ ◊◊◊◊◊ ◊◊◊◊◊ ◊◊◊◊◊ ◊◊◊◊◊	371,200
◊◊◊◊◊ ◊◊◊◊◊ ◊◊◊◊◊ ◊◊◊◊◊ ◊◊◊◊◊ ◊◊◊◊◊ ◊◊◊◊◊ ◊◊◊◊◊ ◊◊◊◊◊ ◊◊◊◊◊	371,250
◊◊◊◊◊ ◊◊◊◊◊ ◊◊◊◊◊ ◊◊◊◊◊ ◊◊◊◊◊ ◊◊◊◊◊ ◊◊◊◊◊ ◊◊◊◊◊ ◊◊◊◊◊ ◊◊◊◊◊	371,300
◊◊◊◊◊ ◊◊◊◊◊ ◊◊◊◊◊ ◊◊◊◊◊ ◊◊◊◊◊ ◊◊◊◊◊ ◊◊◊◊◊ ◊◊◊◊◊ ◊◊◊◊◊ ◊◊◊◊◊	371,350
◊◊◊◊◊ ◊◊◊◊◊ ◊◊◊◊◊ ◊◊◊◊◊ ◊◊◊◊◊ ◊◊◊◊◊ ◊◊◊◊◊ ◊◊◊◊◊ ◊◊◊◊◊ ◊◊◊◊◊	371,400
◊◊◊◊◊ ◊◊◊◊◊ ◊◊◊◊◊ ◊◊◊◊◊ ◊◊◊◊◊ ◊◊◊◊◊ ◊◊◊◊◊ ◊◊◊◊◊ ◊◊◊◊◊ ◊◊◊◊◊	371,450
◊◊◊◊◊ ◊◊◊◊◊ ◊◊◊◊◊ ◊◊◊◊◊ ◊◊◊◊◊ ◊◊◊◊◊ ◊◊◊◊◊ ◊◊◊◊◊ ◊◊◊◊◊ ◊◊◊◊◊	371,500
◊◊◊◊◊ ◊◊◊◊◊ ◊◊◊◊◊ ◊◊◊◊◊ ◊◊◊◊◊ ◊◊◊◊◊ ◊◊◊◊◊ ◊◊◊◊◊ ◊◊◊◊◊ ◊◊◊◊◊	371,550
◊◊◊◊◊ ◊◊◊◊◊ ◊◊◊◊◊ ◊◊◊◊◊ ◊◊◊◊◊ ◊◊◊◊◊ ◊◊◊◊◊ ◊◊◊◊◊ ◊◊◊◊◊ ◊◊◊◊◊	371,600
◊◊◊◊◊ ◊◊◊◊◊ ◊◊◊◊◊ ◊◊◊◊◊ ◊◊◊◊◊ ◊◊◊◊◊ ◊◊◊◊◊ ◊◊◊◊◊ ◊◊◊◊◊ ◊◊◊◊◊	371,650
◊◊◊◊◊ ◊◊◊◊◊ ◊◊◊◊◊ ◊◊◊◊◊ ◊◊◊◊◊ ◊◊◊◊◊ ◊◊◊◊◊ ◊◊◊◊◊ ◊◊◊◊◊ ◊◊◊◊◊	371,700
◊◊◊◊◊ ◊◊◊◊◊ ◊◊◊◊◊ ◊◊◊◊◊ ◊◊◊◊◊ ◊◊◊◊◊ ◊◊◊◊◊ ◊◊◊◊◊ ◊◊◊◊◊ ◊◊◊◊◊	371,750
◊◊◊◊◊ ◊◊◊◊◊ ◊◊◊◊◊ ◊◊◊◊◊ ◊◊◊◊◊ ◊◊◊◊◊ ◊◊◊◊◊ ◊◊◊◊◊ ◊◊◊◊◊ ◊◊◊◊◊	371,800
◊◊◊◊◊ ◊◊◊◊◊ ◊◊◊◊◊ ◊◊◊◊◊ ◊◊◊◊◊ ◊◊◊◊◊ ◊◊◊◊◊ ◊◊◊◊◊ ◊◊◊◊◊ ◊◊◊◊◊	371,850
◊◊◊◊◊ ◊◊◊◊◊ ◊◊◊◊◊ ◊◊◊◊◊ ◊◊◊◊◊ ◊◊◊◊◊ ◊◊◊◊◊ ◊◊◊◊◊ ◊◊◊◊◊ ◊◊◊◊◊	371,900
◊◊◊◊◊ ◊◊◊◊◊ ◊◊◊◊◊ ◊◊◊◊◊ ◊◊◊◊◊ ◊◊◊◊◊ ◊◊◊◊◊ ◊◊◊◊◊ ◊◊◊◊◊ ◊◊◊◊◊	371,950
◊◊◊◊◊ ◊◊◊◊◊ ◊◊◊◊◊ ◊◊◊◊◊ ◊◊◊◊◊ ◊◊◊◊◊ ◊◊◊◊◊ ◊◊◊◊◊ ◊◊◊◊◊ ◊◊◊◊◊	**372,000**

The Number 372,001 to the Number 374,000

Start   Notes

◇◇◇◇◇ ◇◇◇◇◇ ◇◇◇◇◇ ◇◇◇◇◇ ◇◇◇◇◇ ◇◇◇◇◇ ◇◇◇◇◇ ◇◇◇◇◇ ◇◇◇◇◇ ◇◇◇◇◇	372,050
◇◇◇◇◇ ◇◇◇◇◇ ◇◇◇◇◇ ◇◇◇◇◇ ◇◇◇◇◇ ◇◇◇◇◇ ◇◇◇◇◇ ◇◇◇◇◇ ◇◇◇◇◇ ◇◇◇◇◇	372,100
◇◇◇◇◇ ◇◇◇◇◇ ◇◇◇◇◇ ◇◇◇◇◇ ◇◇◇◇◇ ◇◇◇◇◇ ◇◇◇◇◇ ◇◇◇◇◇ ◇◇◇◇◇ ◇◇◇◇◇	372,150
◇◇◇◇◇ ◇◇◇◇◇ ◇◇◇◇◇ ◇◇◇◇◇ ◇◇◇◇◇ ◇◇◇◇◇ ◇◇◇◇◇ ◇◇◇◇◇ ◇◇◇◇◇ ◇◇◇◇◇	372,200
◇◇◇◇◇ ◇◇◇◇◇ ◇◇◇◇◇ ◇◇◇◇◇ ◇◇◇◇◇ ◇◇◇◇◇ ◇◇◇◇◇ ◇◇◇◇◇ ◇◇◇◇◇ ◇◇◇◇◇	372,250
◇◇◇◇◇ ◇◇◇◇◇ ◇◇◇◇◇ ◇◇◇◇◇ ◇◇◇◇◇ ◇◇◇◇◇ ◇◇◇◇◇ ◇◇◇◇◇ ◇◇◇◇◇ ◇◇◇◇◇	372,300
◇◇◇◇◇ ◇◇◇◇◇ ◇◇◇◇◇ ◇◇◇◇◇ ◇◇◇◇◇ ◇◇◇◇◇ ◇◇◇◇◇ ◇◇◇◇◇ ◇◇◇◇◇ ◇◇◇◇◇	372,350
◇◇◇◇◇ ◇◇◇◇◇ ◇◇◇◇◇ ◇◇◇◇◇ ◇◇◇◇◇ ◇◇◇◇◇ ◇◇◇◇◇ ◇◇◇◇◇ ◇◇◇◇◇ ◇◇◇◇◇	372,400
◇◇◇◇◇ ◇◇◇◇◇ ◇◇◇◇◇ ◇◇◇◇◇ ◇◇◇◇◇ ◇◇◇◇◇ ◇◇◇◇◇ ◇◇◇◇◇ ◇◇◇◇◇ ◇◇◇◇◇	372,450
◇◇◇◇◇ ◇◇◇◇◇ ◇◇◇◇◇ ◇◇◇◇◇ ◇◇◇◇◇ ◇◇◇◇◇ ◇◇◇◇◇ ◇◇◇◇◇ ◇◇◇◇◇ ◇◇◇◇◇	372,500
◇◇◇◇◇ ◇◇◇◇◇ ◇◇◇◇◇ ◇◇◇◇◇ ◇◇◇◇◇ ◇◇◇◇◇ ◇◇◇◇◇ ◇◇◇◇◇ ◇◇◇◇◇ ◇◇◇◇◇	372,550
◇◇◇◇◇ ◇◇◇◇◇ ◇◇◇◇◇ ◇◇◇◇◇ ◇◇◇◇◇ ◇◇◇◇◇ ◇◇◇◇◇ ◇◇◇◇◇ ◇◇◇◇◇ ◇◇◇◇◇	372,600
◇◇◇◇◇ ◇◇◇◇◇ ◇◇◇◇◇ ◇◇◇◇◇ ◇◇◇◇◇ ◇◇◇◇◇ ◇◇◇◇◇ ◇◇◇◇◇ ◇◇◇◇◇ ◇◇◇◇◇	372,650
◇◇◇◇◇ ◇◇◇◇◇ ◇◇◇◇◇ ◇◇◇◇◇ ◇◇◇◇◇ ◇◇◇◇◇ ◇◇◇◇◇ ◇◇◇◇◇ ◇◇◇◇◇ ◇◇◇◇◇	372,700
◇◇◇◇◇ ◇◇◇◇◇ ◇◇◇◇◇ ◇◇◇◇◇ ◇◇◇◇◇ ◇◇◇◇◇ ◇◇◇◇◇ ◇◇◇◇◇ ◇◇◇◇◇ ◇◇◇◇◇	372,750
◇◇◇◇◇ ◇◇◇◇◇ ◇◇◇◇◇ ◇◇◇◇◇ ◇◇◇◇◇ ◇◇◇◇◇ ◇◇◇◇◇ ◇◇◇◇◇ ◇◇◇◇◇ ◇◇◇◇◇	372,800
◇◇◇◇◇ ◇◇◇◇◇ ◇◇◇◇◇ ◇◇◇◇◇ ◇◇◇◇◇ ◇◇◇◇◇ ◇◇◇◇◇ ◇◇◇◇◇ ◇◇◇◇◇ ◇◇◇◇◇	372,850
◇◇◇◇◇ ◇◇◇◇◇ ◇◇◇◇◇ ◇◇◇◇◇ ◇◇◇◇◇ ◇◇◇◇◇ ◇◇◇◇◇ ◇◇◇◇◇ ◇◇◇◇◇ ◇◇◇◇◇	372,900
◇◇◇◇◇ ◇◇◇◇◇ ◇◇◇◇◇ ◇◇◇◇◇ ◇◇◇◇◇ ◇◇◇◇◇ ◇◇◇◇◇ ◇◇◇◇◇ ◇◇◇◇◇ ◇◇◇◇◇	372,950
◇◇◇◇◇ ◇◇◇◇◇ ◇◇◇◇◇ ◇◇◇◇◇ ◇◇◇◇◇ ◇◇◇◇◇ ◇◇◇◇◇ ◇◇◇◇◇ ◇◇◇◇◇ ◇◇◇◇◇	**373,000**
◇◇◇◇◇ ◇◇◇◇◇ ◇◇◇◇◇ ◇◇◇◇◇ ◇◇◇◇◇ ◇◇◇◇◇ ◇◇◇◇◇ ◇◇◇◇◇ ◇◇◇◇◇ ◇◇◇◇◇	373,050
◇◇◇◇◇ ◇◇◇◇◇ ◇◇◇◇◇ ◇◇◇◇◇ ◇◇◇◇◇ ◇◇◇◇◇ ◇◇◇◇◇ ◇◇◇◇◇ ◇◇◇◇◇ ◇◇◇◇◇	373,100
◇◇◇◇◇ ◇◇◇◇◇ ◇◇◇◇◇ ◇◇◇◇◇ ◇◇◇◇◇ ◇◇◇◇◇ ◇◇◇◇◇ ◇◇◇◇◇ ◇◇◇◇◇ ◇◇◇◇◇	373,150
◇◇◇◇◇ ◇◇◇◇◇ ◇◇◇◇◇ ◇◇◇◇◇ ◇◇◇◇◇ ◇◇◇◇◇ ◇◇◇◇◇ ◇◇◇◇◇ ◇◇◇◇◇ ◇◇◇◇◇	373,200
◇◇◇◇◇ ◇◇◇◇◇ ◇◇◇◇◇ ◇◇◇◇◇ ◇◇◇◇◇ ◇◇◇◇◇ ◇◇◇◇◇ ◇◇◇◇◇ ◇◇◇◇◇ ◇◇◇◇◇	373,250
◇◇◇◇◇ ◇◇◇◇◇ ◇◇◇◇◇ ◇◇◇◇◇ ◇◇◇◇◇ ◇◇◇◇◇ ◇◇◇◇◇ ◇◇◇◇◇ ◇◇◇◇◇ ◇◇◇◇◇	373,300
◇◇◇◇◇ ◇◇◇◇◇ ◇◇◇◇◇ ◇◇◇◇◇ ◇◇◇◇◇ ◇◇◇◇◇ ◇◇◇◇◇ ◇◇◇◇◇ ◇◇◇◇◇ ◇◇◇◇◇	373,350
◇◇◇◇◇ ◇◇◇◇◇ ◇◇◇◇◇ ◇◇◇◇◇ ◇◇◇◇◇ ◇◇◇◇◇ ◇◇◇◇◇ ◇◇◇◇◇ ◇◇◇◇◇ ◇◇◇◇◇	373,400
◇◇◇◇◇ ◇◇◇◇◇ ◇◇◇◇◇ ◇◇◇◇◇ ◇◇◇◇◇ ◇◇◇◇◇ ◇◇◇◇◇ ◇◇◇◇◇ ◇◇◇◇◇ ◇◇◇◇◇	373,450
◇◇◇◇◇ ◇◇◇◇◇ ◇◇◇◇◇ ◇◇◇◇◇ ◇◇◇◇◇ ◇◇◇◇◇ ◇◇◇◇◇ ◇◇◇◇◇ ◇◇◇◇◇ ◇◇◇◇◇	373,500
◇◇◇◇◇ ◇◇◇◇◇ ◇◇◇◇◇ ◇◇◇◇◇ ◇◇◇◇◇ ◇◇◇◇◇ ◇◇◇◇◇ ◇◇◇◇◇ ◇◇◇◇◇ ◇◇◇◇◇	373,550
◇◇◇◇◇ ◇◇◇◇◇ ◇◇◇◇◇ ◇◇◇◇◇ ◇◇◇◇◇ ◇◇◇◇◇ ◇◇◇◇◇ ◇◇◇◇◇ ◇◇◇◇◇ ◇◇◇◇◇	373,600
◇◇◇◇◇ ◇◇◇◇◇ ◇◇◇◇◇ ◇◇◇◇◇ ◇◇◇◇◇ ◇◇◇◇◇ ◇◇◇◇◇ ◇◇◇◇◇ ◇◇◇◇◇ ◇◇◇◇◇	373,650
◇◇◇◇◇ ◇◇◇◇◇ ◇◇◇◇◇ ◇◇◇◇◇ ◇◇◇◇◇ ◇◇◇◇◇ ◇◇◇◇◇ ◇◇◇◇◇ ◇◇◇◇◇ ◇◇◇◇◇	373,700
◇◇◇◇◇ ◇◇◇◇◇ ◇◇◇◇◇ ◇◇◇◇◇ ◇◇◇◇◇ ◇◇◇◇◇ ◇◇◇◇◇ ◇◇◇◇◇ ◇◇◇◇◇ ◇◇◇◇◇	373,750
◇◇◇◇◇ ◇◇◇◇◇ ◇◇◇◇◇ ◇◇◇◇◇ ◇◇◇◇◇ ◇◇◇◇◇ ◇◇◇◇◇ ◇◇◇◇◇ ◇◇◇◇◇ ◇◇◇◇◇	373,800
◇◇◇◇◇ ◇◇◇◇◇ ◇◇◇◇◇ ◇◇◇◇◇ ◇◇◇◇◇ ◇◇◇◇◇ ◇◇◇◇◇ ◇◇◇◇◇ ◇◇◇◇◇ ◇◇◇◇◇	373,850
◇◇◇◇◇ ◇◇◇◇◇ ◇◇◇◇◇ ◇◇◇◇◇ ◇◇◇◇◇ ◇◇◇◇◇ ◇◇◇◇◇ ◇◇◇◇◇ ◇◇◇◇◇ ◇◇◇◇◇	373,900
◇◇◇◇◇ ◇◇◇◇◇ ◇◇◇◇◇ ◇◇◇◇◇ ◇◇◇◇◇ ◇◇◇◇◇ ◇◇◇◇◇ ◇◇◇◇◇ ◇◇◇◇◇ ◇◇◇◇◇	373,950
◇◇◇◇◇ ◇◇◇◇◇ ◇◇◇◇◇ ◇◇◇◇◇ ◇◇◇◇◇ ◇◇◇◇◇ ◇◇◇◇◇ ◇◇◇◇◇ ◇◇◇◇◇ ◇◇◇◇◇	**374,000**

The Number 374,001 to the Number 376,000

Start   Notes

	374,050
	374,100
	374,150
	374,200
	374,250
	374,300
	374,350
	374,400
	374,450
	374,500
	374,550
	374,600
	374,650
	374,700
	374,750
	374,800
	374,850
	374,900
	374,950
	375,000
	375,050
	375,100
	375,150
	375,200
	375,250
	375,300
	375,350
	375,400
	375,450
	375,500
	375,550
	375,600
	375,650
	375,700
	375,750
	375,800
	375,850
	375,900
	375,950
	376,000

The Number 376,001 to the Number 378,000

Start		Notes

	376,050	
	376,100	
	376,150	
	376,200	
	376,250	
	376,300	
	376,350	
	376,400	
	376,450	
	376,500	
	376,550	
	376,600	
	376,650	
	376,700	
	376,750	
	376,800	
	376,850	
	376,900	
	376,950	
	377,000	
	377,050	
	377,100	
	377,150	
	377,200	
	377,250	
	377,300	
	377,350	
	377,400	
	377,450	
	377,500	
	377,550	
	377,600	
	377,650	
	377,700	
	377,750	
	377,800	
	377,850	
	377,900	
	377,950	
	378,000	

The Number 378,001 to the Number 380,000

Start		Notes
	378,050	
	378,100	
	378,150	
	378,200	
	378,250	
	378,300	
	378,350	
	378,400	
	378,450	
	378,500	
	378,550	
	378,600	
	378,650	
	378,700	
	378,750	
	378,800	
	378,850	
	378,900	
	378,950	
	379,000	
	379,050	
	379,100	
	379,150	
	379,200	
	379,250	
	379,300	
	379,350	
	379,400	
	379,450	
	379,500	
	379,550	
	379,600	
	379,650	
	379,700	
	379,750	
	379,800	
	379,850	
	379,900	
	379,950	
	380,000	

The Number 380,001 to the Number 382,000

Start

	Notes
380,050	
380,100	
380,150	
380,200	
380,250	
380,300	
380,350	
380,400	
380,450	
380,500	
380,550	
380,600	
380,650	
380,700	
380,750	
380,800	
380,850	
380,900	
380,950	
381,000	
381,050	
381,100	
381,150	
381,200	
381,250	
381,300	
381,350	
381,400	
381,450	
381,500	
381,550	
381,600	
381,650	
381,700	
381,750	
381,800	
381,850	
381,900	
381,950	
382,000	

The Number 382,001 to the Number 384,000

Start   Notes

◊◊◊◊◊ ◊◊◊◊◊ ◊◊◊◊◊ ◊◊◊◊◊ ◊◊◊◊◊ ◊◊◊◊◊ ◊◊◊◊◊ ◊◊◊◊◊ ◊◊◊◊◊ ◊◊◊◊◊	382,050
◊◊◊◊◊ ◊◊◊◊◊ ◊◊◊◊◊ ◊◊◊◊◊ ◊◊◊◊◊ ◊◊◊◊◊ ◊◊◊◊◊ ◊◊◊◊◊ ◊◊◊◊◊ ◊◊◊◊◊	382,100
◊◊◊◊◊ ◊◊◊◊◊ ◊◊◊◊◊ ◊◊◊◊◊ ◊◊◊◊◊ ◊◊◊◊◊ ◊◊◊◊◊ ◊◊◊◊◊ ◊◊◊◊◊ ◊◊◊◊◊	382,150
◊◊◊◊◊ ◊◊◊◊◊ ◊◊◊◊◊ ◊◊◊◊◊ ◊◊◊◊◊ ◊◊◊◊◊ ◊◊◊◊◊ ◊◊◊◊◊ ◊◊◊◊◊ ◊◊◊◊◊	382,200
◊◊◊◊◊ ◊◊◊◊◊ ◊◊◊◊◊ ◊◊◊◊◊ ◊◊◊◊◊ ◊◊◊◊◊ ◊◊◊◊◊ ◊◊◊◊◊ ◊◊◊◊◊ ◊◊◊◊◊	382,250
◊◊◊◊◊ ◊◊◊◊◊ ◊◊◊◊◊ ◊◊◊◊◊ ◊◊◊◊◊ ◊◊◊◊◊ ◊◊◊◊◊ ◊◊◊◊◊ ◊◊◊◊◊ ◊◊◊◊◊	382,300
◊◊◊◊◊ ◊◊◊◊◊ ◊◊◊◊◊ ◊◊◊◊◊ ◊◊◊◊◊ ◊◊◊◊◊ ◊◊◊◊◊ ◊◊◊◊◊ ◊◊◊◊◊ ◊◊◊◊◊	382,350
◊◊◊◊◊ ◊◊◊◊◊ ◊◊◊◊◊ ◊◊◊◊◊ ◊◊◊◊◊ ◊◊◊◊◊ ◊◊◊◊◊ ◊◊◊◊◊ ◊◊◊◊◊ ◊◊◊◊◊	382,400
◊◊◊◊◊ ◊◊◊◊◊ ◊◊◊◊◊ ◊◊◊◊◊ ◊◊◊◊◊ ◊◊◊◊◊ ◊◊◊◊◊ ◊◊◊◊◊ ◊◊◊◊◊ ◊◊◊◊◊	382,450
◊◊◊◊◊ ◊◊◊◊◊ ◊◊◊◊◊ ◊◊◊◊◊ ◊◊◊◊◊ ◊◊◊◊◊ ◊◊◊◊◊ ◊◊◊◊◊ ◊◊◊◊◊ ◊◊◊◊◊	382,500
◊◊◊◊◊ ◊◊◊◊◊ ◊◊◊◊◊ ◊◊◊◊◊ ◊◊◊◊◊ ◊◊◊◊◊ ◊◊◊◊◊ ◊◊◊◊◊ ◊◊◊◊◊ ◊◊◊◊◊	382,550
◊◊◊◊◊ ◊◊◊◊◊ ◊◊◊◊◊ ◊◊◊◊◊ ◊◊◊◊◊ ◊◊◊◊◊ ◊◊◊◊◊ ◊◊◊◊◊ ◊◊◊◊◊ ◊◊◊◊◊	382,600
◊◊◊◊◊ ◊◊◊◊◊ ◊◊◊◊◊ ◊◊◊◊◊ ◊◊◊◊◊ ◊◊◊◊◊ ◊◊◊◊◊ ◊◊◊◊◊ ◊◊◊◊◊ ◊◊◊◊◊	382,650
◊◊◊◊◊ ◊◊◊◊◊ ◊◊◊◊◊ ◊◊◊◊◊ ◊◊◊◊◊ ◊◊◊◊◊ ◊◊◊◊◊ ◊◊◊◊◊ ◊◊◊◊◊ ◊◊◊◊◊	382,700
◊◊◊◊◊ ◊◊◊◊◊ ◊◊◊◊◊ ◊◊◊◊◊ ◊◊◊◊◊ ◊◊◊◊◊ ◊◊◊◊◊ ◊◊◊◊◊ ◊◊◊◊◊ ◊◊◊◊◊	382,750
◊◊◊◊◊ ◊◊◊◊◊ ◊◊◊◊◊ ◊◊◊◊◊ ◊◊◊◊◊ ◊◊◊◊◊ ◊◊◊◊◊ ◊◊◊◊◊ ◊◊◊◊◊ ◊◊◊◊◊	382,800
◊◊◊◊◊ ◊◊◊◊◊ ◊◊◊◊◊ ◊◊◊◊◊ ◊◊◊◊◊ ◊◊◊◊◊ ◊◊◊◊◊ ◊◊◊◊◊ ◊◊◊◊◊ ◊◊◊◊◊	382,850
◊◊◊◊◊ ◊◊◊◊◊ ◊◊◊◊◊ ◊◊◊◊◊ ◊◊◊◊◊ ◊◊◊◊◊ ◊◊◊◊◊ ◊◊◊◊◊ ◊◊◊◊◊ ◊◊◊◊◊	382,900
◊◊◊◊◊ ◊◊◊◊◊ ◊◊◊◊◊ ◊◊◊◊◊ ◊◊◊◊◊ ◊◊◊◊◊ ◊◊◊◊◊ ◊◊◊◊◊ ◊◊◊◊◊ ◊◊◊◊◊	382,950
◊◊◊◊◊ ◊◊◊◊◊ ◊◊◊◊◊ ◊◊◊◊◊ ◊◊◊◊◊ ◊◊◊◊◊ ◊◊◊◊◊ ◊◊◊◊◊ ◊◊◊◊◊ ◊◊◊◊◊	**383,000**
◊◊◊◊◊ ◊◊◊◊◊ ◊◊◊◊◊ ◊◊◊◊◊ ◊◊◊◊◊ ◊◊◊◊◊ ◊◊◊◊◊ ◊◊◊◊◊ ◊◊◊◊◊ ◊◊◊◊◊	383,050
◊◊◊◊◊ ◊◊◊◊◊ ◊◊◊◊◊ ◊◊◊◊◊ ◊◊◊◊◊ ◊◊◊◊◊ ◊◊◊◊◊ ◊◊◊◊◊ ◊◊◊◊◊ ◊◊◊◊◊	383,100
◊◊◊◊◊ ◊◊◊◊◊ ◊◊◊◊◊ ◊◊◊◊◊ ◊◊◊◊◊ ◊◊◊◊◊ ◊◊◊◊◊ ◊◊◊◊◊ ◊◊◊◊◊ ◊◊◊◊◊	383,150
◊◊◊◊◊ ◊◊◊◊◊ ◊◊◊◊◊ ◊◊◊◊◊ ◊◊◊◊◊ ◊◊◊◊◊ ◊◊◊◊◊ ◊◊◊◊◊ ◊◊◊◊◊ ◊◊◊◊◊	383,200
◊◊◊◊◊ ◊◊◊◊◊ ◊◊◊◊◊ ◊◊◊◊◊ ◊◊◊◊◊ ◊◊◊◊◊ ◊◊◊◊◊ ◊◊◊◊◊ ◊◊◊◊◊ ◊◊◊◊◊	383,250
◊◊◊◊◊ ◊◊◊◊◊ ◊◊◊◊◊ ◊◊◊◊◊ ◊◊◊◊◊ ◊◊◊◊◊ ◊◊◊◊◊ ◊◊◊◊◊ ◊◊◊◊◊ ◊◊◊◊◊	383,300
◊◊◊◊◊ ◊◊◊◊◊ ◊◊◊◊◊ ◊◊◊◊◊ ◊◊◊◊◊ ◊◊◊◊◊ ◊◊◊◊◊ ◊◊◊◊◊ ◊◊◊◊◊ ◊◊◊◊◊	383,350
◊◊◊◊◊ ◊◊◊◊◊ ◊◊◊◊◊ ◊◊◊◊◊ ◊◊◊◊◊ ◊◊◊◊◊ ◊◊◊◊◊ ◊◊◊◊◊ ◊◊◊◊◊ ◊◊◊◊◊	383,400
◊◊◊◊◊ ◊◊◊◊◊ ◊◊◊◊◊ ◊◊◊◊◊ ◊◊◊◊◊ ◊◊◊◊◊ ◊◊◊◊◊ ◊◊◊◊◊ ◊◊◊◊◊ ◊◊◊◊◊	383,450
◊◊◊◊◊ ◊◊◊◊◊ ◊◊◊◊◊ ◊◊◊◊◊ ◊◊◊◊◊ ◊◊◊◊◊ ◊◊◊◊◊ ◊◊◊◊◊ ◊◊◊◊◊ ◊◊◊◊◊	383,500
◊◊◊◊◊ ◊◊◊◊◊ ◊◊◊◊◊ ◊◊◊◊◊ ◊◊◊◊◊ ◊◊◊◊◊ ◊◊◊◊◊ ◊◊◊◊◊ ◊◊◊◊◊ ◊◊◊◊◊	383,550
◊◊◊◊◊ ◊◊◊◊◊ ◊◊◊◊◊ ◊◊◊◊◊ ◊◊◊◊◊ ◊◊◊◊◊ ◊◊◊◊◊ ◊◊◊◊◊ ◊◊◊◊◊ ◊◊◊◊◊	383,600
◊◊◊◊◊ ◊◊◊◊◊ ◊◊◊◊◊ ◊◊◊◊◊ ◊◊◊◊◊ ◊◊◊◊◊ ◊◊◊◊◊ ◊◊◊◊◊ ◊◊◊◊◊ ◊◊◊◊◊	383,650
◊◊◊◊◊ ◊◊◊◊◊ ◊◊◊◊◊ ◊◊◊◊◊ ◊◊◊◊◊ ◊◊◊◊◊ ◊◊◊◊◊ ◊◊◊◊◊ ◊◊◊◊◊ ◊◊◊◊◊	383,700
◊◊◊◊◊ ◊◊◊◊◊ ◊◊◊◊◊ ◊◊◊◊◊ ◊◊◊◊◊ ◊◊◊◊◊ ◊◊◊◊◊ ◊◊◊◊◊ ◊◊◊◊◊ ◊◊◊◊◊	383,750
◊◊◊◊◊ ◊◊◊◊◊ ◊◊◊◊◊ ◊◊◊◊◊ ◊◊◊◊◊ ◊◊◊◊◊ ◊◊◊◊◊ ◊◊◊◊◊ ◊◊◊◊◊ ◊◊◊◊◊	383,800
◊◊◊◊◊ ◊◊◊◊◊ ◊◊◊◊◊ ◊◊◊◊◊ ◊◊◊◊◊ ◊◊◊◊◊ ◊◊◊◊◊ ◊◊◊◊◊ ◊◊◊◊◊ ◊◊◊◊◊	383,850
◊◊◊◊◊ ◊◊◊◊◊ ◊◊◊◊◊ ◊◊◊◊◊ ◊◊◊◊◊ ◊◊◊◊◊ ◊◊◊◊◊ ◊◊◊◊◊ ◊◊◊◊◊ ◊◊◊◊◊	383,900
◊◊◊◊◊ ◊◊◊◊◊ ◊◊◊◊◊ ◊◊◊◊◊ ◊◊◊◊◊ ◊◊◊◊◊ ◊◊◊◊◊ ◊◊◊◊◊ ◊◊◊◊◊ ◊◊◊◊◊	383,950
◊◊◊◊◊ ◊◊◊◊◊ ◊◊◊◊◊ ◊◊◊◊◊ ◊◊◊◊◊ ◊◊◊◊◊ ◊◊◊◊◊ ◊◊◊◊◊ ◊◊◊◊◊ ◊◊◊◊◊	**384,000**

The Number 384,001 to the Number 386,000

Start   Notes

	384,050
	384,100
	384,150
	384,200
	384,250
	384,300
	384,350
	384,400
	384,450
	384,500
	384,550
	384,600
	384,650
	384,700
	384,750
	384,800
	384,850
	384,900
	384,950
	385,000
	385,050
	385,100
	385,150
	385,200
	385,250
	385,300
	385,350
	385,400
	385,450
	385,500
	385,550
	385,600
	385,650
	385,700
	385,750
	385,800
	385,850
	385,900
	385,950
	386,000

The Number 386,001 to the Number 388,000

Start	Notes
386,050	
386,100	
386,150	
386,200	
386,250	
386,300	
386,350	
386,400	
386,450	
386,500	
386,550	
386,600	
386,650	
386,700	
386,750	
386,800	
386,850	
386,900	
386,950	
387,000	
387,050	
387,100	
387,150	
387,200	
387,250	
387,300	
387,350	
387,400	
387,450	
387,500	
387,550	
387,600	
387,650	
387,700	
387,750	
387,800	
387,850	
387,900	
387,950	
388,000	

The Number 388,001 to the Number 390,000

Start | | Notes

	Number
	388,050
	388,100
	388,150
	388,200
	388,250
	388,300
	388,350
	388,400
	388,450
	388,500
	388,550
	388,600
	388,650
	388,700
	388,750
	388,800
	388,850
	388,900
	388,950
	389,000
	389,050
	389,100
	389,150
	389,200
	389,250
	389,300
	389,350
	389,400
	389,450
	389,500
	389,550
	389,600
	389,650
	389,700
	389,750
	389,800
	389,850
	389,900
	389,950
	390,000

The Number 390,001 to the Number 392,000

Start		Notes
◊◊◊◊◊ ◊◊◊◊◊ ◊◊◊◊◊ ◊◊◊◊◊ ◊◊◊◊◊ ◊◊◊◊◊ ◊◊◊◊◊ ◊◊◊◊◊ ◊◊◊◊◊ ◊◊◊◊◊	390,050	
◊◊◊◊◊ ◊◊◊◊◊ ◊◊◊◊◊ ◊◊◊◊◊ ◊◊◊◊◊ ◊◊◊◊◊ ◊◊◊◊◊ ◊◊◊◊◊ ◊◊◊◊◊ ◊◊◊◊◊	390,100	
◊◊◊◊◊ ◊◊◊◊◊ ◊◊◊◊◊ ◊◊◊◊◊ ◊◊◊◊◊ ◊◊◊◊◊ ◊◊◊◊◊ ◊◊◊◊◊ ◊◊◊◊◊ ◊◊◊◊◊	390,150	
◊◊◊◊◊ ◊◊◊◊◊ ◊◊◊◊◊ ◊◊◊◊◊ ◊◊◊◊◊ ◊◊◊◊◊ ◊◊◊◊◊ ◊◊◊◊◊ ◊◊◊◊◊ ◊◊◊◊◊	390,200	
◊◊◊◊◊ ◊◊◊◊◊ ◊◊◊◊◊ ◊◊◊◊◊ ◊◊◊◊◊ ◊◊◊◊◊ ◊◊◊◊◊ ◊◊◊◊◊ ◊◊◊◊◊ ◊◊◊◊◊	390,250	
◊◊◊◊◊ ◊◊◊◊◊ ◊◊◊◊◊ ◊◊◊◊◊ ◊◊◊◊◊ ◊◊◊◊◊ ◊◊◊◊◊ ◊◊◊◊◊ ◊◊◊◊◊ ◊◊◊◊◊	390,300	
◊◊◊◊◊ ◊◊◊◊◊ ◊◊◊◊◊ ◊◊◊◊◊ ◊◊◊◊◊ ◊◊◊◊◊ ◊◊◊◊◊ ◊◊◊◊◊ ◊◊◊◊◊ ◊◊◊◊◊	390,350	
◊◊◊◊◊ ◊◊◊◊◊ ◊◊◊◊◊ ◊◊◊◊◊ ◊◊◊◊◊ ◊◊◊◊◊ ◊◊◊◊◊ ◊◊◊◊◊ ◊◊◊◊◊ ◊◊◊◊◊	390,400	
◊◊◊◊◊ ◊◊◊◊◊ ◊◊◊◊◊ ◊◊◊◊◊ ◊◊◊◊◊ ◊◊◊◊◊ ◊◊◊◊◊ ◊◊◊◊◊ ◊◊◊◊◊ ◊◊◊◊◊	390,450	
◊◊◊◊◊ ◊◊◊◊◊ ◊◊◊◊◊ ◊◊◊◊◊ ◊◊◊◊◊ ◊◊◊◊◊ ◊◊◊◊◊ ◊◊◊◊◊ ◊◊◊◊◊ ◊◊◊◊◊	390,500	
◊◊◊◊◊ ◊◊◊◊◊ ◊◊◊◊◊ ◊◊◊◊◊ ◊◊◊◊◊ ◊◊◊◊◊ ◊◊◊◊◊ ◊◊◊◊◊ ◊◊◊◊◊ ◊◊◊◊◊	390,550	
◊◊◊◊◊ ◊◊◊◊◊ ◊◊◊◊◊ ◊◊◊◊◊ ◊◊◊◊◊ ◊◊◊◊◊ ◊◊◊◊◊ ◊◊◊◊◊ ◊◊◊◊◊ ◊◊◊◊◊	390,600	
◊◊◊◊◊ ◊◊◊◊◊ ◊◊◊◊◊ ◊◊◊◊◊ ◊◊◊◊◊ ◊◊◊◊◊ ◊◊◊◊◊ ◊◊◊◊◊ ◊◊◊◊◊ ◊◊◊◊◊	390,650	
◊◊◊◊◊ ◊◊◊◊◊ ◊◊◊◊◊ ◊◊◊◊◊ ◊◊◊◊◊ ◊◊◊◊◊ ◊◊◊◊◊ ◊◊◊◊◊ ◊◊◊◊◊ ◊◊◊◊◊	390,700	
◊◊◊◊◊ ◊◊◊◊◊ ◊◊◊◊◊ ◊◊◊◊◊ ◊◊◊◊◊ ◊◊◊◊◊ ◊◊◊◊◊ ◊◊◊◊◊ ◊◊◊◊◊ ◊◊◊◊◊	390,750	
◊◊◊◊◊ ◊◊◊◊◊ ◊◊◊◊◊ ◊◊◊◊◊ ◊◊◊◊◊ ◊◊◊◊◊ ◊◊◊◊◊ ◊◊◊◊◊ ◊◊◊◊◊ ◊◊◊◊◊	390,800	
◊◊◊◊◊ ◊◊◊◊◊ ◊◊◊◊◊ ◊◊◊◊◊ ◊◊◊◊◊ ◊◊◊◊◊ ◊◊◊◊◊ ◊◊◊◊◊ ◊◊◊◊◊ ◊◊◊◊◊	390,850	
◊◊◊◊◊ ◊◊◊◊◊ ◊◊◊◊◊ ◊◊◊◊◊ ◊◊◊◊◊ ◊◊◊◊◊ ◊◊◊◊◊ ◊◊◊◊◊ ◊◊◊◊◊ ◊◊◊◊◊	390,900	
◊◊◊◊◊ ◊◊◊◊◊ ◊◊◊◊◊ ◊◊◊◊◊ ◊◊◊◊◊ ◊◊◊◊◊ ◊◊◊◊◊ ◊◊◊◊◊ ◊◊◊◊◊ ◊◊◊◊◊	390,950	
◊◊◊◊◊ ◊◊◊◊◊ ◊◊◊◊◊ ◊◊◊◊◊ ◊◊◊◊◊ ◊◊◊◊◊ ◊◊◊◊◊ ◊◊◊◊◊ ◊◊◊◊◊ ◊◊◊◊◊	**391,000**	
◊◊◊◊◊ ◊◊◊◊◊ ◊◊◊◊◊ ◊◊◊◊◊ ◊◊◊◊◊ ◊◊◊◊◊ ◊◊◊◊◊ ◊◊◊◊◊ ◊◊◊◊◊ ◊◊◊◊◊	391,050	
◊◊◊◊◊ ◊◊◊◊◊ ◊◊◊◊◊ ◊◊◊◊◊ ◊◊◊◊◊ ◊◊◊◊◊ ◊◊◊◊◊ ◊◊◊◊◊ ◊◊◊◊◊ ◊◊◊◊◊	391,100	
◊◊◊◊◊ ◊◊◊◊◊ ◊◊◊◊◊ ◊◊◊◊◊ ◊◊◊◊◊ ◊◊◊◊◊ ◊◊◊◊◊ ◊◊◊◊◊ ◊◊◊◊◊ ◊◊◊◊◊	391,150	
◊◊◊◊◊ ◊◊◊◊◊ ◊◊◊◊◊ ◊◊◊◊◊ ◊◊◊◊◊ ◊◊◊◊◊ ◊◊◊◊◊ ◊◊◊◊◊ ◊◊◊◊◊ ◊◊◊◊◊	391,200	
◊◊◊◊◊ ◊◊◊◊◊ ◊◊◊◊◊ ◊◊◊◊◊ ◊◊◊◊◊ ◊◊◊◊◊ ◊◊◊◊◊ ◊◊◊◊◊ ◊◊◊◊◊ ◊◊◊◊◊	391,250	
◊◊◊◊◊ ◊◊◊◊◊ ◊◊◊◊◊ ◊◊◊◊◊ ◊◊◊◊◊ ◊◊◊◊◊ ◊◊◊◊◊ ◊◊◊◊◊ ◊◊◊◊◊ ◊◊◊◊◊	391,300	
◊◊◊◊◊ ◊◊◊◊◊ ◊◊◊◊◊ ◊◊◊◊◊ ◊◊◊◊◊ ◊◊◊◊◊ ◊◊◊◊◊ ◊◊◊◊◊ ◊◊◊◊◊ ◊◊◊◊◊	391,350	
◊◊◊◊◊ ◊◊◊◊◊ ◊◊◊◊◊ ◊◊◊◊◊ ◊◊◊◊◊ ◊◊◊◊◊ ◊◊◊◊◊ ◊◊◊◊◊ ◊◊◊◊◊ ◊◊◊◊◊	391,400	
◊◊◊◊◊ ◊◊◊◊◊ ◊◊◊◊◊ ◊◊◊◊◊ ◊◊◊◊◊ ◊◊◊◊◊ ◊◊◊◊◊ ◊◊◊◊◊ ◊◊◊◊◊ ◊◊◊◊◊	391,450	
◊◊◊◊◊ ◊◊◊◊◊ ◊◊◊◊◊ ◊◊◊◊◊ ◊◊◊◊◊ ◊◊◊◊◊ ◊◊◊◊◊ ◊◊◊◊◊ ◊◊◊◊◊ ◊◊◊◊◊	391,500	
◊◊◊◊◊ ◊◊◊◊◊ ◊◊◊◊◊ ◊◊◊◊◊ ◊◊◊◊◊ ◊◊◊◊◊ ◊◊◊◊◊ ◊◊◊◊◊ ◊◊◊◊◊ ◊◊◊◊◊	391,550	
◊◊◊◊◊ ◊◊◊◊◊ ◊◊◊◊◊ ◊◊◊◊◊ ◊◊◊◊◊ ◊◊◊◊◊ ◊◊◊◊◊ ◊◊◊◊◊ ◊◊◊◊◊ ◊◊◊◊◊	391,600	
◊◊◊◊◊ ◊◊◊◊◊ ◊◊◊◊◊ ◊◊◊◊◊ ◊◊◊◊◊ ◊◊◊◊◊ ◊◊◊◊◊ ◊◊◊◊◊ ◊◊◊◊◊ ◊◊◊◊◊	391,650	
◊◊◊◊◊ ◊◊◊◊◊ ◊◊◊◊◊ ◊◊◊◊◊ ◊◊◊◊◊ ◊◊◊◊◊ ◊◊◊◊◊ ◊◊◊◊◊ ◊◊◊◊◊ ◊◊◊◊◊	391,700	
◊◊◊◊◊ ◊◊◊◊◊ ◊◊◊◊◊ ◊◊◊◊◊ ◊◊◊◊◊ ◊◊◊◊◊ ◊◊◊◊◊ ◊◊◊◊◊ ◊◊◊◊◊ ◊◊◊◊◊	391,750	
◊◊◊◊◊ ◊◊◊◊◊ ◊◊◊◊◊ ◊◊◊◊◊ ◊◊◊◊◊ ◊◊◊◊◊ ◊◊◊◊◊ ◊◊◊◊◊ ◊◊◊◊◊ ◊◊◊◊◊	391,800	
◊◊◊◊◊ ◊◊◊◊◊ ◊◊◊◊◊ ◊◊◊◊◊ ◊◊◊◊◊ ◊◊◊◊◊ ◊◊◊◊◊ ◊◊◊◊◊ ◊◊◊◊◊ ◊◊◊◊◊	391,850	
◊◊◊◊◊ ◊◊◊◊◊ ◊◊◊◊◊ ◊◊◊◊◊ ◊◊◊◊◊ ◊◊◊◊◊ ◊◊◊◊◊ ◊◊◊◊◊ ◊◊◊◊◊ ◊◊◊◊◊	391,900	
◊◊◊◊◊ ◊◊◊◊◊ ◊◊◊◊◊ ◊◊◊◊◊ ◊◊◊◊◊ ◊◊◊◊◊ ◊◊◊◊◊ ◊◊◊◊◊ ◊◊◊◊◊ ◊◊◊◊◊	391,950	
◊◊◊◊◊ ◊◊◊◊◊ ◊◊◊◊◊ ◊◊◊◊◊ ◊◊◊◊◊ ◊◊◊◊◊ ◊◊◊◊◊ ◊◊◊◊◊ ◊◊◊◊◊ ◊◊◊◊◊	**392,000**	

The Number 392,001 to the Number 394,000

Start   Notes

◊◊◊◊◊ ◊◊◊◊◊ ◊◊◊◊◊ ◊◊◊◊◊ ◊◊◊◊◊ ◊◊◊◊◊ ◊◊◊◊◊ ◊◊◊◊◊ ◊◊◊◊◊ ◊◊◊◊◊	392,050
◊◊◊◊◊ ◊◊◊◊◊ ◊◊◊◊◊ ◊◊◊◊◊ ◊◊◊◊◊ ◊◊◊◊◊ ◊◊◊◊◊ ◊◊◊◊◊ ◊◊◊◊◊ ◊◊◊◊◊	392,100
◊◊◊◊◊ ◊◊◊◊◊ ◊◊◊◊◊ ◊◊◊◊◊ ◊◊◊◊◊ ◊◊◊◊◊ ◊◊◊◊◊ ◊◊◊◊◊ ◊◊◊◊◊ ◊◊◊◊◊	392,150
◊◊◊◊◊ ◊◊◊◊◊ ◊◊◊◊◊ ◊◊◊◊◊ ◊◊◊◊◊ ◊◊◊◊◊ ◊◊◊◊◊ ◊◊◊◊◊ ◊◊◊◊◊ ◊◊◊◊◊	392,200
◊◊◊◊◊ ◊◊◊◊◊ ◊◊◊◊◊ ◊◊◊◊◊ ◊◊◊◊◊ ◊◊◊◊◊ ◊◊◊◊◊ ◊◊◊◊◊ ◊◊◊◊◊ ◊◊◊◊◊	392,250
◊◊◊◊◊ ◊◊◊◊◊ ◊◊◊◊◊ ◊◊◊◊◊ ◊◊◊◊◊ ◊◊◊◊◊ ◊◊◊◊◊ ◊◊◊◊◊ ◊◊◊◊◊ ◊◊◊◊◊	392,300
◊◊◊◊◊ ◊◊◊◊◊ ◊◊◊◊◊ ◊◊◊◊◊ ◊◊◊◊◊ ◊◊◊◊◊ ◊◊◊◊◊ ◊◊◊◊◊ ◊◊◊◊◊ ◊◊◊◊◊	392,350
◊◊◊◊◊ ◊◊◊◊◊ ◊◊◊◊◊ ◊◊◊◊◊ ◊◊◊◊◊ ◊◊◊◊◊ ◊◊◊◊◊ ◊◊◊◊◊ ◊◊◊◊◊ ◊◊◊◊◊	392,400
◊◊◊◊◊ ◊◊◊◊◊ ◊◊◊◊◊ ◊◊◊◊◊ ◊◊◊◊◊ ◊◊◊◊◊ ◊◊◊◊◊ ◊◊◊◊◊ ◊◊◊◊◊ ◊◊◊◊◊	392,450
◊◊◊◊◊ ◊◊◊◊◊ ◊◊◊◊◊ ◊◊◊◊◊ ◊◊◊◊◊ ◊◊◊◊◊ ◊◊◊◊◊ ◊◊◊◊◊ ◊◊◊◊◊ ◊◊◊◊◊	392,500
◊◊◊◊◊ ◊◊◊◊◊ ◊◊◊◊◊ ◊◊◊◊◊ ◊◊◊◊◊ ◊◊◊◊◊ ◊◊◊◊◊ ◊◊◊◊◊ ◊◊◊◊◊ ◊◊◊◊◊	392,550
◊◊◊◊◊ ◊◊◊◊◊ ◊◊◊◊◊ ◊◊◊◊◊ ◊◊◊◊◊ ◊◊◊◊◊ ◊◊◊◊◊ ◊◊◊◊◊ ◊◊◊◊◊ ◊◊◊◊◊	392,600
◊◊◊◊◊ ◊◊◊◊◊ ◊◊◊◊◊ ◊◊◊◊◊ ◊◊◊◊◊ ◊◊◊◊◊ ◊◊◊◊◊ ◊◊◊◊◊ ◊◊◊◊◊ ◊◊◊◊◊	392,650
◊◊◊◊◊ ◊◊◊◊◊ ◊◊◊◊◊ ◊◊◊◊◊ ◊◊◊◊◊ ◊◊◊◊◊ ◊◊◊◊◊ ◊◊◊◊◊ ◊◊◊◊◊ ◊◊◊◊◊	392,700
◊◊◊◊◊ ◊◊◊◊◊ ◊◊◊◊◊ ◊◊◊◊◊ ◊◊◊◊◊ ◊◊◊◊◊ ◊◊◊◊◊ ◊◊◊◊◊ ◊◊◊◊◊ ◊◊◊◊◊	392,750
◊◊◊◊◊ ◊◊◊◊◊ ◊◊◊◊◊ ◊◊◊◊◊ ◊◊◊◊◊ ◊◊◊◊◊ ◊◊◊◊◊ ◊◊◊◊◊ ◊◊◊◊◊ ◊◊◊◊◊	392,800
◊◊◊◊◊ ◊◊◊◊◊ ◊◊◊◊◊ ◊◊◊◊◊ ◊◊◊◊◊ ◊◊◊◊◊ ◊◊◊◊◊ ◊◊◊◊◊ ◊◊◊◊◊ ◊◊◊◊◊	392,850
◊◊◊◊◊ ◊◊◊◊◊ ◊◊◊◊◊ ◊◊◊◊◊ ◊◊◊◊◊ ◊◊◊◊◊ ◊◊◊◊◊ ◊◊◊◊◊ ◊◊◊◊◊ ◊◊◊◊◊	392,900
◊◊◊◊◊ ◊◊◊◊◊ ◊◊◊◊◊ ◊◊◊◊◊ ◊◊◊◊◊ ◊◊◊◊◊ ◊◊◊◊◊ ◊◊◊◊◊ ◊◊◊◊◊ ◊◊◊◊◊	392,950
◊◊◊◊◊ ◊◊◊◊◊ ◊◊◊◊◊ ◊◊◊◊◊ ◊◊◊◊◊ ◊◊◊◊◊ ◊◊◊◊◊ ◊◊◊◊◊ ◊◊◊◊◊ ◊◊◊◊◊	**393,000**
◊◊◊◊◊ ◊◊◊◊◊ ◊◊◊◊◊ ◊◊◊◊◊ ◊◊◊◊◊ ◊◊◊◊◊ ◊◊◊◊◊ ◊◊◊◊◊ ◊◊◊◊◊ ◊◊◊◊◊	393,050
◊◊◊◊◊ ◊◊◊◊◊ ◊◊◊◊◊ ◊◊◊◊◊ ◊◊◊◊◊ ◊◊◊◊◊ ◊◊◊◊◊ ◊◊◊◊◊ ◊◊◊◊◊ ◊◊◊◊◊	393,100
◊◊◊◊◊ ◊◊◊◊◊ ◊◊◊◊◊ ◊◊◊◊◊ ◊◊◊◊◊ ◊◊◊◊◊ ◊◊◊◊◊ ◊◊◊◊◊ ◊◊◊◊◊ ◊◊◊◊◊	393,150
◊◊◊◊◊ ◊◊◊◊◊ ◊◊◊◊◊ ◊◊◊◊◊ ◊◊◊◊◊ ◊◊◊◊◊ ◊◊◊◊◊ ◊◊◊◊◊ ◊◊◊◊◊ ◊◊◊◊◊	393,200
◊◊◊◊◊ ◊◊◊◊◊ ◊◊◊◊◊ ◊◊◊◊◊ ◊◊◊◊◊ ◊◊◊◊◊ ◊◊◊◊◊ ◊◊◊◊◊ ◊◊◊◊◊ ◊◊◊◊◊	393,250
◊◊◊◊◊ ◊◊◊◊◊ ◊◊◊◊◊ ◊◊◊◊◊ ◊◊◊◊◊ ◊◊◊◊◊ ◊◊◊◊◊ ◊◊◊◊◊ ◊◊◊◊◊ ◊◊◊◊◊	393,300
◊◊◊◊◊ ◊◊◊◊◊ ◊◊◊◊◊ ◊◊◊◊◊ ◊◊◊◊◊ ◊◊◊◊◊ ◊◊◊◊◊ ◊◊◊◊◊ ◊◊◊◊◊ ◊◊◊◊◊	393,350
◊◊◊◊◊ ◊◊◊◊◊ ◊◊◊◊◊ ◊◊◊◊◊ ◊◊◊◊◊ ◊◊◊◊◊ ◊◊◊◊◊ ◊◊◊◊◊ ◊◊◊◊◊ ◊◊◊◊◊	393,400
◊◊◊◊◊ ◊◊◊◊◊ ◊◊◊◊◊ ◊◊◊◊◊ ◊◊◊◊◊ ◊◊◊◊◊ ◊◊◊◊◊ ◊◊◊◊◊ ◊◊◊◊◊ ◊◊◊◊◊	393,450
◊◊◊◊◊ ◊◊◊◊◊ ◊◊◊◊◊ ◊◊◊◊◊ ◊◊◊◊◊ ◊◊◊◊◊ ◊◊◊◊◊ ◊◊◊◊◊ ◊◊◊◊◊ ◊◊◊◊◊	393,500
◊◊◊◊◊ ◊◊◊◊◊ ◊◊◊◊◊ ◊◊◊◊◊ ◊◊◊◊◊ ◊◊◊◊◊ ◊◊◊◊◊ ◊◊◊◊◊ ◊◊◊◊◊ ◊◊◊◊◊	393,550
◊◊◊◊◊ ◊◊◊◊◊ ◊◊◊◊◊ ◊◊◊◊◊ ◊◊◊◊◊ ◊◊◊◊◊ ◊◊◊◊◊ ◊◊◊◊◊ ◊◊◊◊◊ ◊◊◊◊◊	393,600
◊◊◊◊◊ ◊◊◊◊◊ ◊◊◊◊◊ ◊◊◊◊◊ ◊◊◊◊◊ ◊◊◊◊◊ ◊◊◊◊◊ ◊◊◊◊◊ ◊◊◊◊◊ ◊◊◊◊◊	393,650
◊◊◊◊◊ ◊◊◊◊◊ ◊◊◊◊◊ ◊◊◊◊◊ ◊◊◊◊◊ ◊◊◊◊◊ ◊◊◊◊◊ ◊◊◊◊◊ ◊◊◊◊◊ ◊◊◊◊◊	393,700
◊◊◊◊◊ ◊◊◊◊◊ ◊◊◊◊◊ ◊◊◊◊◊ ◊◊◊◊◊ ◊◊◊◊◊ ◊◊◊◊◊ ◊◊◊◊◊ ◊◊◊◊◊ ◊◊◊◊◊	393,750
◊◊◊◊◊ ◊◊◊◊◊ ◊◊◊◊◊ ◊◊◊◊◊ ◊◊◊◊◊ ◊◊◊◊◊ ◊◊◊◊◊ ◊◊◊◊◊ ◊◊◊◊◊ ◊◊◊◊◊	393,800
◊◊◊◊◊ ◊◊◊◊◊ ◊◊◊◊◊ ◊◊◊◊◊ ◊◊◊◊◊ ◊◊◊◊◊ ◊◊◊◊◊ ◊◊◊◊◊ ◊◊◊◊◊ ◊◊◊◊◊	393,850
◊◊◊◊◊ ◊◊◊◊◊ ◊◊◊◊◊ ◊◊◊◊◊ ◊◊◊◊◊ ◊◊◊◊◊ ◊◊◊◊◊ ◊◊◊◊◊ ◊◊◊◊◊ ◊◊◊◊◊	393,900
◊◊◊◊◊ ◊◊◊◊◊ ◊◊◊◊◊ ◊◊◊◊◊ ◊◊◊◊◊ ◊◊◊◊◊ ◊◊◊◊◊ ◊◊◊◊◊ ◊◊◊◊◊ ◊◊◊◊◊	393,950
◊◊◊◊◊ ◊◊◊◊◊ ◊◊◊◊◊ ◊◊◊◊◊ ◊◊◊◊◊ ◊◊◊◊◊ ◊◊◊◊◊ ◊◊◊◊◊ ◊◊◊◊◊ ◊◊◊◊◊	**394,000**

The Number 394,001 to the Number 396,000

Start	Notes
◊◊◊◊◊ ◊◊◊◊◊ ◊◊◊◊◊ ◊◊◊◊◊ ◊◊◊◊◊ ◊◊◊◊◊ ◊◊◊◊◊ ◊◊◊◊◊ ◊◊◊◊◊ ◊◊◊◊◊	394,050
◊◊◊◊◊ ◊◊◊◊◊ ◊◊◊◊◊ ◊◊◊◊◊ ◊◊◊◊◊ ◊◊◊◊◊ ◊◊◊◊◊ ◊◊◊◊◊ ◊◊◊◊◊ ◊◊◊◊◊	394,100
◊◊◊◊◊ ◊◊◊◊◊ ◊◊◊◊◊ ◊◊◊◊◊ ◊◊◊◊◊ ◊◊◊◊◊ ◊◊◊◊◊ ◊◊◊◊◊ ◊◊◊◊◊ ◊◊◊◊◊	394,150
◊◊◊◊◊ ◊◊◊◊◊ ◊◊◊◊◊ ◊◊◊◊◊ ◊◊◊◊◊ ◊◊◊◊◊ ◊◊◊◊◊ ◊◊◊◊◊ ◊◊◊◊◊ ◊◊◊◊◊	394,200
◊◊◊◊◊ ◊◊◊◊◊ ◊◊◊◊◊ ◊◊◊◊◊ ◊◊◊◊◊ ◊◊◊◊◊ ◊◊◊◊◊ ◊◊◊◊◊ ◊◊◊◊◊ ◊◊◊◊◊	394,250
◊◊◊◊◊ ◊◊◊◊◊ ◊◊◊◊◊ ◊◊◊◊◊ ◊◊◊◊◊ ◊◊◊◊◊ ◊◊◊◊◊ ◊◊◊◊◊ ◊◊◊◊◊ ◊◊◊◊◊	394,300
◊◊◊◊◊ ◊◊◊◊◊ ◊◊◊◊◊ ◊◊◊◊◊ ◊◊◊◊◊ ◊◊◊◊◊ ◊◊◊◊◊ ◊◊◊◊◊ ◊◊◊◊◊ ◊◊◊◊◊	394,350
◊◊◊◊◊ ◊◊◊◊◊ ◊◊◊◊◊ ◊◊◊◊◊ ◊◊◊◊◊ ◊◊◊◊◊ ◊◊◊◊◊ ◊◊◊◊◊ ◊◊◊◊◊ ◊◊◊◊◊	394,400
◊◊◊◊◊ ◊◊◊◊◊ ◊◊◊◊◊ ◊◊◊◊◊ ◊◊◊◊◊ ◊◊◊◊◊ ◊◊◊◊◊ ◊◊◊◊◊ ◊◊◊◊◊ ◊◊◊◊◊	394,450
◊◊◊◊◊ ◊◊◊◊◊ ◊◊◊◊◊ ◊◊◊◊◊ ◊◊◊◊◊ ◊◊◊◊◊ ◊◊◊◊◊ ◊◊◊◊◊ ◊◊◊◊◊ ◊◊◊◊◊	394,500
◊◊◊◊◊ ◊◊◊◊◊ ◊◊◊◊◊ ◊◊◊◊◊ ◊◊◊◊◊ ◊◊◊◊◊ ◊◊◊◊◊ ◊◊◊◊◊ ◊◊◊◊◊ ◊◊◊◊◊	394,550
◊◊◊◊◊ ◊◊◊◊◊ ◊◊◊◊◊ ◊◊◊◊◊ ◊◊◊◊◊ ◊◊◊◊◊ ◊◊◊◊◊ ◊◊◊◊◊ ◊◊◊◊◊ ◊◊◊◊◊	394,600
◊◊◊◊◊ ◊◊◊◊◊ ◊◊◊◊◊ ◊◊◊◊◊ ◊◊◊◊◊ ◊◊◊◊◊ ◊◊◊◊◊ ◊◊◊◊◊ ◊◊◊◊◊ ◊◊◊◊◊	394,650
◊◊◊◊◊ ◊◊◊◊◊ ◊◊◊◊◊ ◊◊◊◊◊ ◊◊◊◊◊ ◊◊◊◊◊ ◊◊◊◊◊ ◊◊◊◊◊ ◊◊◊◊◊ ◊◊◊◊◊	394,700
◊◊◊◊◊ ◊◊◊◊◊ ◊◊◊◊◊ ◊◊◊◊◊ ◊◊◊◊◊ ◊◊◊◊◊ ◊◊◊◊◊ ◊◊◊◊◊ ◊◊◊◊◊ ◊◊◊◊◊	394,750
◊◊◊◊◊ ◊◊◊◊◊ ◊◊◊◊◊ ◊◊◊◊◊ ◊◊◊◊◊ ◊◊◊◊◊ ◊◊◊◊◊ ◊◊◊◊◊ ◊◊◊◊◊ ◊◊◊◊◊	394,800
◊◊◊◊◊ ◊◊◊◊◊ ◊◊◊◊◊ ◊◊◊◊◊ ◊◊◊◊◊ ◊◊◊◊◊ ◊◊◊◊◊ ◊◊◊◊◊ ◊◊◊◊◊ ◊◊◊◊◊	394,850
◊◊◊◊◊ ◊◊◊◊◊ ◊◊◊◊◊ ◊◊◊◊◊ ◊◊◊◊◊ ◊◊◊◊◊ ◊◊◊◊◊ ◊◊◊◊◊ ◊◊◊◊◊ ◊◊◊◊◊	394,900
◊◊◊◊◊ ◊◊◊◊◊ ◊◊◊◊◊ ◊◊◊◊◊ ◊◊◊◊◊ ◊◊◊◊◊ ◊◊◊◊◊ ◊◊◊◊◊ ◊◊◊◊◊ ◊◊◊◊◊	394,950
◊◊◊◊◊ ◊◊◊◊◊ ◊◊◊◊◊ ◊◊◊◊◊ ◊◊◊◊◊ ◊◊◊◊◊ ◊◊◊◊◊ ◊◊◊◊◊ ◊◊◊◊◊ ◊◊◊◊◊	**395,000**
◊◊◊◊◊ ◊◊◊◊◊ ◊◊◊◊◊ ◊◊◊◊◊ ◊◊◊◊◊ ◊◊◊◊◊ ◊◊◊◊◊ ◊◊◊◊◊ ◊◊◊◊◊ ◊◊◊◊◊	395,050
◊◊◊◊◊ ◊◊◊◊◊ ◊◊◊◊◊ ◊◊◊◊◊ ◊◊◊◊◊ ◊◊◊◊◊ ◊◊◊◊◊ ◊◊◊◊◊ ◊◊◊◊◊ ◊◊◊◊◊	395,100
◊◊◊◊◊ ◊◊◊◊◊ ◊◊◊◊◊ ◊◊◊◊◊ ◊◊◊◊◊ ◊◊◊◊◊ ◊◊◊◊◊ ◊◊◊◊◊ ◊◊◊◊◊ ◊◊◊◊◊	395,150
◊◊◊◊◊ ◊◊◊◊◊ ◊◊◊◊◊ ◊◊◊◊◊ ◊◊◊◊◊ ◊◊◊◊◊ ◊◊◊◊◊ ◊◊◊◊◊ ◊◊◊◊◊ ◊◊◊◊◊	395,200
◊◊◊◊◊ ◊◊◊◊◊ ◊◊◊◊◊ ◊◊◊◊◊ ◊◊◊◊◊ ◊◊◊◊◊ ◊◊◊◊◊ ◊◊◊◊◊ ◊◊◊◊◊ ◊◊◊◊◊	395,250
◊◊◊◊◊ ◊◊◊◊◊ ◊◊◊◊◊ ◊◊◊◊◊ ◊◊◊◊◊ ◊◊◊◊◊ ◊◊◊◊◊ ◊◊◊◊◊ ◊◊◊◊◊ ◊◊◊◊◊	395,300
◊◊◊◊◊ ◊◊◊◊◊ ◊◊◊◊◊ ◊◊◊◊◊ ◊◊◊◊◊ ◊◊◊◊◊ ◊◊◊◊◊ ◊◊◊◊◊ ◊◊◊◊◊ ◊◊◊◊◊	395,350
◊◊◊◊◊ ◊◊◊◊◊ ◊◊◊◊◊ ◊◊◊◊◊ ◊◊◊◊◊ ◊◊◊◊◊ ◊◊◊◊◊ ◊◊◊◊◊ ◊◊◊◊◊ ◊◊◊◊◊	395,400
◊◊◊◊◊ ◊◊◊◊◊ ◊◊◊◊◊ ◊◊◊◊◊ ◊◊◊◊◊ ◊◊◊◊◊ ◊◊◊◊◊ ◊◊◊◊◊ ◊◊◊◊◊ ◊◊◊◊◊	395,450
◊◊◊◊◊ ◊◊◊◊◊ ◊◊◊◊◊ ◊◊◊◊◊ ◊◊◊◊◊ ◊◊◊◊◊ ◊◊◊◊◊ ◊◊◊◊◊ ◊◊◊◊◊ ◊◊◊◊◊	395,500
◊◊◊◊◊ ◊◊◊◊◊ ◊◊◊◊◊ ◊◊◊◊◊ ◊◊◊◊◊ ◊◊◊◊◊ ◊◊◊◊◊ ◊◊◊◊◊ ◊◊◊◊◊ ◊◊◊◊◊	395,550
◊◊◊◊◊ ◊◊◊◊◊ ◊◊◊◊◊ ◊◊◊◊◊ ◊◊◊◊◊ ◊◊◊◊◊ ◊◊◊◊◊ ◊◊◊◊◊ ◊◊◊◊◊ ◊◊◊◊◊	395,600
◊◊◊◊◊ ◊◊◊◊◊ ◊◊◊◊◊ ◊◊◊◊◊ ◊◊◊◊◊ ◊◊◊◊◊ ◊◊◊◊◊ ◊◊◊◊◊ ◊◊◊◊◊ ◊◊◊◊◊	395,650
◊◊◊◊◊ ◊◊◊◊◊ ◊◊◊◊◊ ◊◊◊◊◊ ◊◊◊◊◊ ◊◊◊◊◊ ◊◊◊◊◊ ◊◊◊◊◊ ◊◊◊◊◊ ◊◊◊◊◊	395,700
◊◊◊◊◊ ◊◊◊◊◊ ◊◊◊◊◊ ◊◊◊◊◊ ◊◊◊◊◊ ◊◊◊◊◊ ◊◊◊◊◊ ◊◊◊◊◊ ◊◊◊◊◊ ◊◊◊◊◊	395,750
◊◊◊◊◊ ◊◊◊◊◊ ◊◊◊◊◊ ◊◊◊◊◊ ◊◊◊◊◊ ◊◊◊◊◊ ◊◊◊◊◊ ◊◊◊◊◊ ◊◊◊◊◊ ◊◊◊◊◊	395,800
◊◊◊◊◊ ◊◊◊◊◊ ◊◊◊◊◊ ◊◊◊◊◊ ◊◊◊◊◊ ◊◊◊◊◊ ◊◊◊◊◊ ◊◊◊◊◊ ◊◊◊◊◊ ◊◊◊◊◊	395,850
◊◊◊◊◊ ◊◊◊◊◊ ◊◊◊◊◊ ◊◊◊◊◊ ◊◊◊◊◊ ◊◊◊◊◊ ◊◊◊◊◊ ◊◊◊◊◊ ◊◊◊◊◊ ◊◊◊◊◊	395,900
◊◊◊◊◊ ◊◊◊◊◊ ◊◊◊◊◊ ◊◊◊◊◊ ◊◊◊◊◊ ◊◊◊◊◊ ◊◊◊◊◊ ◊◊◊◊◊ ◊◊◊◊◊ ◊◊◊◊◊	395,950
◊◊◊◊◊ ◊◊◊◊◊ ◊◊◊◊◊ ◊◊◊◊◊ ◊◊◊◊◊ ◊◊◊◊◊ ◊◊◊◊◊ ◊◊◊◊◊ ◊◊◊◊◊ ◊◊◊◊◊	**396,000**

The Number 396,001 to the Number 398,000

Start | | Notes

	396,050
	396,100
	396,150
	396,200
	396,250
	396,300
	396,350
	396,400
	396,450
	396,500
	396,550
	396,600
	396,650
	396,700
	396,750
	396,800
	396,850
	396,900
	396,950
	397,000
	397,050
	397,100
	397,150
	397,200
	397,250
	397,300
	397,350
	397,400
	397,450
	397,500
	397,550
	397,600
	397,650
	397,700
	397,750
	397,800
	397,850
	397,900
	397,950
	398,000

The Number 398,001 to the Number 400,000

Start

◊◊◊◊◊ ◊◊◊◊◊ ◊◊◊◊◊ ◊◊◊◊◊ ◊◊◊◊◊ ◊◊◊◊◊ ◊◊◊◊◊ ◊◊◊◊◊ ◊◊◊◊◊ ◊◊◊◊◊	398,050
◊◊◊◊◊ ◊◊◊◊◊ ◊◊◊◊◊ ◊◊◊◊◊ ◊◊◊◊◊ ◊◊◊◊◊ ◊◊◊◊◊ ◊◊◊◊◊ ◊◊◊◊◊ ◊◊◊◊◊	398,100
◊◊◊◊◊ ◊◊◊◊◊ ◊◊◊◊◊ ◊◊◊◊◊ ◊◊◊◊◊ ◊◊◊◊◊ ◊◊◊◊◊ ◊◊◊◊◊ ◊◊◊◊◊ ◊◊◊◊◊	398,150
◊◊◊◊◊ ◊◊◊◊◊ ◊◊◊◊◊ ◊◊◊◊◊ ◊◊◊◊◊ ◊◊◊◊◊ ◊◊◊◊◊ ◊◊◊◊◊ ◊◊◊◊◊ ◊◊◊◊◊	398,200
◊◊◊◊◊ ◊◊◊◊◊ ◊◊◊◊◊ ◊◊◊◊◊ ◊◊◊◊◊ ◊◊◊◊◊ ◊◊◊◊◊ ◊◊◊◊◊ ◊◊◊◊◊ ◊◊◊◊◊	398,250
◊◊◊◊◊ ◊◊◊◊◊ ◊◊◊◊◊ ◊◊◊◊◊ ◊◊◊◊◊ ◊◊◊◊◊ ◊◊◊◊◊ ◊◊◊◊◊ ◊◊◊◊◊ ◊◊◊◊◊	398,300
◊◊◊◊◊ ◊◊◊◊◊ ◊◊◊◊◊ ◊◊◊◊◊ ◊◊◊◊◊ ◊◊◊◊◊ ◊◊◊◊◊ ◊◊◊◊◊ ◊◊◊◊◊ ◊◊◊◊◊	398,350
◊◊◊◊◊ ◊◊◊◊◊ ◊◊◊◊◊ ◊◊◊◊◊ ◊◊◊◊◊ ◊◊◊◊◊ ◊◊◊◊◊ ◊◊◊◊◊ ◊◊◊◊◊ ◊◊◊◊◊	398,400
◊◊◊◊◊ ◊◊◊◊◊ ◊◊◊◊◊ ◊◊◊◊◊ ◊◊◊◊◊ ◊◊◊◊◊ ◊◊◊◊◊ ◊◊◊◊◊ ◊◊◊◊◊ ◊◊◊◊◊	398,450
◊◊◊◊◊ ◊◊◊◊◊ ◊◊◊◊◊ ◊◊◊◊◊ ◊◊◊◊◊ ◊◊◊◊◊ ◊◊◊◊◊ ◊◊◊◊◊ ◊◊◊◊◊ ◊◊◊◊◊	398,500
◊◊◊◊◊ ◊◊◊◊◊ ◊◊◊◊◊ ◊◊◊◊◊ ◊◊◊◊◊ ◊◊◊◊◊ ◊◊◊◊◊ ◊◊◊◊◊ ◊◊◊◊◊ ◊◊◊◊◊	398,550
◊◊◊◊◊ ◊◊◊◊◊ ◊◊◊◊◊ ◊◊◊◊◊ ◊◊◊◊◊ ◊◊◊◊◊ ◊◊◊◊◊ ◊◊◊◊◊ ◊◊◊◊◊ ◊◊◊◊◊	398,600
◊◊◊◊◊ ◊◊◊◊◊ ◊◊◊◊◊ ◊◊◊◊◊ ◊◊◊◊◊ ◊◊◊◊◊ ◊◊◊◊◊ ◊◊◊◊◊ ◊◊◊◊◊ ◊◊◊◊◊	398,650
◊◊◊◊◊ ◊◊◊◊◊ ◊◊◊◊◊ ◊◊◊◊◊ ◊◊◊◊◊ ◊◊◊◊◊ ◊◊◊◊◊ ◊◊◊◊◊ ◊◊◊◊◊ ◊◊◊◊◊	398,700
◊◊◊◊◊ ◊◊◊◊◊ ◊◊◊◊◊ ◊◊◊◊◊ ◊◊◊◊◊ ◊◊◊◊◊ ◊◊◊◊◊ ◊◊◊◊◊ ◊◊◊◊◊ ◊◊◊◊◊	398,750
◊◊◊◊◊ ◊◊◊◊◊ ◊◊◊◊◊ ◊◊◊◊◊ ◊◊◊◊◊ ◊◊◊◊◊ ◊◊◊◊◊ ◊◊◊◊◊ ◊◊◊◊◊ ◊◊◊◊◊	398,800
◊◊◊◊◊ ◊◊◊◊◊ ◊◊◊◊◊ ◊◊◊◊◊ ◊◊◊◊◊ ◊◊◊◊◊ ◊◊◊◊◊ ◊◊◊◊◊ ◊◊◊◊◊ ◊◊◊◊◊	398,850
◊◊◊◊◊ ◊◊◊◊◊ ◊◊◊◊◊ ◊◊◊◊◊ ◊◊◊◊◊ ◊◊◊◊◊ ◊◊◊◊◊ ◊◊◊◊◊ ◊◊◊◊◊ ◊◊◊◊◊	398,900
◊◊◊◊◊ ◊◊◊◊◊ ◊◊◊◊◊ ◊◊◊◊◊ ◊◊◊◊◊ ◊◊◊◊◊ ◊◊◊◊◊ ◊◊◊◊◊ ◊◊◊◊◊ ◊◊◊◊◊	398,950
◊◊◊◊◊ ◊◊◊◊◊ ◊◊◊◊◊ ◊◊◊◊◊ ◊◊◊◊◊ ◊◊◊◊◊ ◊◊◊◊◊ ◊◊◊◊◊ ◊◊◊◊◊ ◊◊◊◊◊	**399,000**
◊◊◊◊◊ ◊◊◊◊◊ ◊◊◊◊◊ ◊◊◊◊◊ ◊◊◊◊◊ ◊◊◊◊◊ ◊◊◊◊◊ ◊◊◊◊◊ ◊◊◊◊◊ ◊◊◊◊◊	399,050
◊◊◊◊◊ ◊◊◊◊◊ ◊◊◊◊◊ ◊◊◊◊◊ ◊◊◊◊◊ ◊◊◊◊◊ ◊◊◊◊◊ ◊◊◊◊◊ ◊◊◊◊◊ ◊◊◊◊◊	399,100
◊◊◊◊◊ ◊◊◊◊◊ ◊◊◊◊◊ ◊◊◊◊◊ ◊◊◊◊◊ ◊◊◊◊◊ ◊◊◊◊◊ ◊◊◊◊◊ ◊◊◊◊◊ ◊◊◊◊◊	399,150
◊◊◊◊◊ ◊◊◊◊◊ ◊◊◊◊◊ ◊◊◊◊◊ ◊◊◊◊◊ ◊◊◊◊◊ ◊◊◊◊◊ ◊◊◊◊◊ ◊◊◊◊◊ ◊◊◊◊◊	399,200
◊◊◊◊◊ ◊◊◊◊◊ ◊◊◊◊◊ ◊◊◊◊◊ ◊◊◊◊◊ ◊◊◊◊◊ ◊◊◊◊◊ ◊◊◊◊◊ ◊◊◊◊◊ ◊◊◊◊◊	399,250
◊◊◊◊◊ ◊◊◊◊◊ ◊◊◊◊◊ ◊◊◊◊◊ ◊◊◊◊◊ ◊◊◊◊◊ ◊◊◊◊◊ ◊◊◊◊◊ ◊◊◊◊◊ ◊◊◊◊◊	399,300
◊◊◊◊◊ ◊◊◊◊◊ ◊◊◊◊◊ ◊◊◊◊◊ ◊◊◊◊◊ ◊◊◊◊◊ ◊◊◊◊◊ ◊◊◊◊◊ ◊◊◊◊◊ ◊◊◊◊◊	399,350
◊◊◊◊◊ ◊◊◊◊◊ ◊◊◊◊◊ ◊◊◊◊◊ ◊◊◊◊◊ ◊◊◊◊◊ ◊◊◊◊◊ ◊◊◊◊◊ ◊◊◊◊◊ ◊◊◊◊◊	399,400
◊◊◊◊◊ ◊◊◊◊◊ ◊◊◊◊◊ ◊◊◊◊◊ ◊◊◊◊◊ ◊◊◊◊◊ ◊◊◊◊◊ ◊◊◊◊◊ ◊◊◊◊◊ ◊◊◊◊◊	399,450
◊◊◊◊◊ ◊◊◊◊◊ ◊◊◊◊◊ ◊◊◊◊◊ ◊◊◊◊◊ ◊◊◊◊◊ ◊◊◊◊◊ ◊◊◊◊◊ ◊◊◊◊◊ ◊◊◊◊◊	399,500
◊◊◊◊◊ ◊◊◊◊◊ ◊◊◊◊◊ ◊◊◊◊◊ ◊◊◊◊◊ ◊◊◊◊◊ ◊◊◊◊◊ ◊◊◊◊◊ ◊◊◊◊◊ ◊◊◊◊◊	399,550
◊◊◊◊◊ ◊◊◊◊◊ ◊◊◊◊◊ ◊◊◊◊◊ ◊◊◊◊◊ ◊◊◊◊◊ ◊◊◊◊◊ ◊◊◊◊◊ ◊◊◊◊◊ ◊◊◊◊◊	399,600
◊◊◊◊◊ ◊◊◊◊◊ ◊◊◊◊◊ ◊◊◊◊◊ ◊◊◊◊◊ ◊◊◊◊◊ ◊◊◊◊◊ ◊◊◊◊◊ ◊◊◊◊◊ ◊◊◊◊◊	399,650
◊◊◊◊◊ ◊◊◊◊◊ ◊◊◊◊◊ ◊◊◊◊◊ ◊◊◊◊◊ ◊◊◊◊◊ ◊◊◊◊◊ ◊◊◊◊◊ ◊◊◊◊◊ ◊◊◊◊◊	399,700
◊◊◊◊◊ ◊◊◊◊◊ ◊◊◊◊◊ ◊◊◊◊◊ ◊◊◊◊◊ ◊◊◊◊◊ ◊◊◊◊◊ ◊◊◊◊◊ ◊◊◊◊◊ ◊◊◊◊◊	399,750
◊◊◊◊◊ ◊◊◊◊◊ ◊◊◊◊◊ ◊◊◊◊◊ ◊◊◊◊◊ ◊◊◊◊◊ ◊◊◊◊◊ ◊◊◊◊◊ ◊◊◊◊◊ ◊◊◊◊◊	399,800
◊◊◊◊◊ ◊◊◊◊◊ ◊◊◊◊◊ ◊◊◊◊◊ ◊◊◊◊◊ ◊◊◊◊◊ ◊◊◊◊◊ ◊◊◊◊◊ ◊◊◊◊◊ ◊◊◊◊◊	399,850
◊◊◊◊◊ ◊◊◊◊◊ ◊◊◊◊◊ ◊◊◊◊◊ ◊◊◊◊◊ ◊◊◊◊◊ ◊◊◊◊◊ ◊◊◊◊◊ ◊◊◊◊◊ ◊◊◊◊◊	399,900
◊◊◊◊◊ ◊◊◊◊◊ ◊◊◊◊◊ ◊◊◊◊◊ ◊◊◊◊◊ ◊◊◊◊◊ ◊◊◊◊◊ ◊◊◊◊◊ ◊◊◊◊◊ ◊◊◊◊◊	399,950
◊◊◊◊◊ ◊◊◊◊◊ ◊◊◊◊◊ ◊◊◊◊◊ ◊◊◊◊◊ ◊◊◊◊◊ ◊◊◊◊◊ ◊◊◊◊◊ ◊◊◊◊◊ ◊◊◊◊◊	**400,000**

The Number 400,001 to the Number 402,000

Start

Notes

	400,050
	400,100
	400,150
	400,200
	400,250
	400,300
	400,350
	400,400
	400,450
	400,500
	400,550
	400,600
	400,650
	400,700
	400,750
	400,800
	400,850
	400,900
	400,950
	401,000
	401,050
	401,100
	401,150
	401,200
	401,250
	401,300
	401,350
	401,400
	401,450
	401,500
	401,550
	401,600
	401,650
	401,700
	401,750
	401,800
	401,850
	401,900
	401,950
	402,000

The Number 402,001 to the Number 404,000

Start	Notes
	402,050
	402,100
	402,150
	402,200
	402,250
	402,300
	402,350
	402,400
	402,450
	402,500
	402,550
	402,600
	402,650
	402,700
	402,750
	402,800
	402,850
	402,900
	402,950
	403,000
	403,050
	403,100
	403,150
	403,200
	403,250
	403,300
	403,350
	403,400
	403,450
	403,500
	403,550
	403,600
	403,650
	403,700
	403,750
	403,800
	403,850
	403,900
	403,950
	404,000

The Number 404,001 to the Number 406,000

Start   Notes

	404,050
	404,100
	404,150
	404,200
	404,250
	404,300
	404,350
	404,400
	404,450
	404,500
	404,550
	404,600
	404,650
	404,700
	404,750
	404,800
	404,850
	404,900
	404,950
	405,000
	405,050
	405,100
	405,150
	405,200
	405,250
	405,300
	405,350
	405,400
	405,450
	405,500
	405,550
	405,600
	405,650
	405,700
	405,750
	405,800
	405,850
	405,900
	405,950
	406,000

The Number 406,001 to the Number 408,000

Start

	Notes
406,050	
406,100	
406,150	
406,200	
406,250	
406,300	
406,350	
406,400	
406,450	
406,500	
406,550	
406,600	
406,650	
406,700	
406,750	
406,800	
406,850	
406,900	
406,950	
407,000	
407,050	
407,100	
407,150	
407,200	
407,250	
407,300	
407,350	
407,400	
407,450	
407,500	
407,550	
407,600	
407,650	
407,700	
407,750	
407,800	
407,850	
407,900	
407,950	
408,000	

The Number 408,001 to the Number 410,000

Start | | Notes

	408,050
	408,100
	408,150
	408,200
	408,250
	408,300
	408,350
	408,400
	408,450
	408,500
	408,550
	408,600
	408,650
	408,700
	408,750
	408,800
	408,850
	408,900
	408,950
	409,000
	409,050
	409,100
	409,150
	409,200
	409,250
	409,300
	409,350
	409,400
	409,450
	409,500
	409,550
	409,600
	409,650
	409,700
	409,750
	409,800
	409,850
	409,900
	409,950
	410,000

The Number 410,001 to the Number 412,000

Start	Notes
◊◊◊◊◊ ◊◊◊◊◊ ◊◊◊◊◊ ◊◊◊◊◊ ◊◊◊◊◊ ◊◊◊◊◊ ◊◊◊◊◊ ◊◊◊◊◊ ◊◊◊◊◊ ◊◊◊◊◊	410,050
◊◊◊◊◊ ◊◊◊◊◊ ◊◊◊◊◊ ◊◊◊◊◊ ◊◊◊◊◊ ◊◊◊◊◊ ◊◊◊◊◊ ◊◊◊◊◊ ◊◊◊◊◊ ◊◊◊◊◊	410,100
◊◊◊◊◊ ◊◊◊◊◊ ◊◊◊◊◊ ◊◊◊◊◊ ◊◊◊◊◊ ◊◊◊◊◊ ◊◊◊◊◊ ◊◊◊◊◊ ◊◊◊◊◊ ◊◊◊◊◊	410,150
◊◊◊◊◊ ◊◊◊◊◊ ◊◊◊◊◊ ◊◊◊◊◊ ◊◊◊◊◊ ◊◊◊◊◊ ◊◊◊◊◊ ◊◊◊◊◊ ◊◊◊◊◊ ◊◊◊◊◊	410,200
◊◊◊◊◊ ◊◊◊◊◊ ◊◊◊◊◊ ◊◊◊◊◊ ◊◊◊◊◊ ◊◊◊◊◊ ◊◊◊◊◊ ◊◊◊◊◊ ◊◊◊◊◊ ◊◊◊◊◊	410,250
◊◊◊◊◊ ◊◊◊◊◊ ◊◊◊◊◊ ◊◊◊◊◊ ◊◊◊◊◊ ◊◊◊◊◊ ◊◊◊◊◊ ◊◊◊◊◊ ◊◊◊◊◊ ◊◊◊◊◊	410,300
◊◊◊◊◊ ◊◊◊◊◊ ◊◊◊◊◊ ◊◊◊◊◊ ◊◊◊◊◊ ◊◊◊◊◊ ◊◊◊◊◊ ◊◊◊◊◊ ◊◊◊◊◊ ◊◊◊◊◊	410,350
◊◊◊◊◊ ◊◊◊◊◊ ◊◊◊◊◊ ◊◊◊◊◊ ◊◊◊◊◊ ◊◊◊◊◊ ◊◊◊◊◊ ◊◊◊◊◊ ◊◊◊◊◊ ◊◊◊◊◊	410,400
◊◊◊◊◊ ◊◊◊◊◊ ◊◊◊◊◊ ◊◊◊◊◊ ◊◊◊◊◊ ◊◊◊◊◊ ◊◊◊◊◊ ◊◊◊◊◊ ◊◊◊◊◊ ◊◊◊◊◊	410,450
◊◊◊◊◊ ◊◊◊◊◊ ◊◊◊◊◊ ◊◊◊◊◊ ◊◊◊◊◊ ◊◊◊◊◊ ◊◊◊◊◊ ◊◊◊◊◊ ◊◊◊◊◊ ◊◊◊◊◊	410,500
◊◊◊◊◊ ◊◊◊◊◊ ◊◊◊◊◊ ◊◊◊◊◊ ◊◊◊◊◊ ◊◊◊◊◊ ◊◊◊◊◊ ◊◊◊◊◊ ◊◊◊◊◊ ◊◊◊◊◊	410,550
◊◊◊◊◊ ◊◊◊◊◊ ◊◊◊◊◊ ◊◊◊◊◊ ◊◊◊◊◊ ◊◊◊◊◊ ◊◊◊◊◊ ◊◊◊◊◊ ◊◊◊◊◊ ◊◊◊◊◊	410,600
◊◊◊◊◊ ◊◊◊◊◊ ◊◊◊◊◊ ◊◊◊◊◊ ◊◊◊◊◊ ◊◊◊◊◊ ◊◊◊◊◊ ◊◊◊◊◊ ◊◊◊◊◊ ◊◊◊◊◊	410,650
◊◊◊◊◊ ◊◊◊◊◊ ◊◊◊◊◊ ◊◊◊◊◊ ◊◊◊◊◊ ◊◊◊◊◊ ◊◊◊◊◊ ◊◊◊◊◊ ◊◊◊◊◊ ◊◊◊◊◊	410,700
◊◊◊◊◊ ◊◊◊◊◊ ◊◊◊◊◊ ◊◊◊◊◊ ◊◊◊◊◊ ◊◊◊◊◊ ◊◊◊◊◊ ◊◊◊◊◊ ◊◊◊◊◊ ◊◊◊◊◊	410,750
◊◊◊◊◊ ◊◊◊◊◊ ◊◊◊◊◊ ◊◊◊◊◊ ◊◊◊◊◊ ◊◊◊◊◊ ◊◊◊◊◊ ◊◊◊◊◊ ◊◊◊◊◊ ◊◊◊◊◊	410,800
◊◊◊◊◊ ◊◊◊◊◊ ◊◊◊◊◊ ◊◊◊◊◊ ◊◊◊◊◊ ◊◊◊◊◊ ◊◊◊◊◊ ◊◊◊◊◊ ◊◊◊◊◊ ◊◊◊◊◊	410,850
◊◊◊◊◊ ◊◊◊◊◊ ◊◊◊◊◊ ◊◊◊◊◊ ◊◊◊◊◊ ◊◊◊◊◊ ◊◊◊◊◊ ◊◊◊◊◊ ◊◊◊◊◊ ◊◊◊◊◊	410,900
◊◊◊◊◊ ◊◊◊◊◊ ◊◊◊◊◊ ◊◊◊◊◊ ◊◊◊◊◊ ◊◊◊◊◊ ◊◊◊◊◊ ◊◊◊◊◊ ◊◊◊◊◊ ◊◊◊◊◊	410,950
◊◊◊◊◊ ◊◊◊◊◊ ◊◊◊◊◊ ◊◊◊◊◊ ◊◊◊◊◊ ◊◊◊◊◊ ◊◊◊◊◊ ◊◊◊◊◊ ◊◊◊◊◊ ◊◊◊◊◊	**411,000**
◊◊◊◊◊ ◊◊◊◊◊ ◊◊◊◊◊ ◊◊◊◊◊ ◊◊◊◊◊ ◊◊◊◊◊ ◊◊◊◊◊ ◊◊◊◊◊ ◊◊◊◊◊ ◊◊◊◊◊	411,050
◊◊◊◊◊ ◊◊◊◊◊ ◊◊◊◊◊ ◊◊◊◊◊ ◊◊◊◊◊ ◊◊◊◊◊ ◊◊◊◊◊ ◊◊◊◊◊ ◊◊◊◊◊ ◊◊◊◊◊	411,100
◊◊◊◊◊ ◊◊◊◊◊ ◊◊◊◊◊ ◊◊◊◊◊ ◊◊◊◊◊ ◊◊◊◊◊ ◊◊◊◊◊ ◊◊◊◊◊ ◊◊◊◊◊ ◊◊◊◊◊	411,150
◊◊◊◊◊ ◊◊◊◊◊ ◊◊◊◊◊ ◊◊◊◊◊ ◊◊◊◊◊ ◊◊◊◊◊ ◊◊◊◊◊ ◊◊◊◊◊ ◊◊◊◊◊ ◊◊◊◊◊	411,200
◊◊◊◊◊ ◊◊◊◊◊ ◊◊◊◊◊ ◊◊◊◊◊ ◊◊◊◊◊ ◊◊◊◊◊ ◊◊◊◊◊ ◊◊◊◊◊ ◊◊◊◊◊ ◊◊◊◊◊	411,250
◊◊◊◊◊ ◊◊◊◊◊ ◊◊◊◊◊ ◊◊◊◊◊ ◊◊◊◊◊ ◊◊◊◊◊ ◊◊◊◊◊ ◊◊◊◊◊ ◊◊◊◊◊ ◊◊◊◊◊	411,300
◊◊◊◊◊ ◊◊◊◊◊ ◊◊◊◊◊ ◊◊◊◊◊ ◊◊◊◊◊ ◊◊◊◊◊ ◊◊◊◊◊ ◊◊◊◊◊ ◊◊◊◊◊ ◊◊◊◊◊	411,350
◊◊◊◊◊ ◊◊◊◊◊ ◊◊◊◊◊ ◊◊◊◊◊ ◊◊◊◊◊ ◊◊◊◊◊ ◊◊◊◊◊ ◊◊◊◊◊ ◊◊◊◊◊ ◊◊◊◊◊	411,400
◊◊◊◊◊ ◊◊◊◊◊ ◊◊◊◊◊ ◊◊◊◊◊ ◊◊◊◊◊ ◊◊◊◊◊ ◊◊◊◊◊ ◊◊◊◊◊ ◊◊◊◊◊ ◊◊◊◊◊	411,450
◊◊◊◊◊ ◊◊◊◊◊ ◊◊◊◊◊ ◊◊◊◊◊ ◊◊◊◊◊ ◊◊◊◊◊ ◊◊◊◊◊ ◊◊◊◊◊ ◊◊◊◊◊ ◊◊◊◊◊	411,500
◊◊◊◊◊ ◊◊◊◊◊ ◊◊◊◊◊ ◊◊◊◊◊ ◊◊◊◊◊ ◊◊◊◊◊ ◊◊◊◊◊ ◊◊◊◊◊ ◊◊◊◊◊ ◊◊◊◊◊	411,550
◊◊◊◊◊ ◊◊◊◊◊ ◊◊◊◊◊ ◊◊◊◊◊ ◊◊◊◊◊ ◊◊◊◊◊ ◊◊◊◊◊ ◊◊◊◊◊ ◊◊◊◊◊ ◊◊◊◊◊	411,600
◊◊◊◊◊ ◊◊◊◊◊ ◊◊◊◊◊ ◊◊◊◊◊ ◊◊◊◊◊ ◊◊◊◊◊ ◊◊◊◊◊ ◊◊◊◊◊ ◊◊◊◊◊ ◊◊◊◊◊	411,650
◊◊◊◊◊ ◊◊◊◊◊ ◊◊◊◊◊ ◊◊◊◊◊ ◊◊◊◊◊ ◊◊◊◊◊ ◊◊◊◊◊ ◊◊◊◊◊ ◊◊◊◊◊ ◊◊◊◊◊	411,700
◊◊◊◊◊ ◊◊◊◊◊ ◊◊◊◊◊ ◊◊◊◊◊ ◊◊◊◊◊ ◊◊◊◊◊ ◊◊◊◊◊ ◊◊◊◊◊ ◊◊◊◊◊ ◊◊◊◊◊	411,750
◊◊◊◊◊ ◊◊◊◊◊ ◊◊◊◊◊ ◊◊◊◊◊ ◊◊◊◊◊ ◊◊◊◊◊ ◊◊◊◊◊ ◊◊◊◊◊ ◊◊◊◊◊ ◊◊◊◊◊	411,800
◊◊◊◊◊ ◊◊◊◊◊ ◊◊◊◊◊ ◊◊◊◊◊ ◊◊◊◊◊ ◊◊◊◊◊ ◊◊◊◊◊ ◊◊◊◊◊ ◊◊◊◊◊ ◊◊◊◊◊	411,850
◊◊◊◊◊ ◊◊◊◊◊ ◊◊◊◊◊ ◊◊◊◊◊ ◊◊◊◊◊ ◊◊◊◊◊ ◊◊◊◊◊ ◊◊◊◊◊ ◊◊◊◊◊ ◊◊◊◊◊	411,900
◊◊◊◊◊ ◊◊◊◊◊ ◊◊◊◊◊ ◊◊◊◊◊ ◊◊◊◊◊ ◊◊◊◊◊ ◊◊◊◊◊ ◊◊◊◊◊ ◊◊◊◊◊ ◊◊◊◊◊	411,950
◊◊◊◊◊ ◊◊◊◊◊ ◊◊◊◊◊ ◊◊◊◊◊ ◊◊◊◊◊ ◊◊◊◊◊ ◊◊◊◊◊ ◊◊◊◊◊ ◊◊◊◊◊ ◊◊◊◊◊	**412,000**

The Number 412,001 to the Number 414,000

Start   Notes

	412,050
	412,100
	412,150
	412,200
	412,250
	412,300
	412,350
	412,400
	412,450
	412,500
	412,550
	412,600
	412,650
	412,700
	412,750
	412,800
	412,850
	412,900
	412,950
	413,000
	413,050
	413,100
	413,150
	413,200
	413,250
	413,300
	413,350
	413,400
	413,450
	413,500
	413,550
	413,600
	413,650
	413,700
	413,750
	413,800
	413,850
	413,900
	413,950
	414,000

The Number 414,001 to the Number 416,000

Start

	Notes
414,050	
414,100	
414,150	
414,200	
414,250	
414,300	
414,350	
414,400	
414,450	
414,500	
414,550	
414,600	
414,650	
414,700	
414,750	
414,800	
414,850	
414,900	
414,950	
415,000	
415,050	
415,100	
415,150	
415,200	
415,250	
415,300	
415,350	
415,400	
415,450	
415,500	
415,550	
415,600	
415,650	
415,700	
415,750	
415,800	
415,850	
415,900	
415,950	
416,000	

The Number 416,001 to the Number 418,000

Start   Notes

	416,050
	416,100
	416,150
	416,200
	416,250
	416,300
	416,350
	416,400
	416,450
	416,500
	416,550
	416,600
	416,650
	416,700
	416,750
	416,800
	416,850
	416,900
	416,950
	417,000
	417,050
	417,100
	417,150
	417,200
	417,250
	417,300
	417,350
	417,400
	417,450
	417,500
	417,550
	417,600
	417,650
	417,700
	417,750
	417,800
	417,850
	417,900
	417,950
	418,000

The Number 418,001 to the Number 420,000

Start	Notes
	418,050
	418,100
	418,150
	418,200
	418,250
	418,300
	418,350
	418,400
	418,450
	418,500
	418,550
	418,600
	418,650
	418,700
	418,750
	418,800
	418,850
	418,900
	418,950
	419,000
	419,050
	419,100
	419,150
	419,200
	419,250
	419,300
	419,350
	419,400
	419,450
	419,500
	419,550
	419,600
	419,650
	419,700
	419,750
	419,800
	419,850
	419,900
	419,950
	420,000

The Number 420,001 to the Number 422,000

Start		Notes
	420,050	
	420,100	
	420,150	
	420,200	
	420,250	
	420,300	
	420,350	
	420,400	
	420,450	
	420,500	
	420,550	
	420,600	
	420,650	
	420,700	
	420,750	
	420,800	
	420,850	
	420,900	
	420,950	
	421,000	
	421,050	
	421,100	
	421,150	
	421,200	
	421,250	
	421,300	
	421,350	
	421,400	
	421,450	
	421,500	
	421,550	
	421,600	
	421,650	
	421,700	
	421,750	
	421,800	
	421,850	
	421,900	
	421,950	
	422,000	

The Number 422,001 to the Number 424,000

Start

Notes

	422,050
	422,100
	422,150
	422,200
	422,250
	422,300
	422,350
	422,400
	422,450
	422,500
	422,550
	422,600
	422,650
	422,700
	422,750
	422,800
	422,850
	422,900
	422,950
	423,000
	423,050
	423,100
	423,150
	423,200
	423,250
	423,300
	423,350
	423,400
	423,450
	423,500
	423,550
	423,600
	423,650
	423,700
	423,750
	423,800
	423,850
	423,900
	423,950
	424,000

The Number 424,001 to the Number 426,000

Start   Notes

	424,050
	424,100
	424,150
	424,200
	424,250
	424,300
	424,350
	424,400
	424,450
	424,500
	424,550
	424,600
	424,650
	424,700
	424,750
	424,800
	424,850
	424,900
	424,950
	425,000
	425,050
	425,100
	425,150
	425,200
	425,250
	425,300
	425,350
	425,400
	425,450
	425,500
	425,550
	425,600
	425,650
	425,700
	425,750
	425,800
	425,850
	425,900
	425,950
	426,000

The Number 426,001 to the Number 428,000

Start		Notes
	426,050	
	426,100	
	426,150	
	426,200	
	426,250	
	426,300	
	426,350	
	426,400	
	426,450	
	426,500	
	426,550	
	426,600	
	426,650	
	426,700	
	426,750	
	426,800	
	426,850	
	426,900	
	426,950	
	427,000	
	427,050	
	427,100	
	427,150	
	427,200	
	427,250	
	427,300	
	427,350	
	427,400	
	427,450	
	427,500	
	427,550	
	427,600	
	427,650	
	427,700	
	427,750	
	427,800	
	427,850	
	427,900	
	427,950	
	428,000	

The Number 428,001 to the Number 430,000

Start   **Notes**

	428,050
	428,100
	428,150
	428,200
	428,250
	428,300
	428,350
	428,400
	428,450
	428,500
	428,550
	428,600
	428,650
	428,700
	428,750
	428,800
	428,850
	428,900
	428,950
	429,000
	429,050
	429,100
	429,150
	429,200
	429,250
	429,300
	429,350
	429,400
	429,450
	429,500
	429,550
	429,600
	429,650
	429,700
	429,750
	429,800
	429,850
	429,900
	429,950
	430,000

The Number 430,001 to the Number 432,000

Start · · · Notes

	Number	Notes
	430,050	
	430,100	
	430,150	
	430,200	
	430,250	
	430,300	
	430,350	
	430,400	
	430,450	
	430,500	
	430,550	
	430,600	
	430,650	
	430,700	
	430,750	
	430,800	
	430,850	
	430,900	
	430,950	
	431,000	
	431,050	
	431,100	
	431,150	
	431,200	
	431,250	
	431,300	
	431,350	
	431,400	
	431,450	
	431,500	
	431,550	
	431,600	
	431,650	
	431,700	
	431,750	
	431,800	
	431,850	
	431,900	
	431,950	
	432,000	

The Number 432,001 to the Number 434,000

Start		Notes
	432,050	
	432,100	
	432,150	
	432,200	
	432,250	
	432,300	
	432,350	
	432,400	
	432,450	
	432,500	
	432,550	
	432,600	
	432,650	
	432,700	
	432,750	
	432,800	
	432,850	
	432,900	
	432,950	
	433,000	
	433,050	
	433,100	
	433,150	
	433,200	
	433,250	
	433,300	
	433,350	
	433,400	
	433,450	
	433,500	
	433,550	
	433,600	
	433,650	
	433,700	
	433,750	
	433,800	
	433,850	
	433,900	
	433,950	
	434,000	

The Number 434,001 to the Number 436,000

Start	Notes
	434,050
	434,100
	434,150
	434,200
	434,250
	434,300
	434,350
	434,400
	434,450
	434,500
	434,550
	434,600
	434,650
	434,700
	434,750
	434,800
	434,850
	434,900
	434,950
	435,000
	435,050
	435,100
	435,150
	435,200
	435,250
	435,300
	435,350
	435,400
	435,450
	435,500
	435,550
	435,600
	435,650
	435,700
	435,750
	435,800
	435,850
	435,900
	435,950
	436,000

The Number 436,001 to the Number 438,000

Start | | Notes

436,050	
436,100	
436,150	
436,200	
436,250	
436,300	
436,350	
436,400	
436,450	
436,500	
436,550	
436,600	
436,650	
436,700	
436,750	
436,800	
436,850	
436,900	
436,950	
437,000	
437,050	
437,100	
437,150	
437,200	
437,250	
437,300	
437,350	
437,400	
437,450	
437,500	
437,550	
437,600	
437,650	
437,700	
437,750	
437,800	
437,850	
437,900	
437,950	
438,000	

The Number 438,001 to the Number 440,000

Start | | **Notes**

	438,050
	438,100
	438,150
	438,200
	438,250
	438,300
	438,350
	438,400
	438,450
	438,500
	438,550
	438,600
	438,650
	438,700
	438,750
	438,800
	438,850
	438,900
	438,950
	439,000
	439,050
	439,100
	439,150
	439,200
	439,250
	439,300
	439,350
	439,400
	439,450
	439,500
	439,550
	439,600
	439,650
	439,700
	439,750
	439,800
	439,850
	439,900
	439,950
	440,000

The Number 440,001 to the Number 442,000

Start | | **Notes**

	440,050
	440,100
	440,150
	440,200
	440,250
	440,300
	440,350
	440,400
	440,450
	440,500
	440,550
	440,600
	440,650
	440,700
	440,750
	440,800
	440,850
	440,900
	440,950
	441,000
	441,050
	441,100
	441,150
	441,200
	441,250
	441,300
	441,350
	441,400
	441,450
	441,500
	441,550
	441,600
	441,650
	441,700
	441,750
	441,800
	441,850
	441,900
	441,950
	442,000

The Number 442,001 to the Number 444,000

Start		Notes
	442,050	
	442,100	
	442,150	
	442,200	
	442,250	
	442,300	
	442,350	
	442,400	
	442,450	
	442,500	
	442,550	
	442,600	
	442,650	
	442,700	
	442,750	
	442,800	
	442,850	
	442,900	
	442,950	
	443,000	
	443,050	
	443,100	
	443,150	
	443,200	
	443,250	
	443,300	
	443,350	
	443,400	
	443,450	
	443,500	
	443,550	
	443,600	
	443,650	
	443,700	
	443,750	
	443,800	
	443,850	
	443,900	
	443,950	
	444,000	

The Number 444,001 to the Number 446,000

Start

Notes

◊◊◊◊◊ ◊◊◊◊◊ ◊◊◊◊◊ ◊◊◊◊◊ ◊◊◊◊◊ ◊◊◊◊◊ ◊◊◊◊◊ ◊◊◊◊◊ ◊◊◊◊◊ ◊◊◊◊◊	444,050
◊◊◊◊◊ ◊◊◊◊◊ ◊◊◊◊◊ ◊◊◊◊◊ ◊◊◊◊◊ ◊◊◊◊◊ ◊◊◊◊◊ ◊◊◊◊◊ ◊◊◊◊◊ ◊◊◊◊◊	444,100
◊◊◊◊◊ ◊◊◊◊◊ ◊◊◊◊◊ ◊◊◊◊◊ ◊◊◊◊◊ ◊◊◊◊◊ ◊◊◊◊◊ ◊◊◊◊◊ ◊◊◊◊◊ ◊◊◊◊◊	444,150
◊◊◊◊◊ ◊◊◊◊◊ ◊◊◊◊◊ ◊◊◊◊◊ ◊◊◊◊◊ ◊◊◊◊◊ ◊◊◊◊◊ ◊◊◊◊◊ ◊◊◊◊◊ ◊◊◊◊◊	444,200
◊◊◊◊◊ ◊◊◊◊◊ ◊◊◊◊◊ ◊◊◊◊◊ ◊◊◊◊◊ ◊◊◊◊◊ ◊◊◊◊◊ ◊◊◊◊◊ ◊◊◊◊◊ ◊◊◊◊◊	444,250
◊◊◊◊◊ ◊◊◊◊◊ ◊◊◊◊◊ ◊◊◊◊◊ ◊◊◊◊◊ ◊◊◊◊◊ ◊◊◊◊◊ ◊◊◊◊◊ ◊◊◊◊◊ ◊◊◊◊◊	444,300
◊◊◊◊◊ ◊◊◊◊◊ ◊◊◊◊◊ ◊◊◊◊◊ ◊◊◊◊◊ ◊◊◊◊◊ ◊◊◊◊◊ ◊◊◊◊◊ ◊◊◊◊◊ ◊◊◊◊◊	444,350
◊◊◊◊◊ ◊◊◊◊◊ ◊◊◊◊◊ ◊◊◊◊◊ ◊◊◊◊◊ ◊◊◊◊◊ ◊◊◊◊◊ ◊◊◊◊◊ ◊◊◊◊◊ ◊◊◊◊◊	444,400
◊◊◊◊◊ ◊◊◊◊◊ ◊◊◊◊◊ ◊◊◊◊◊ ◊◊◊◊◊ ◊◊◊◊◊ ◊◊◊◊◊ ◊◊◊◊◊ ◊◊◊◊◊ ◊◊◊◊◊	444,450
◊◊◊◊◊ ◊◊◊◊◊ ◊◊◊◊◊ ◊◊◊◊◊ ◊◊◊◊◊ ◊◊◊◊◊ ◊◊◊◊◊ ◊◊◊◊◊ ◊◊◊◊◊ ◊◊◊◊◊	444,500
◊◊◊◊◊ ◊◊◊◊◊ ◊◊◊◊◊ ◊◊◊◊◊ ◊◊◊◊◊ ◊◊◊◊◊ ◊◊◊◊◊ ◊◊◊◊◊ ◊◊◊◊◊ ◊◊◊◊◊	444,550
◊◊◊◊◊ ◊◊◊◊◊ ◊◊◊◊◊ ◊◊◊◊◊ ◊◊◊◊◊ ◊◊◊◊◊ ◊◊◊◊◊ ◊◊◊◊◊ ◊◊◊◊◊ ◊◊◊◊◊	444,600
◊◊◊◊◊ ◊◊◊◊◊ ◊◊◊◊◊ ◊◊◊◊◊ ◊◊◊◊◊ ◊◊◊◊◊ ◊◊◊◊◊ ◊◊◊◊◊ ◊◊◊◊◊ ◊◊◊◊◊	444,650
◊◊◊◊◊ ◊◊◊◊◊ ◊◊◊◊◊ ◊◊◊◊◊ ◊◊◊◊◊ ◊◊◊◊◊ ◊◊◊◊◊ ◊◊◊◊◊ ◊◊◊◊◊ ◊◊◊◊◊	444,700
◊◊◊◊◊ ◊◊◊◊◊ ◊◊◊◊◊ ◊◊◊◊◊ ◊◊◊◊◊ ◊◊◊◊◊ ◊◊◊◊◊ ◊◊◊◊◊ ◊◊◊◊◊ ◊◊◊◊◊	444,750
◊◊◊◊◊ ◊◊◊◊◊ ◊◊◊◊◊ ◊◊◊◊◊ ◊◊◊◊◊ ◊◊◊◊◊ ◊◊◊◊◊ ◊◊◊◊◊ ◊◊◊◊◊ ◊◊◊◊◊	444,800
◊◊◊◊◊ ◊◊◊◊◊ ◊◊◊◊◊ ◊◊◊◊◊ ◊◊◊◊◊ ◊◊◊◊◊ ◊◊◊◊◊ ◊◊◊◊◊ ◊◊◊◊◊ ◊◊◊◊◊	444,850
◊◊◊◊◊ ◊◊◊◊◊ ◊◊◊◊◊ ◊◊◊◊◊ ◊◊◊◊◊ ◊◊◊◊◊ ◊◊◊◊◊ ◊◊◊◊◊ ◊◊◊◊◊ ◊◊◊◊◊	444,900
◊◊◊◊◊ ◊◊◊◊◊ ◊◊◊◊◊ ◊◊◊◊◊ ◊◊◊◊◊ ◊◊◊◊◊ ◊◊◊◊◊ ◊◊◊◊◊ ◊◊◊◊◊ ◊◊◊◊◊	444,950
◊◊◊◊◊ ◊◊◊◊◊ ◊◊◊◊◊ ◊◊◊◊◊ ◊◊◊◊◊ ◊◊◊◊◊ ◊◊◊◊◊ ◊◊◊◊◊ ◊◊◊◊◊ ◊◊◊◊◊	**445,000**
◊◊◊◊◊ ◊◊◊◊◊ ◊◊◊◊◊ ◊◊◊◊◊ ◊◊◊◊◊ ◊◊◊◊◊ ◊◊◊◊◊ ◊◊◊◊◊ ◊◊◊◊◊ ◊◊◊◊◊	445,050
◊◊◊◊◊ ◊◊◊◊◊ ◊◊◊◊◊ ◊◊◊◊◊ ◊◊◊◊◊ ◊◊◊◊◊ ◊◊◊◊◊ ◊◊◊◊◊ ◊◊◊◊◊ ◊◊◊◊◊	445,100
◊◊◊◊◊ ◊◊◊◊◊ ◊◊◊◊◊ ◊◊◊◊◊ ◊◊◊◊◊ ◊◊◊◊◊ ◊◊◊◊◊ ◊◊◊◊◊ ◊◊◊◊◊ ◊◊◊◊◊	445,150
◊◊◊◊◊ ◊◊◊◊◊ ◊◊◊◊◊ ◊◊◊◊◊ ◊◊◊◊◊ ◊◊◊◊◊ ◊◊◊◊◊ ◊◊◊◊◊ ◊◊◊◊◊ ◊◊◊◊◊	445,200
◊◊◊◊◊ ◊◊◊◊◊ ◊◊◊◊◊ ◊◊◊◊◊ ◊◊◊◊◊ ◊◊◊◊◊ ◊◊◊◊◊ ◊◊◊◊◊ ◊◊◊◊◊ ◊◊◊◊◊	445,250
◊◊◊◊◊ ◊◊◊◊◊ ◊◊◊◊◊ ◊◊◊◊◊ ◊◊◊◊◊ ◊◊◊◊◊ ◊◊◊◊◊ ◊◊◊◊◊ ◊◊◊◊◊ ◊◊◊◊◊	445,300
◊◊◊◊◊ ◊◊◊◊◊ ◊◊◊◊◊ ◊◊◊◊◊ ◊◊◊◊◊ ◊◊◊◊◊ ◊◊◊◊◊ ◊◊◊◊◊ ◊◊◊◊◊ ◊◊◊◊◊	445,350
◊◊◊◊◊ ◊◊◊◊◊ ◊◊◊◊◊ ◊◊◊◊◊ ◊◊◊◊◊ ◊◊◊◊◊ ◊◊◊◊◊ ◊◊◊◊◊ ◊◊◊◊◊ ◊◊◊◊◊	445,400
◊◊◊◊◊ ◊◊◊◊◊ ◊◊◊◊◊ ◊◊◊◊◊ ◊◊◊◊◊ ◊◊◊◊◊ ◊◊◊◊◊ ◊◊◊◊◊ ◊◊◊◊◊ ◊◊◊◊◊	445,450
◊◊◊◊◊ ◊◊◊◊◊ ◊◊◊◊◊ ◊◊◊◊◊ ◊◊◊◊◊ ◊◊◊◊◊ ◊◊◊◊◊ ◊◊◊◊◊ ◊◊◊◊◊ ◊◊◊◊◊	445,500
◊◊◊◊◊ ◊◊◊◊◊ ◊◊◊◊◊ ◊◊◊◊◊ ◊◊◊◊◊ ◊◊◊◊◊ ◊◊◊◊◊ ◊◊◊◊◊ ◊◊◊◊◊ ◊◊◊◊◊	445,550
◊◊◊◊◊ ◊◊◊◊◊ ◊◊◊◊◊ ◊◊◊◊◊ ◊◊◊◊◊ ◊◊◊◊◊ ◊◊◊◊◊ ◊◊◊◊◊ ◊◊◊◊◊ ◊◊◊◊◊	445,600
◊◊◊◊◊ ◊◊◊◊◊ ◊◊◊◊◊ ◊◊◊◊◊ ◊◊◊◊◊ ◊◊◊◊◊ ◊◊◊◊◊ ◊◊◊◊◊ ◊◊◊◊◊ ◊◊◊◊◊	445,650
◊◊◊◊◊ ◊◊◊◊◊ ◊◊◊◊◊ ◊◊◊◊◊ ◊◊◊◊◊ ◊◊◊◊◊ ◊◊◊◊◊ ◊◊◊◊◊ ◊◊◊◊◊ ◊◊◊◊◊	445,700
◊◊◊◊◊ ◊◊◊◊◊ ◊◊◊◊◊ ◊◊◊◊◊ ◊◊◊◊◊ ◊◊◊◊◊ ◊◊◊◊◊ ◊◊◊◊◊ ◊◊◊◊◊ ◊◊◊◊◊	445,750
◊◊◊◊◊ ◊◊◊◊◊ ◊◊◊◊◊ ◊◊◊◊◊ ◊◊◊◊◊ ◊◊◊◊◊ ◊◊◊◊◊ ◊◊◊◊◊ ◊◊◊◊◊ ◊◊◊◊◊	445,800
◊◊◊◊◊ ◊◊◊◊◊ ◊◊◊◊◊ ◊◊◊◊◊ ◊◊◊◊◊ ◊◊◊◊◊ ◊◊◊◊◊ ◊◊◊◊◊ ◊◊◊◊◊ ◊◊◊◊◊	445,850
◊◊◊◊◊ ◊◊◊◊◊ ◊◊◊◊◊ ◊◊◊◊◊ ◊◊◊◊◊ ◊◊◊◊◊ ◊◊◊◊◊ ◊◊◊◊◊ ◊◊◊◊◊ ◊◊◊◊◊	445,900
◊◊◊◊◊ ◊◊◊◊◊ ◊◊◊◊◊ ◊◊◊◊◊ ◊◊◊◊◊ ◊◊◊◊◊ ◊◊◊◊◊ ◊◊◊◊◊ ◊◊◊◊◊ ◊◊◊◊◊	445,950
◊◊◊◊◊ ◊◊◊◊◊ ◊◊◊◊◊ ◊◊◊◊◊ ◊◊◊◊◊ ◊◊◊◊◊ ◊◊◊◊◊ ◊◊◊◊◊ ◊◊◊◊◊ ◊◊◊◊	**446,000**

The Number 446,001 to the Number 448,000

Start		Notes
	446,050	
	446,100	
	446,150	
	446,200	
	446,250	
	446,300	
	446,350	
	446,400	
	446,450	
	446,500	
	446,550	
	446,600	
	446,650	
	446,700	
	446,750	
	446,800	
	446,850	
	446,900	
	446,950	
	447,000	
	447,050	
	447,100	
	447,150	
	447,200	
	447,250	
	447,300	
	447,350	
	447,400	
	447,450	
	447,500	
	447,550	
	447,600	
	447,650	
	447,700	
	447,750	
	447,800	
	447,850	
	447,900	
	447,950	
	448,000	

The Number 448,001 to the Number 450,000

Start | | Notes

	Number	Notes
	448,050	
	448,100	
	448,150	
	448,200	
	448,250	
	448,300	
	448,350	
	448,400	
	448,450	
	448,500	
	448,550	
	448,600	
	448,650	
	448,700	
	448,750	
	448,800	
	448,850	
	448,900	
	448,950	
	449,000	
	449,050	
	449,100	
	449,150	
	449,200	
	449,250	
	449,300	
	449,350	
	449,400	
	449,450	
	449,500	
	449,550	
	449,600	
	449,650	
	449,700	
	449,750	
	449,800	
	449,850	
	449,900	
	449,950	
	450,000	

◊ The Number 450,001 to the Number 452,000

Start		Notes
	450,050	
	450,100	
	450,150	
	450,200	
	450,250	
	450,300	
	450,350	
	450,400	
	450,450	
	450,500	
	450,550	
	450,600	
	450,650	
	450,700	
	450,750	
	450,800	
	450,850	
	450,900	
	450,950	
	451,000	
	451,050	
	451,100	
	451,150	
	451,200	
	451,250	
	451,300	
	451,350	
	451,400	
	451,450	
	451,500	
	451,550	
	451,600	
	451,650	
	451,700	
	451,750	
	451,800	
	451,850	
	451,900	
	451,950	
	452,000	

The Number 452,001 to the Number 454,000

Start	Notes
	452,050
	452,100
	452,150
	452,200
	452,250
	452,300
	452,350
	452,400
	452,450
	452,500
	452,550
	452,600
	452,650
	452,700
	452,750
	452,800
	452,850
	452,900
	452,950
	453,000
	453,050
	453,100
	453,150
	453,200
	453,250
	453,300
	453,350
	453,400
	453,450
	453,500
	453,550
	453,600
	453,650
	453,700
	453,750
	453,800
	453,850
	453,900
	453,950
	454,000

The Number 454,001 to the Number 456,000

Start

Notes

	454,050
	454,100
	454,150
	454,200
	454,250
	454,300
	454,350
	454,400
	454,450
	454,500
	454,550
	454,600
	454,650
	454,700
	454,750
	454,800
	454,850
	454,900
	454,950
	455,000
	455,050
	455,100
	455,150
	455,200
	455,250
	455,300
	455,350
	455,400
	455,450
	455,500
	455,550
	455,600
	455,650
	455,700
	455,750
	455,800
	455,850
	455,900
	455,950
	456,000

The Number 456,001 to the Number 458,000

Start   Notes

	456,050
	456,100
	456,150
	456,200
	456,250
	456,300
	456,350
	456,400
	456,450
	456,500
	456,550
	456,600
	456,650
	456,700
	456,750
	456,800
	456,850
	456,900
	456,950
	457,000
	457,050
	457,100
	457,150
	457,200
	457,250
	457,300
	457,350
	457,400
	457,450
	457,500
	457,550
	457,600
	457,650
	457,700
	457,750
	457,800
	457,850
	457,900
	457,950
	458,000

The Number 458,001 to the Number 460,000

Start		Notes
◊◊◊◊◊ ◊◊◊◊◊ ◊◊◊◊◊ ◊◊◊◊◊ ◊◊◊◊◊ ◊◊◊◊◊ ◊◊◊◊◊ ◊◊◊◊◊ ◊◊◊◊◊ ◊◊◊◊◊	458,050
◊◊◊◊◊ ◊◊◊◊◊ ◊◊◊◊◊ ◊◊◊◊◊ ◊◊◊◊◊ ◊◊◊◊◊ ◊◊◊◊◊ ◊◊◊◊◊ ◊◊◊◊◊ ◊◊◊◊◊	458,100
◊◊◊◊◊ ◊◊◊◊◊ ◊◊◊◊◊ ◊◊◊◊◊ ◊◊◊◊◊ ◊◊◊◊◊ ◊◊◊◊◊ ◊◊◊◊◊ ◊◊◊◊◊ ◊◊◊◊◊	458,150
◊◊◊◊◊ ◊◊◊◊◊ ◊◊◊◊◊ ◊◊◊◊◊ ◊◊◊◊◊ ◊◊◊◊◊ ◊◊◊◊◊ ◊◊◊◊◊ ◊◊◊◊◊ ◊◊◊◊◊	458,200
◊◊◊◊◊ ◊◊◊◊◊ ◊◊◊◊◊ ◊◊◊◊◊ ◊◊◊◊◊ ◊◊◊◊◊ ◊◊◊◊◊ ◊◊◊◊◊ ◊◊◊◊◊ ◊◊◊◊◊	458,250
◊◊◊◊◊ ◊◊◊◊◊ ◊◊◊◊◊ ◊◊◊◊◊ ◊◊◊◊◊ ◊◊◊◊◊ ◊◊◊◊◊ ◊◊◊◊◊ ◊◊◊◊◊ ◊◊◊◊◊	458,300
◊◊◊◊◊ ◊◊◊◊◊ ◊◊◊◊◊ ◊◊◊◊◊ ◊◊◊◊◊ ◊◊◊◊◊ ◊◊◊◊◊ ◊◊◊◊◊ ◊◊◊◊◊ ◊◊◊◊◊	458,350
◊◊◊◊◊ ◊◊◊◊◊ ◊◊◊◊◊ ◊◊◊◊◊ ◊◊◊◊◊ ◊◊◊◊◊ ◊◊◊◊◊ ◊◊◊◊◊ ◊◊◊◊◊ ◊◊◊◊◊	458,400
◊◊◊◊◊ ◊◊◊◊◊ ◊◊◊◊◊ ◊◊◊◊◊ ◊◊◊◊◊ ◊◊◊◊◊ ◊◊◊◊◊ ◊◊◊◊◊ ◊◊◊◊◊ ◊◊◊◊◊	458,450
◊◊◊◊◊ ◊◊◊◊◊ ◊◊◊◊◊ ◊◊◊◊◊ ◊◊◊◊◊ ◊◊◊◊◊ ◊◊◊◊◊ ◊◊◊◊◊ ◊◊◊◊◊ ◊◊◊◊◊	458,500
◊◊◊◊◊ ◊◊◊◊◊ ◊◊◊◊◊ ◊◊◊◊◊ ◊◊◊◊◊ ◊◊◊◊◊ ◊◊◊◊◊ ◊◊◊◊◊ ◊◊◊◊◊ ◊◊◊◊◊	458,550
◊◊◊◊◊ ◊◊◊◊◊ ◊◊◊◊◊ ◊◊◊◊◊ ◊◊◊◊◊ ◊◊◊◊◊ ◊◊◊◊◊ ◊◊◊◊◊ ◊◊◊◊◊ ◊◊◊◊◊	458,600
◊◊◊◊◊ ◊◊◊◊◊ ◊◊◊◊◊ ◊◊◊◊◊ ◊◊◊◊◊ ◊◊◊◊◊ ◊◊◊◊◊ ◊◊◊◊◊ ◊◊◊◊◊ ◊◊◊◊◊	458,650
◊◊◊◊◊ ◊◊◊◊◊ ◊◊◊◊◊ ◊◊◊◊◊ ◊◊◊◊◊ ◊◊◊◊◊ ◊◊◊◊◊ ◊◊◊◊◊ ◊◊◊◊◊ ◊◊◊◊◊	458,700
◊◊◊◊◊ ◊◊◊◊◊ ◊◊◊◊◊ ◊◊◊◊◊ ◊◊◊◊◊ ◊◊◊◊◊ ◊◊◊◊◊ ◊◊◊◊◊ ◊◊◊◊◊ ◊◊◊◊◊	458,750
◊◊◊◊◊ ◊◊◊◊◊ ◊◊◊◊◊ ◊◊◊◊◊ ◊◊◊◊◊ ◊◊◊◊◊ ◊◊◊◊◊ ◊◊◊◊◊ ◊◊◊◊◊ ◊◊◊◊◊	458,800
◊◊◊◊◊ ◊◊◊◊◊ ◊◊◊◊◊ ◊◊◊◊◊ ◊◊◊◊◊ ◊◊◊◊◊ ◊◊◊◊◊ ◊◊◊◊◊ ◊◊◊◊◊ ◊◊◊◊◊	458,850
◊◊◊◊◊ ◊◊◊◊◊ ◊◊◊◊◊ ◊◊◊◊◊ ◊◊◊◊◊ ◊◊◊◊◊ ◊◊◊◊◊ ◊◊◊◊◊ ◊◊◊◊◊ ◊◊◊◊◊	458,900
◊◊◊◊◊ ◊◊◊◊◊ ◊◊◊◊◊ ◊◊◊◊◊ ◊◊◊◊◊ ◊◊◊◊◊ ◊◊◊◊◊ ◊◊◊◊◊ ◊◊◊◊◊ ◊◊◊◊◊	458,950
◊◊◊◊◊ ◊◊◊◊◊ ◊◊◊◊◊ ◊◊◊◊◊ ◊◊◊◊◊ ◊◊◊◊◊ ◊◊◊◊◊ ◊◊◊◊◊ ◊◊◊◊◊ ◊◊◊◊◊	**459,000**
◊◊◊◊◊ ◊◊◊◊◊ ◊◊◊◊◊ ◊◊◊◊◊ ◊◊◊◊◊ ◊◊◊◊◊ ◊◊◊◊◊ ◊◊◊◊◊ ◊◊◊◊◊ ◊◊◊◊◊	459,050
◊◊◊◊◊ ◊◊◊◊◊ ◊◊◊◊◊ ◊◊◊◊◊ ◊◊◊◊◊ ◊◊◊◊◊ ◊◊◊◊◊ ◊◊◊◊◊ ◊◊◊◊◊ ◊◊◊◊◊	459,100
◊◊◊◊◊ ◊◊◊◊◊ ◊◊◊◊◊ ◊◊◊◊◊ ◊◊◊◊◊ ◊◊◊◊◊ ◊◊◊◊◊ ◊◊◊◊◊ ◊◊◊◊◊ ◊◊◊◊◊	459,150
◊◊◊◊◊ ◊◊◊◊◊ ◊◊◊◊◊ ◊◊◊◊◊ ◊◊◊◊◊ ◊◊◊◊◊ ◊◊◊◊◊ ◊◊◊◊◊ ◊◊◊◊◊ ◊◊◊◊◊	459,200
◊◊◊◊◊ ◊◊◊◊◊ ◊◊◊◊◊ ◊◊◊◊◊ ◊◊◊◊◊ ◊◊◊◊◊ ◊◊◊◊◊ ◊◊◊◊◊ ◊◊◊◊◊ ◊◊◊◊◊	459,250
◊◊◊◊◊ ◊◊◊◊◊ ◊◊◊◊◊ ◊◊◊◊◊ ◊◊◊◊◊ ◊◊◊◊◊ ◊◊◊◊◊ ◊◊◊◊◊ ◊◊◊◊◊ ◊◊◊◊◊	459,300
◊◊◊◊◊ ◊◊◊◊◊ ◊◊◊◊◊ ◊◊◊◊◊ ◊◊◊◊◊ ◊◊◊◊◊ ◊◊◊◊◊ ◊◊◊◊◊ ◊◊◊◊◊ ◊◊◊◊◊	459,350
◊◊◊◊◊ ◊◊◊◊◊ ◊◊◊◊◊ ◊◊◊◊◊ ◊◊◊◊◊ ◊◊◊◊◊ ◊◊◊◊◊ ◊◊◊◊◊ ◊◊◊◊◊ ◊◊◊◊◊	459,400
◊◊◊◊◊ ◊◊◊◊◊ ◊◊◊◊◊ ◊◊◊◊◊ ◊◊◊◊◊ ◊◊◊◊◊ ◊◊◊◊◊ ◊◊◊◊◊ ◊◊◊◊◊ ◊◊◊◊◊	459,450
◊◊◊◊◊ ◊◊◊◊◊ ◊◊◊◊◊ ◊◊◊◊◊ ◊◊◊◊◊ ◊◊◊◊◊ ◊◊◊◊◊ ◊◊◊◊◊ ◊◊◊◊◊ ◊◊◊◊◊	459,500
◊◊◊◊◊ ◊◊◊◊◊ ◊◊◊◊◊ ◊◊◊◊◊ ◊◊◊◊◊ ◊◊◊◊◊ ◊◊◊◊◊ ◊◊◊◊◊ ◊◊◊◊◊ ◊◊◊◊◊	459,550
◊◊◊◊◊ ◊◊◊◊◊ ◊◊◊◊◊ ◊◊◊◊◊ ◊◊◊◊◊ ◊◊◊◊◊ ◊◊◊◊◊ ◊◊◊◊◊ ◊◊◊◊◊ ◊◊◊◊◊	459,600
◊◊◊◊◊ ◊◊◊◊◊ ◊◊◊◊◊ ◊◊◊◊◊ ◊◊◊◊◊ ◊◊◊◊◊ ◊◊◊◊◊ ◊◊◊◊◊ ◊◊◊◊◊ ◊◊◊◊◊	459,650
◊◊◊◊◊ ◊◊◊◊◊ ◊◊◊◊◊ ◊◊◊◊◊ ◊◊◊◊◊ ◊◊◊◊◊ ◊◊◊◊◊ ◊◊◊◊◊ ◊◊◊◊◊ ◊◊◊◊◊	459,700
◊◊◊◊◊ ◊◊◊◊◊ ◊◊◊◊◊ ◊◊◊◊◊ ◊◊◊◊◊ ◊◊◊◊◊ ◊◊◊◊◊ ◊◊◊◊◊ ◊◊◊◊◊ ◊◊◊◊◊	459,750
◊◊◊◊◊ ◊◊◊◊◊ ◊◊◊◊◊ ◊◊◊◊◊ ◊◊◊◊◊ ◊◊◊◊◊ ◊◊◊◊◊ ◊◊◊◊◊ ◊◊◊◊◊ ◊◊◊◊◊	459,800
◊◊◊◊◊ ◊◊◊◊◊ ◊◊◊◊◊ ◊◊◊◊◊ ◊◊◊◊◊ ◊◊◊◊◊ ◊◊◊◊◊ ◊◊◊◊◊ ◊◊◊◊◊ ◊◊◊◊◊	459,850
◊◊◊◊◊ ◊◊◊◊◊ ◊◊◊◊◊ ◊◊◊◊◊ ◊◊◊◊◊ ◊◊◊◊◊ ◊◊◊◊◊ ◊◊◊◊◊ ◊◊◊◊◊ ◊◊◊◊◊	459,900
◊◊◊◊◊ ◊◊◊◊◊ ◊◊◊◊◊ ◊◊◊◊◊ ◊◊◊◊◊ ◊◊◊◊◊ ◊◊◊◊◊ ◊◊◊◊◊ ◊◊◊◊◊ ◊◊◊◊◊	459,950
◊◊◊◊◊ ◊◊◊◊◊ ◊◊◊◊◊ ◊◊◊◊◊ ◊◊◊◊◊ ◊◊◊◊◊ ◊◊◊◊◊ ◊◊◊◊◊ ◊◊◊◊◊ ◊◊◊◊◊	**460,000**

The Number 460,001 to the Number 462,000

Start		Notes

	460,050
	460,100
	460,150
	460,200
	460,250
	460,300
	460,350
	460,400
	460,450
	460,500
	460,550
	460,600
	460,650
	460,700
	460,750
	460,800
	460,850
	460,900
	460,950
	461,000
	461,050
	461,100
	461,150
	461,200
	461,250
	461,300
	461,350
	461,400
	461,450
	461,500
	461,550
	461,600
	461,650
	461,700
	461,750
	461,800
	461,850
	461,900
	461,950
	462,000

The Number 462,001 to the Number 464,000

Start   Notes

◊◊◊◊◊ ◊◊◊◊◊ ◊◊◊◊◊ ◊◊◊◊◊ ◊◊◊◊◊ ◊◊◊◊◊ ◊◊◊◊◊ ◊◊◊◊◊ ◊◊◊◊◊ ◊◊◊◊◊	462,050
◊◊◊◊◊ ◊◊◊◊◊ ◊◊◊◊◊ ◊◊◊◊◊ ◊◊◊◊◊ ◊◊◊◊◊ ◊◊◊◊◊ ◊◊◊◊◊ ◊◊◊◊◊ ◊◊◊◊◊	462,100
◊◊◊◊◊ ◊◊◊◊◊ ◊◊◊◊◊ ◊◊◊◊◊ ◊◊◊◊◊ ◊◊◊◊◊ ◊◊◊◊◊ ◊◊◊◊◊ ◊◊◊◊◊ ◊◊◊◊◊	462,150
◊◊◊◊◊ ◊◊◊◊◊ ◊◊◊◊◊ ◊◊◊◊◊ ◊◊◊◊◊ ◊◊◊◊◊ ◊◊◊◊◊ ◊◊◊◊◊ ◊◊◊◊◊ ◊◊◊◊◊	462,200
◊◊◊◊◊ ◊◊◊◊◊ ◊◊◊◊◊ ◊◊◊◊◊ ◊◊◊◊◊ ◊◊◊◊◊ ◊◊◊◊◊ ◊◊◊◊◊ ◊◊◊◊◊ ◊◊◊◊◊	462,250
◊◊◊◊◊ ◊◊◊◊◊ ◊◊◊◊◊ ◊◊◊◊◊ ◊◊◊◊◊ ◊◊◊◊◊ ◊◊◊◊◊ ◊◊◊◊◊ ◊◊◊◊◊ ◊◊◊◊◊	462,300
◊◊◊◊◊ ◊◊◊◊◊ ◊◊◊◊◊ ◊◊◊◊◊ ◊◊◊◊◊ ◊◊◊◊◊ ◊◊◊◊◊ ◊◊◊◊◊ ◊◊◊◊◊ ◊◊◊◊◊	462,350
◊◊◊◊◊ ◊◊◊◊◊ ◊◊◊◊◊ ◊◊◊◊◊ ◊◊◊◊◊ ◊◊◊◊◊ ◊◊◊◊◊ ◊◊◊◊◊ ◊◊◊◊◊ ◊◊◊◊◊	462,400
◊◊◊◊◊ ◊◊◊◊◊ ◊◊◊◊◊ ◊◊◊◊◊ ◊◊◊◊◊ ◊◊◊◊◊ ◊◊◊◊◊ ◊◊◊◊◊ ◊◊◊◊◊ ◊◊◊◊◊	462,450
◊◊◊◊◊ ◊◊◊◊◊ ◊◊◊◊◊ ◊◊◊◊◊ ◊◊◊◊◊ ◊◊◊◊◊ ◊◊◊◊◊ ◊◊◊◊◊ ◊◊◊◊◊ ◊◊◊◊◊	462,500
◊◊◊◊◊ ◊◊◊◊◊ ◊◊◊◊◊ ◊◊◊◊◊ ◊◊◊◊◊ ◊◊◊◊◊ ◊◊◊◊◊ ◊◊◊◊◊ ◊◊◊◊◊ ◊◊◊◊◊	462,550
◊◊◊◊◊ ◊◊◊◊◊ ◊◊◊◊◊ ◊◊◊◊◊ ◊◊◊◊◊ ◊◊◊◊◊ ◊◊◊◊◊ ◊◊◊◊◊ ◊◊◊◊◊ ◊◊◊◊◊	462,600
◊◊◊◊◊ ◊◊◊◊◊ ◊◊◊◊◊ ◊◊◊◊◊ ◊◊◊◊◊ ◊◊◊◊◊ ◊◊◊◊◊ ◊◊◊◊◊ ◊◊◊◊◊ ◊◊◊◊◊	462,650
◊◊◊◊◊ ◊◊◊◊◊ ◊◊◊◊◊ ◊◊◊◊◊ ◊◊◊◊◊ ◊◊◊◊◊ ◊◊◊◊◊ ◊◊◊◊◊ ◊◊◊◊◊ ◊◊◊◊◊	462,700
◊◊◊◊◊ ◊◊◊◊◊ ◊◊◊◊◊ ◊◊◊◊◊ ◊◊◊◊◊ ◊◊◊◊◊ ◊◊◊◊◊ ◊◊◊◊◊ ◊◊◊◊◊ ◊◊◊◊◊	462,750
◊◊◊◊◊ ◊◊◊◊◊ ◊◊◊◊◊ ◊◊◊◊◊ ◊◊◊◊◊ ◊◊◊◊◊ ◊◊◊◊◊ ◊◊◊◊◊ ◊◊◊◊◊ ◊◊◊◊◊	462,800
◊◊◊◊◊ ◊◊◊◊◊ ◊◊◊◊◊ ◊◊◊◊◊ ◊◊◊◊◊ ◊◊◊◊◊ ◊◊◊◊◊ ◊◊◊◊◊ ◊◊◊◊◊ ◊◊◊◊◊	462,850
◊◊◊◊◊ ◊◊◊◊◊ ◊◊◊◊◊ ◊◊◊◊◊ ◊◊◊◊◊ ◊◊◊◊◊ ◊◊◊◊◊ ◊◊◊◊◊ ◊◊◊◊◊ ◊◊◊◊◊	462,900
◊◊◊◊◊ ◊◊◊◊◊ ◊◊◊◊◊ ◊◊◊◊◊ ◊◊◊◊◊ ◊◊◊◊◊ ◊◊◊◊◊ ◊◊◊◊◊ ◊◊◊◊◊ ◊◊◊◊◊	462,950
◊◊◊◊◊ ◊◊◊◊◊ ◊◊◊◊◊ ◊◊◊◊◊ ◊◊◊◊◊ ◊◊◊◊◊ ◊◊◊◊◊ ◊◊◊◊◊ ◊◊◊◊◊ ◊◊◊◊◊	**463,000**
◊◊◊◊◊ ◊◊◊◊◊ ◊◊◊◊◊ ◊◊◊◊◊ ◊◊◊◊◊ ◊◊◊◊◊ ◊◊◊◊◊ ◊◊◊◊◊ ◊◊◊◊◊ ◊◊◊◊◊	463,050
◊◊◊◊◊ ◊◊◊◊◊ ◊◊◊◊◊ ◊◊◊◊◊ ◊◊◊◊◊ ◊◊◊◊◊ ◊◊◊◊◊ ◊◊◊◊◊ ◊◊◊◊◊ ◊◊◊◊◊	463,100
◊◊◊◊◊ ◊◊◊◊◊ ◊◊◊◊◊ ◊◊◊◊◊ ◊◊◊◊◊ ◊◊◊◊◊ ◊◊◊◊◊ ◊◊◊◊◊ ◊◊◊◊◊ ◊◊◊◊◊	463,150
◊◊◊◊◊ ◊◊◊◊◊ ◊◊◊◊◊ ◊◊◊◊◊ ◊◊◊◊◊ ◊◊◊◊◊ ◊◊◊◊◊ ◊◊◊◊◊ ◊◊◊◊◊ ◊◊◊◊◊	463,200
◊◊◊◊◊ ◊◊◊◊◊ ◊◊◊◊◊ ◊◊◊◊◊ ◊◊◊◊◊ ◊◊◊◊◊ ◊◊◊◊◊ ◊◊◊◊◊ ◊◊◊◊◊ ◊◊◊◊◊	463,250
◊◊◊◊◊ ◊◊◊◊◊ ◊◊◊◊◊ ◊◊◊◊◊ ◊◊◊◊◊ ◊◊◊◊◊ ◊◊◊◊◊ ◊◊◊◊◊ ◊◊◊◊◊ ◊◊◊◊◊	463,300
◊◊◊◊◊ ◊◊◊◊◊ ◊◊◊◊◊ ◊◊◊◊◊ ◊◊◊◊◊ ◊◊◊◊◊ ◊◊◊◊◊ ◊◊◊◊◊ ◊◊◊◊◊ ◊◊◊◊◊	463,350
◊◊◊◊◊ ◊◊◊◊◊ ◊◊◊◊◊ ◊◊◊◊◊ ◊◊◊◊◊ ◊◊◊◊◊ ◊◊◊◊◊ ◊◊◊◊◊ ◊◊◊◊◊ ◊◊◊◊◊	463,400
◊◊◊◊◊ ◊◊◊◊◊ ◊◊◊◊◊ ◊◊◊◊◊ ◊◊◊◊◊ ◊◊◊◊◊ ◊◊◊◊◊ ◊◊◊◊◊ ◊◊◊◊◊ ◊◊◊◊◊	463,450
◊◊◊◊◊ ◊◊◊◊◊ ◊◊◊◊◊ ◊◊◊◊◊ ◊◊◊◊◊ ◊◊◊◊◊ ◊◊◊◊◊ ◊◊◊◊◊ ◊◊◊◊◊ ◊◊◊◊◊	463,500
◊◊◊◊◊ ◊◊◊◊◊ ◊◊◊◊◊ ◊◊◊◊◊ ◊◊◊◊◊ ◊◊◊◊◊ ◊◊◊◊◊ ◊◊◊◊◊ ◊◊◊◊◊ ◊◊◊◊◊	463,550
◊◊◊◊◊ ◊◊◊◊◊ ◊◊◊◊◊ ◊◊◊◊◊ ◊◊◊◊◊ ◊◊◊◊◊ ◊◊◊◊◊ ◊◊◊◊◊ ◊◊◊◊◊ ◊◊◊◊◊	463,600
◊◊◊◊◊ ◊◊◊◊◊ ◊◊◊◊◊ ◊◊◊◊◊ ◊◊◊◊◊ ◊◊◊◊◊ ◊◊◊◊◊ ◊◊◊◊◊ ◊◊◊◊◊ ◊◊◊◊◊	463,650
◊◊◊◊◊ ◊◊◊◊◊ ◊◊◊◊◊ ◊◊◊◊◊ ◊◊◊◊◊ ◊◊◊◊◊ ◊◊◊◊◊ ◊◊◊◊◊ ◊◊◊◊◊ ◊◊◊◊◊	463,700
◊◊◊◊◊ ◊◊◊◊◊ ◊◊◊◊◊ ◊◊◊◊◊ ◊◊◊◊◊ ◊◊◊◊◊ ◊◊◊◊◊ ◊◊◊◊◊ ◊◊◊◊◊ ◊◊◊◊◊	463,750
◊◊◊◊◊ ◊◊◊◊◊ ◊◊◊◊◊ ◊◊◊◊◊ ◊◊◊◊◊ ◊◊◊◊◊ ◊◊◊◊◊ ◊◊◊◊◊ ◊◊◊◊◊ ◊◊◊◊◊	463,800
◊◊◊◊◊ ◊◊◊◊◊ ◊◊◊◊◊ ◊◊◊◊◊ ◊◊◊◊◊ ◊◊◊◊◊ ◊◊◊◊◊ ◊◊◊◊◊ ◊◊◊◊◊ ◊◊◊◊◊	463,850
◊◊◊◊◊ ◊◊◊◊◊ ◊◊◊◊◊ ◊◊◊◊◊ ◊◊◊◊◊ ◊◊◊◊◊ ◊◊◊◊◊ ◊◊◊◊◊ ◊◊◊◊◊ ◊◊◊◊◊	463,900
◊◊◊◊◊ ◊◊◊◊◊ ◊◊◊◊◊ ◊◊◊◊◊ ◊◊◊◊◊ ◊◊◊◊◊ ◊◊◊◊◊ ◊◊◊◊◊ ◊◊◊◊◊ ◊◊◊◊◊	463,950
◊◊◊◊◊ ◊◊◊◊◊ ◊◊◊◊◊ ◊◊◊◊◊ ◊◊◊◊◊ ◊◊◊◊◊ ◊◊◊◊◊ ◊◊◊◊◊ ◊◊◊◊◊ ◊◊◊◊◊	**464,000**

◊ The Number 464,001 to the Number 466,000

Start | **Notes**

	464,050
	464,100
	464,150
	464,200
	464,250
	464,300
	464,350
	464,400
	464,450
	464,500
	464,550
	464,600
	464,650
	464,700
	464,750
	464,800
	464,850
	464,900
	464,950
	465,000
	465,050
	465,100
	465,150
	465,200
	465,250
	465,300
	465,350
	465,400
	465,450
	465,500
	465,550
	465,600
	465,650
	465,700
	465,750
	465,800
	465,850
	465,900
	465,950
	466,000

The Number 466,001 to the Number 468,000

Start	Notes
	466,050
	466,100
	466,150
	466,200
	466,250
	466,300
	466,350
	466,400
	466,450
	466,500
	466,550
	466,600
	466,650
	466,700
	466,750
	466,800
	466,850
	466,900
	466,950
	467,000
	467,050
	467,100
	467,150
	467,200
	467,250
	467,300
	467,350
	467,400
	467,450
	467,500
	467,550
	467,600
	467,650
	467,700
	467,750
	467,800
	467,850
	467,900
	467,950
	468,000

The Number 468,001 to the Number 470,000

Start	Notes
◊◊◊◊◊ ◊◊◊◊◊ ◊◊◊◊◊ ◊◊◊◊◊ ◊◊◊◊◊ ◊◊◊◊◊ ◊◊◊◊◊ ◊◊◊◊◊ ◊◊◊◊◊ ◊◊◊◊◊	468,050
◊◊◊◊◊ ◊◊◊◊◊ ◊◊◊◊◊ ◊◊◊◊◊ ◊◊◊◊◊ ◊◊◊◊◊ ◊◊◊◊◊ ◊◊◊◊◊ ◊◊◊◊◊ ◊◊◊◊◊	468,100
◊◊◊◊◊ ◊◊◊◊◊ ◊◊◊◊◊ ◊◊◊◊◊ ◊◊◊◊◊ ◊◊◊◊◊ ◊◊◊◊◊ ◊◊◊◊◊ ◊◊◊◊◊ ◊◊◊◊◊	468,150
◊◊◊◊◊ ◊◊◊◊◊ ◊◊◊◊◊ ◊◊◊◊◊ ◊◊◊◊◊ ◊◊◊◊◊ ◊◊◊◊◊ ◊◊◊◊◊ ◊◊◊◊◊ ◊◊◊◊◊	468,200
◊◊◊◊◊ ◊◊◊◊◊ ◊◊◊◊◊ ◊◊◊◊◊ ◊◊◊◊◊ ◊◊◊◊◊ ◊◊◊◊◊ ◊◊◊◊◊ ◊◊◊◊◊ ◊◊◊◊◊	468,250
◊◊◊◊◊ ◊◊◊◊◊ ◊◊◊◊◊ ◊◊◊◊◊ ◊◊◊◊◊ ◊◊◊◊◊ ◊◊◊◊◊ ◊◊◊◊◊ ◊◊◊◊◊ ◊◊◊◊◊	468,300
◊◊◊◊◊ ◊◊◊◊◊ ◊◊◊◊◊ ◊◊◊◊◊ ◊◊◊◊◊ ◊◊◊◊◊ ◊◊◊◊◊ ◊◊◊◊◊ ◊◊◊◊◊ ◊◊◊◊◊	468,350
◊◊◊◊◊ ◊◊◊◊◊ ◊◊◊◊◊ ◊◊◊◊◊ ◊◊◊◊◊ ◊◊◊◊◊ ◊◊◊◊◊ ◊◊◊◊◊ ◊◊◊◊◊ ◊◊◊◊◊	468,400
◊◊◊◊◊ ◊◊◊◊◊ ◊◊◊◊◊ ◊◊◊◊◊ ◊◊◊◊◊ ◊◊◊◊◊ ◊◊◊◊◊ ◊◊◊◊◊ ◊◊◊◊◊ ◊◊◊◊◊	468,450
◊◊◊◊◊ ◊◊◊◊◊ ◊◊◊◊◊ ◊◊◊◊◊ ◊◊◊◊◊ ◊◊◊◊◊ ◊◊◊◊◊ ◊◊◊◊◊ ◊◊◊◊◊ ◊◊◊◊◊	468,500
◊◊◊◊◊ ◊◊◊◊◊ ◊◊◊◊◊ ◊◊◊◊◊ ◊◊◊◊◊ ◊◊◊◊◊ ◊◊◊◊◊ ◊◊◊◊◊ ◊◊◊◊◊ ◊◊◊◊◊	468,550
◊◊◊◊◊ ◊◊◊◊◊ ◊◊◊◊◊ ◊◊◊◊◊ ◊◊◊◊◊ ◊◊◊◊◊ ◊◊◊◊◊ ◊◊◊◊◊ ◊◊◊◊◊ ◊◊◊◊◊	468,600
◊◊◊◊◊ ◊◊◊◊◊ ◊◊◊◊◊ ◊◊◊◊◊ ◊◊◊◊◊ ◊◊◊◊◊ ◊◊◊◊◊ ◊◊◊◊◊ ◊◊◊◊◊ ◊◊◊◊◊	468,650
◊◊◊◊◊ ◊◊◊◊◊ ◊◊◊◊◊ ◊◊◊◊◊ ◊◊◊◊◊ ◊◊◊◊◊ ◊◊◊◊◊ ◊◊◊◊◊ ◊◊◊◊◊ ◊◊◊◊◊	468,700
◊◊◊◊◊ ◊◊◊◊◊ ◊◊◊◊◊ ◊◊◊◊◊ ◊◊◊◊◊ ◊◊◊◊◊ ◊◊◊◊◊ ◊◊◊◊◊ ◊◊◊◊◊ ◊◊◊◊◊	468,750
◊◊◊◊◊ ◊◊◊◊◊ ◊◊◊◊◊ ◊◊◊◊◊ ◊◊◊◊◊ ◊◊◊◊◊ ◊◊◊◊◊ ◊◊◊◊◊ ◊◊◊◊◊ ◊◊◊◊◊	468,800
◊◊◊◊◊ ◊◊◊◊◊ ◊◊◊◊◊ ◊◊◊◊◊ ◊◊◊◊◊ ◊◊◊◊◊ ◊◊◊◊◊ ◊◊◊◊◊ ◊◊◊◊◊ ◊◊◊◊◊	468,850
◊◊◊◊◊ ◊◊◊◊◊ ◊◊◊◊◊ ◊◊◊◊◊ ◊◊◊◊◊ ◊◊◊◊◊ ◊◊◊◊◊ ◊◊◊◊◊ ◊◊◊◊◊ ◊◊◊◊◊	468,900
◊◊◊◊◊ ◊◊◊◊◊ ◊◊◊◊◊ ◊◊◊◊◊ ◊◊◊◊◊ ◊◊◊◊◊ ◊◊◊◊◊ ◊◊◊◊◊ ◊◊◊◊◊ ◊◊◊◊◊	468,950
◊◊◊◊◊ ◊◊◊◊◊ ◊◊◊◊◊ ◊◊◊◊◊ ◊◊◊◊◊ ◊◊◊◊◊ ◊◊◊◊◊ ◊◊◊◊◊ ◊◊◊◊◊ ◊◊◊◊◊	**469,000**
◊◊◊◊◊ ◊◊◊◊◊ ◊◊◊◊◊ ◊◊◊◊◊ ◊◊◊◊◊ ◊◊◊◊◊ ◊◊◊◊◊ ◊◊◊◊◊ ◊◊◊◊◊ ◊◊◊◊◊	469,050
◊◊◊◊◊ ◊◊◊◊◊ ◊◊◊◊◊ ◊◊◊◊◊ ◊◊◊◊◊ ◊◊◊◊◊ ◊◊◊◊◊ ◊◊◊◊◊ ◊◊◊◊◊ ◊◊◊◊◊	469,100
◊◊◊◊◊ ◊◊◊◊◊ ◊◊◊◊◊ ◊◊◊◊◊ ◊◊◊◊◊ ◊◊◊◊◊ ◊◊◊◊◊ ◊◊◊◊◊ ◊◊◊◊◊ ◊◊◊◊◊	469,150
◊◊◊◊◊ ◊◊◊◊◊ ◊◊◊◊◊ ◊◊◊◊◊ ◊◊◊◊◊ ◊◊◊◊◊ ◊◊◊◊◊ ◊◊◊◊◊ ◊◊◊◊◊ ◊◊◊◊◊	469,200
◊◊◊◊◊ ◊◊◊◊◊ ◊◊◊◊◊ ◊◊◊◊◊ ◊◊◊◊◊ ◊◊◊◊◊ ◊◊◊◊◊ ◊◊◊◊◊ ◊◊◊◊◊ ◊◊◊◊◊	469,250
◊◊◊◊◊ ◊◊◊◊◊ ◊◊◊◊◊ ◊◊◊◊◊ ◊◊◊◊◊ ◊◊◊◊◊ ◊◊◊◊◊ ◊◊◊◊◊ ◊◊◊◊◊ ◊◊◊◊◊	469,300
◊◊◊◊◊ ◊◊◊◊◊ ◊◊◊◊◊ ◊◊◊◊◊ ◊◊◊◊◊ ◊◊◊◊◊ ◊◊◊◊◊ ◊◊◊◊◊ ◊◊◊◊◊ ◊◊◊◊◊	469,350
◊◊◊◊◊ ◊◊◊◊◊ ◊◊◊◊◊ ◊◊◊◊◊ ◊◊◊◊◊ ◊◊◊◊◊ ◊◊◊◊◊ ◊◊◊◊◊ ◊◊◊◊◊ ◊◊◊◊◊	469,400
◊◊◊◊◊ ◊◊◊◊◊ ◊◊◊◊◊ ◊◊◊◊◊ ◊◊◊◊◊ ◊◊◊◊◊ ◊◊◊◊◊ ◊◊◊◊◊ ◊◊◊◊◊ ◊◊◊◊◊	469,450
◊◊◊◊◊ ◊◊◊◊◊ ◊◊◊◊◊ ◊◊◊◊◊ ◊◊◊◊◊ ◊◊◊◊◊ ◊◊◊◊◊ ◊◊◊◊◊ ◊◊◊◊◊ ◊◊◊◊◊	469,500
◊◊◊◊◊ ◊◊◊◊◊ ◊◊◊◊◊ ◊◊◊◊◊ ◊◊◊◊◊ ◊◊◊◊◊ ◊◊◊◊◊ ◊◊◊◊◊ ◊◊◊◊◊ ◊◊◊◊◊	469,550
◊◊◊◊◊ ◊◊◊◊◊ ◊◊◊◊◊ ◊◊◊◊◊ ◊◊◊◊◊ ◊◊◊◊◊ ◊◊◊◊◊ ◊◊◊◊◊ ◊◊◊◊◊ ◊◊◊◊◊	469,600
◊◊◊◊◊ ◊◊◊◊◊ ◊◊◊◊◊ ◊◊◊◊◊ ◊◊◊◊◊ ◊◊◊◊◊ ◊◊◊◊◊ ◊◊◊◊◊ ◊◊◊◊◊ ◊◊◊◊◊	469,650
◊◊◊◊◊ ◊◊◊◊◊ ◊◊◊◊◊ ◊◊◊◊◊ ◊◊◊◊◊ ◊◊◊◊◊ ◊◊◊◊◊ ◊◊◊◊◊ ◊◊◊◊◊ ◊◊◊◊◊	469,700
◊◊◊◊◊ ◊◊◊◊◊ ◊◊◊◊◊ ◊◊◊◊◊ ◊◊◊◊◊ ◊◊◊◊◊ ◊◊◊◊◊ ◊◊◊◊◊ ◊◊◊◊◊ ◊◊◊◊◊	469,750
◊◊◊◊◊ ◊◊◊◊◊ ◊◊◊◊◊ ◊◊◊◊◊ ◊◊◊◊◊ ◊◊◊◊◊ ◊◊◊◊◊ ◊◊◊◊◊ ◊◊◊◊◊ ◊◊◊◊◊	469,800
◊◊◊◊◊ ◊◊◊◊◊ ◊◊◊◊◊ ◊◊◊◊◊ ◊◊◊◊◊ ◊◊◊◊◊ ◊◊◊◊◊ ◊◊◊◊◊ ◊◊◊◊◊ ◊◊◊◊◊	469,850
◊◊◊◊◊ ◊◊◊◊◊ ◊◊◊◊◊ ◊◊◊◊◊ ◊◊◊◊◊ ◊◊◊◊◊ ◊◊◊◊◊ ◊◊◊◊◊ ◊◊◊◊◊ ◊◊◊◊◊	469,900
◊◊◊◊◊ ◊◊◊◊◊ ◊◊◊◊◊ ◊◊◊◊◊ ◊◊◊◊◊ ◊◊◊◊◊ ◊◊◊◊◊ ◊◊◊◊◊ ◊◊◊◊◊ ◊◊◊◊◊	469,950
◊◊◊◊◊ ◊◊◊◊◊ ◊◊◊◊◊ ◊◊◊◊◊ ◊◊◊◊◊ ◊◊◊◊◊ ◊◊◊◊◊ ◊◊◊◊◊ ◊◊◊◊◊ ◊◊◊◊◊	**470,000**

The Number 470,001 to the Number 472,000

Start

Notes

	470,050
	470,100
	470,150
	470,200
	470,250
	470,300
	470,350
	470,400
	470,450
	470,500
	470,550
	470,600
	470,650
	470,700
	470,750
	470,800
	470,850
	470,900
	470,950
	471,000
	471,050
	471,100
	471,150
	471,200
	471,250
	471,300
	471,350
	471,400
	471,450
	471,500
	471,550
	471,600
	471,650
	471,700
	471,750
	471,800
	471,850
	471,900
	471,950
	472,000

The Number 472,001 to the Number 474,000

Start

Notes

	472,050
	472,100
	472,150
	472,200
	472,250
	472,300
	472,350
	472,400
	472,450
	472,500
	472,550
	472,600
	472,650
	472,700
	472,750
	472,800
	472,850
	472,900
	472,950
	473,000
	473,050
	473,100
	473,150
	473,200
	473,250
	473,300
	473,350
	473,400
	473,450
	473,500
	473,550
	473,600
	473,650
	473,700
	473,750
	473,800
	473,850
	473,900
	473,950
	474,000

The Number 474,001 to the Number 476,000

Start		Notes
	474,050	
	474,100	
	474,150	
	474,200	
	474,250	
	474,300	
	474,350	
	474,400	
	474,450	
	474,500	
	474,550	
	474,600	
	474,650	
	474,700	
	474,750	
	474,800	
	474,850	
	474,900	
	474,950	
	475,000	
	475,050	
	475,100	
	475,150	
	475,200	
	475,250	
	475,300	
	475,350	
	475,400	
	475,450	
	475,500	
	475,550	
	475,600	
	475,650	
	475,700	
	475,750	
	475,800	
	475,850	
	475,900	
	475,950	
	476,000	

The Number 476,001 to the Number 478,000

Start   Notes

	476,050
	476,100
	476,150
	476,200
	476,250
	476,300
	476,350
	476,400
	476,450
	476,500
	476,550
	476,600
	476,650
	476,700
	476,750
	476,800
	476,850
	476,900
	476,950
	477,000
	477,050
	477,100
	477,150
	477,200
	477,250
	477,300
	477,350
	477,400
	477,450
	477,500
	477,550
	477,600
	477,650
	477,700
	477,750
	477,800
	477,850
	477,900
	477,950
	478,000

The Number 478,001 to the Number 480,000

Start	Notes
478,050	
478,100	
478,150	
478,200	
478,250	
478,300	
478,350	
478,400	
478,450	
478,500	
478,550	
478,600	
478,650	
478,700	
478,750	
478,800	
478,850	
478,900	
478,950	
479,000	
479,050	
479,100	
479,150	
479,200	
479,250	
479,300	
479,350	
479,400	
479,450	
479,500	
479,550	
479,600	
479,650	
479,700	
479,750	
479,800	
479,850	
479,900	
479,950	
480,000	

The Number 480,001 to the Number 482,000

Start   Notes

	480,050
	480,100
	480,150
	480,200
	480,250
	480,300
	480,350
	480,400
	480,450
	480,500
	480,550
	480,600
	480,650
	480,700
	480,750
	480,800
	480,850
	480,900
	480,950
	481,000
	481,050
	481,100
	481,150
	481,200
	481,250
	481,300
	481,350
	481,400
	481,450
	481,500
	481,550
	481,600
	481,650
	481,700
	481,750
	481,800
	481,850
	481,900
	481,950
	482,000

The Number 482,001 to the Number 484,000

Start		Notes
	482,050	
	482,100	
	482,150	
	482,200	
	482,250	
	482,300	
	482,350	
	482,400	
	482,450	
	482,500	
	482,550	
	482,600	
	482,650	
	482,700	
	482,750	
	482,800	
	482,850	
	482,900	
	482,950	
	483,000	
	483,050	
	483,100	
	483,150	
	483,200	
	483,250	
	483,300	
	483,350	
	483,400	
	483,450	
	483,500	
	483,550	
	483,600	
	483,650	
	483,700	
	483,750	
	483,800	
	483,850	
	483,900	
	483,950	
	484,000	

The Number 484,001 to the Number 486,000

Start

Notes

	484,050
	484,100
	484,150
	484,200
	484,250
	484,300
	484,350
	484,400
	484,450
	484,500
	484,550
	484,600
	484,650
	484,700
	484,750
	484,800
	484,850
	484,900
	484,950
	485,000
	485,050
	485,100
	485,150
	485,200
	485,250
	485,300
	485,350
	485,400
	485,450
	485,500
	485,550
	485,600
	485,650
	485,700
	485,750
	485,800
	485,850
	485,900
	485,950
	486,000

The Number 486,001 to the Number 488,000

Start   Notes

	486,050
	486,100
	486,150
	486,200
	486,250
	486,300
	486,350
	486,400
	486,450
	486,500
	486,550
	486,600
	486,650
	486,700
	486,750
	486,800
	486,850
	486,900
	486,950
	487,000
	487,050
	487,100
	487,150
	487,200
	487,250
	487,300
	487,350
	487,400
	487,450
	487,500
	487,550
	487,600
	487,650
	487,700
	487,750
	487,800
	487,850
	487,900
	487,950
	488,000

The Number 488,001 to the Number 490,000

Start

Notes

	Notes
488,050	
488,100	
488,150	
488,200	
488,250	
488,300	
488,350	
488,400	
488,450	
488,500	
488,550	
488,600	
488,650	
488,700	
488,750	
488,800	
488,850	
488,900	
488,950	
489,000	
489,050	
489,100	
489,150	
489,200	
489,250	
489,300	
489,350	
489,400	
489,450	
489,500	
489,550	
489,600	
489,650	
489,700	
489,750	
489,800	
489,850	
489,900	
489,950	
490,000	

The Number 490,001 to the Number 492,000

Start		Notes
	490,050	
	490,100	
	490,150	
	490,200	
	490,250	
	490,300	
	490,350	
	490,400	
	490,450	
	490,500	
	490,550	
	490,600	
	490,650	
	490,700	
	490,750	
	490,800	
	490,850	
	490,900	
	490,950	
	491,000	
	491,050	
	491,100	
	491,150	
	491,200	
	491,250	
	491,300	
	491,350	
	491,400	
	491,450	
	491,500	
	491,550	
	491,600	
	491,650	
	491,700	
	491,750	
	491,800	
	491,850	
	491,900	
	491,950	
	492,000	

The Number 492,001 to the Number 494,000

Start										Number	Notes
										492,050	
										492,100	
										492,150	
										492,200	
										492,250	
										492,300	
										492,350	
										492,400	
										492,450	
										492,500	
										492,550	
										492,600	
										492,650	
										492,700	
										492,750	
										492,800	
										492,850	
										492,900	
										492,950	
										493,000	
										493,050	
										493,100	
										493,150	
										493,200	
										493,250	
										493,300	
										493,350	
										493,400	
										493,450	
										493,500	
										493,550	
										493,600	
										493,650	
										493,700	
										493,750	
										493,800	
										493,850	
										493,900	
										493,950	
										494,000	

The Number 494,001 to the Number 496,000

Start	Notes
494,050	
494,100	
494,150	
494,200	
494,250	
494,300	
494,350	
494,400	
494,450	
494,500	
494,550	
494,600	
494,650	
494,700	
494,750	
494,800	
494,850	
494,900	
494,950	
495,000	
495,050	
495,100	
495,150	
495,200	
495,250	
495,300	
495,350	
495,400	
495,450	
495,500	
495,550	
495,600	
495,650	
495,700	
495,750	
495,800	
495,850	
495,900	
495,950	
496,000	

The Number 496,001 to the Number 498,000

Start | | Notes

	496,050
	496,100
	496,150
	496,200
	496,250
	496,300
	496,350
	496,400
	496,450
	496,500
	496,550
	496,600
	496,650
	496,700
	496,750
	496,800
	496,850
	496,900
	496,950
	497,000
	497,050
	497,100
	497,150
	497,200
	497,250
	497,300
	497,350
	497,400
	497,450
	497,500
	497,550
	497,600
	497,650
	497,700
	497,750
	497,800
	497,850
	497,900
	497,950
	498,000

◊ The Number 498,001 to the Number 500,000

Start		Notes
◊◊◊◊◊ ◊◊◊◊◊ ◊◊◊◊◊ ◊◊◊◊◊ ◊◊◊◊◊ ◊◊◊◊◊ ◊◊◊◊◊ ◊◊◊◊◊ ◊◊◊◊◊ ◊◊◊◊◊	498,050	
◊◊◊◊◊ ◊◊◊◊◊ ◊◊◊◊◊ ◊◊◊◊◊ ◊◊◊◊◊ ◊◊◊◊◊ ◊◊◊◊◊ ◊◊◊◊◊ ◊◊◊◊◊ ◊◊◊◊◊	498,100	
◊◊◊◊◊ ◊◊◊◊◊ ◊◊◊◊◊ ◊◊◊◊◊ ◊◊◊◊◊ ◊◊◊◊◊ ◊◊◊◊◊ ◊◊◊◊◊ ◊◊◊◊◊ ◊◊◊◊◊	498,150	
◊◊◊◊◊ ◊◊◊◊◊ ◊◊◊◊◊ ◊◊◊◊◊ ◊◊◊◊◊ ◊◊◊◊◊ ◊◊◊◊◊ ◊◊◊◊◊ ◊◊◊◊◊ ◊◊◊◊◊	498,200	
◊◊◊◊◊ ◊◊◊◊◊ ◊◊◊◊◊ ◊◊◊◊◊ ◊◊◊◊◊ ◊◊◊◊◊ ◊◊◊◊◊ ◊◊◊◊◊ ◊◊◊◊◊ ◊◊◊◊◊	498,250	
◊◊◊◊◊ ◊◊◊◊◊ ◊◊◊◊◊ ◊◊◊◊◊ ◊◊◊◊◊ ◊◊◊◊◊ ◊◊◊◊◊ ◊◊◊◊◊ ◊◊◊◊◊ ◊◊◊◊◊	498,300	
◊◊◊◊◊ ◊◊◊◊◊ ◊◊◊◊◊ ◊◊◊◊◊ ◊◊◊◊◊ ◊◊◊◊◊ ◊◊◊◊◊ ◊◊◊◊◊ ◊◊◊◊◊ ◊◊◊◊◊	498,350	
◊◊◊◊◊ ◊◊◊◊◊ ◊◊◊◊◊ ◊◊◊◊◊ ◊◊◊◊◊ ◊◊◊◊◊ ◊◊◊◊◊ ◊◊◊◊◊ ◊◊◊◊◊ ◊◊◊◊◊	498,400	
◊◊◊◊◊ ◊◊◊◊◊ ◊◊◊◊◊ ◊◊◊◊◊ ◊◊◊◊◊ ◊◊◊◊◊ ◊◊◊◊◊ ◊◊◊◊◊ ◊◊◊◊◊ ◊◊◊◊◊	498,450	
◊◊◊◊◊ ◊◊◊◊◊ ◊◊◊◊◊ ◊◊◊◊◊ ◊◊◊◊◊ ◊◊◊◊◊ ◊◊◊◊◊ ◊◊◊◊◊ ◊◊◊◊◊ ◊◊◊◊◊	498,500	
◊◊◊◊◊ ◊◊◊◊◊ ◊◊◊◊◊ ◊◊◊◊◊ ◊◊◊◊◊ ◊◊◊◊◊ ◊◊◊◊◊ ◊◊◊◊◊ ◊◊◊◊◊ ◊◊◊◊◊	498,550	
◊◊◊◊◊ ◊◊◊◊◊ ◊◊◊◊◊ ◊◊◊◊◊ ◊◊◊◊◊ ◊◊◊◊◊ ◊◊◊◊◊ ◊◊◊◊◊ ◊◊◊◊◊ ◊◊◊◊◊	498,600	
◊◊◊◊◊ ◊◊◊◊◊ ◊◊◊◊◊ ◊◊◊◊◊ ◊◊◊◊◊ ◊◊◊◊◊ ◊◊◊◊◊ ◊◊◊◊◊ ◊◊◊◊◊ ◊◊◊◊◊	498,650	
◊◊◊◊◊ ◊◊◊◊◊ ◊◊◊◊◊ ◊◊◊◊◊ ◊◊◊◊◊ ◊◊◊◊◊ ◊◊◊◊◊ ◊◊◊◊◊ ◊◊◊◊◊ ◊◊◊◊◊	498,700	
◊◊◊◊◊ ◊◊◊◊◊ ◊◊◊◊◊ ◊◊◊◊◊ ◊◊◊◊◊ ◊◊◊◊◊ ◊◊◊◊◊ ◊◊◊◊◊ ◊◊◊◊◊ ◊◊◊◊◊	498,750	
◊◊◊◊◊ ◊◊◊◊◊ ◊◊◊◊◊ ◊◊◊◊◊ ◊◊◊◊◊ ◊◊◊◊◊ ◊◊◊◊◊ ◊◊◊◊◊ ◊◊◊◊◊ ◊◊◊◊◊	498,800	
◊◊◊◊◊ ◊◊◊◊◊ ◊◊◊◊◊ ◊◊◊◊◊ ◊◊◊◊◊ ◊◊◊◊◊ ◊◊◊◊◊ ◊◊◊◊◊ ◊◊◊◊◊ ◊◊◊◊◊	498,850	
◊◊◊◊◊ ◊◊◊◊◊ ◊◊◊◊◊ ◊◊◊◊◊ ◊◊◊◊◊ ◊◊◊◊◊ ◊◊◊◊◊ ◊◊◊◊◊ ◊◊◊◊◊ ◊◊◊◊◊	498,900	
◊◊◊◊◊ ◊◊◊◊◊ ◊◊◊◊◊ ◊◊◊◊◊ ◊◊◊◊◊ ◊◊◊◊◊ ◊◊◊◊◊ ◊◊◊◊◊ ◊◊◊◊◊ ◊◊◊◊◊	498,950	
◊◊◊◊◊ ◊◊◊◊◊ ◊◊◊◊◊ ◊◊◊◊◊ ◊◊◊◊◊ ◊◊◊◊◊ ◊◊◊◊◊ ◊◊◊◊◊ ◊◊◊◊◊ ◊◊◊◊◊	**499,000**	
◊◊◊◊◊ ◊◊◊◊◊ ◊◊◊◊◊ ◊◊◊◊◊ ◊◊◊◊◊ ◊◊◊◊◊ ◊◊◊◊◊ ◊◊◊◊◊ ◊◊◊◊◊ ◊◊◊◊◊	499,050	
◊◊◊◊◊ ◊◊◊◊◊ ◊◊◊◊◊ ◊◊◊◊◊ ◊◊◊◊◊ ◊◊◊◊◊ ◊◊◊◊◊ ◊◊◊◊◊ ◊◊◊◊◊ ◊◊◊◊◊	499,100	
◊◊◊◊◊ ◊◊◊◊◊ ◊◊◊◊◊ ◊◊◊◊◊ ◊◊◊◊◊ ◊◊◊◊◊ ◊◊◊◊◊ ◊◊◊◊◊ ◊◊◊◊◊ ◊◊◊◊◊	499,150	
◊◊◊◊◊ ◊◊◊◊◊ ◊◊◊◊◊ ◊◊◊◊◊ ◊◊◊◊◊ ◊◊◊◊◊ ◊◊◊◊◊ ◊◊◊◊◊ ◊◊◊◊◊ ◊◊◊◊◊	499,200	
◊◊◊◊◊ ◊◊◊◊◊ ◊◊◊◊◊ ◊◊◊◊◊ ◊◊◊◊◊ ◊◊◊◊◊ ◊◊◊◊◊ ◊◊◊◊◊ ◊◊◊◊◊ ◊◊◊◊◊	499,250	
◊◊◊◊◊ ◊◊◊◊◊ ◊◊◊◊◊ ◊◊◊◊◊ ◊◊◊◊◊ ◊◊◊◊◊ ◊◊◊◊◊ ◊◊◊◊◊ ◊◊◊◊◊ ◊◊◊◊◊	499,300	
◊◊◊◊◊ ◊◊◊◊◊ ◊◊◊◊◊ ◊◊◊◊◊ ◊◊◊◊◊ ◊◊◊◊◊ ◊◊◊◊◊ ◊◊◊◊◊ ◊◊◊◊◊ ◊◊◊◊◊	499,350	
◊◊◊◊◊ ◊◊◊◊◊ ◊◊◊◊◊ ◊◊◊◊◊ ◊◊◊◊◊ ◊◊◊◊◊ ◊◊◊◊◊ ◊◊◊◊◊ ◊◊◊◊◊ ◊◊◊◊◊	499,400	
◊◊◊◊◊ ◊◊◊◊◊ ◊◊◊◊◊ ◊◊◊◊◊ ◊◊◊◊◊ ◊◊◊◊◊ ◊◊◊◊◊ ◊◊◊◊◊ ◊◊◊◊◊ ◊◊◊◊◊	499,450	
◊◊◊◊◊ ◊◊◊◊◊ ◊◊◊◊◊ ◊◊◊◊◊ ◊◊◊◊◊ ◊◊◊◊◊ ◊◊◊◊◊ ◊◊◊◊◊ ◊◊◊◊◊ ◊◊◊◊◊	499,500	
◊◊◊◊◊ ◊◊◊◊◊ ◊◊◊◊◊ ◊◊◊◊◊ ◊◊◊◊◊ ◊◊◊◊◊ ◊◊◊◊◊ ◊◊◊◊◊ ◊◊◊◊◊ ◊◊◊◊◊	499,550	
◊◊◊◊◊ ◊◊◊◊◊ ◊◊◊◊◊ ◊◊◊◊◊ ◊◊◊◊◊ ◊◊◊◊◊ ◊◊◊◊◊ ◊◊◊◊◊ ◊◊◊◊◊ ◊◊◊◊◊	499,600	
◊◊◊◊◊ ◊◊◊◊◊ ◊◊◊◊◊ ◊◊◊◊◊ ◊◊◊◊◊ ◊◊◊◊◊ ◊◊◊◊◊ ◊◊◊◊◊ ◊◊◊◊◊ ◊◊◊◊◊	499,650	
◊◊◊◊◊ ◊◊◊◊◊ ◊◊◊◊◊ ◊◊◊◊◊ ◊◊◊◊◊ ◊◊◊◊◊ ◊◊◊◊◊ ◊◊◊◊◊ ◊◊◊◊◊ ◊◊◊◊◊	499,700	
◊◊◊◊◊ ◊◊◊◊◊ ◊◊◊◊◊ ◊◊◊◊◊ ◊◊◊◊◊ ◊◊◊◊◊ ◊◊◊◊◊ ◊◊◊◊◊ ◊◊◊◊◊ ◊◊◊◊◊	499,750	
◊◊◊◊◊ ◊◊◊◊◊ ◊◊◊◊◊ ◊◊◊◊◊ ◊◊◊◊◊ ◊◊◊◊◊ ◊◊◊◊◊ ◊◊◊◊◊ ◊◊◊◊◊ ◊◊◊◊◊	499,800	
◊◊◊◊◊ ◊◊◊◊◊ ◊◊◊◊◊ ◊◊◊◊◊ ◊◊◊◊◊ ◊◊◊◊◊ ◊◊◊◊◊ ◊◊◊◊◊ ◊◊◊◊◊ ◊◊◊◊◊	499,850	
◊◊◊◊◊ ◊◊◊◊◊ ◊◊◊◊◊ ◊◊◊◊◊ ◊◊◊◊◊ ◊◊◊◊◊ ◊◊◊◊◊ ◊◊◊◊◊ ◊◊◊◊◊ ◊◊◊◊◊	499,900	
◊◊◊◊◊ ◊◊◊◊◊ ◊◊◊◊◊ ◊◊◊◊◊ ◊◊◊◊◊ ◊◊◊◊◊ ◊◊◊◊◊ ◊◊◊◊◊ ◊◊◊◊◊ ◊◊◊◊◊	499,950	
◊◊◊◊◊ ◊◊◊◊◊ ◊◊◊◊◊ ◊◊◊◊◊ ◊◊◊◊◊ ◊◊◊◊◊ ◊◊◊◊◊ ◊◊◊◊◊ ◊◊◊◊◊ ◊◊◊◊◊	**500,000**	

The Number 500,001 to the Number 502,000

Start		Notes

	Number	Notes
	500,050	
	500,100	
	500,150	
	500,200	
	500,250	
	500,300	
	500,350	
	500,400	
	500,450	
	500,500	
	500,550	
	500,600	
	500,650	
	500,700	
	500,750	
	500,800	
	500,850	
	500,900	
	500,950	
	501,000	
	501,050	
	501,100	
	501,150	
	501,200	
	501,250	
	501,300	
	501,350	
	501,400	
	501,450	
	501,500	
	501,550	
	501,600	
	501,650	
	501,700	
	501,750	
	501,800	
	501,850	
	501,900	
	501,950	
	502,000	

The Number 502,001 to the Number 504,000

	Notes
Start	
	502,050
	502,100
	502,150
	502,200
	502,250
	502,300
	502,350
	502,400
	502,450
	502,500
	502,550
	502,600
	502,650
	502,700
	502,750
	502,800
	502,850
	502,900
	502,950
	503,000
	503,050
	503,100
	503,150
	503,200
	503,250
	503,300
	503,350
	503,400
	503,450
	503,500
	503,550
	503,600
	503,650
	503,700
	503,750
	503,800
	503,850
	503,900
	503,950
	504,000

The Number 504,001 to the Number 506,000

Start		Notes
	504,050	
	504,100	
	504,150	
	504,200	
	504,250	
	504,300	
	504,350	
	504,400	
	504,450	
	504,500	
	504,550	
	504,600	
	504,650	
	504,700	
	504,750	
	504,800	
	504,850	
	504,900	
	504,950	
	505,000	
	505,050	
	505,100	
	505,150	
	505,200	
	505,250	
	505,300	
	505,350	
	505,400	
	505,450	
	505,500	
	505,550	
	505,600	
	505,650	
	505,700	
	505,750	
	505,800	
	505,850	
	505,900	
	505,950	
	506,000	

The Number 506,001 to the Number 508,000

Start		Notes
	506,050	
	506,100	
	506,150	
	506,200	
	506,250	
	506,300	
	506,350	
	506,400	
	506,450	
	506,500	
	506,550	
	506,600	
	506,650	
	506,700	
	506,750	
	506,800	
	506,850	
	506,900	
	506,950	
	507,000	
	507,050	
	507,100	
	507,150	
	507,200	
	507,250	
	507,300	
	507,350	
	507,400	
	507,450	
	507,500	
	507,550	
	507,600	
	507,650	
	507,700	
	507,750	
	507,800	
	507,850	
	507,900	
	507,950	
	508,000	

The Number 508,001 to the Number 510,000

Start

Notes

	Number
	508,050
	508,100
	508,150
	508,200
	508,250
	508,300
	508,350
	508,400
	508,450
	508,500
	508,550
	508,600
	508,650
	508,700
	508,750
	508,800
	508,850
	508,900
	508,950
	509,000
	509,050
	509,100
	509,150
	509,200
	509,250
	509,300
	509,350
	509,400
	509,450
	509,500
	509,550
	509,600
	509,650
	509,700
	509,750
	509,800
	509,850
	509,900
	509,950
	510,000

The Number 510,001 to the Number 512,000

Start		Notes
	510,050	
	510,100	
	510,150	
	510,200	
	510,250	
	510,300	
	510,350	
	510,400	
	510,450	
	510,500	
	510,550	
	510,600	
	510,650	
	510,700	
	510,750	
	510,800	
	510,850	
	510,900	
	510,950	
	511,000	
	511,050	
	511,100	
	511,150	
	511,200	
	511,250	
	511,300	
	511,350	
	511,400	
	511,450	
	511,500	
	511,550	
	511,600	
	511,650	
	511,700	
	511,750	
	511,800	
	511,850	
	511,900	
	511,950	
	512,000	

The Number 512,001 to the Number 514,000

Start   **Notes**

512,050	
512,100	
512,150	
512,200	
512,250	
512,300	
512,350	
512,400	
512,450	
512,500	
512,550	
512,600	
512,650	
512,700	
512,750	
512,800	
512,850	
512,900	
512,950	
513,000	
513,050	
513,100	
513,150	
513,200	
513,250	
513,300	
513,350	
513,400	
513,450	
513,500	
513,550	
513,600	
513,650	
513,700	
513,750	
513,800	
513,850	
513,900	
513,950	
514,000	

The Number 514,001 to the Number 516,000

Start	Notes
	514,050
	514,100
	514,150
	514,200
	514,250
	514,300
	514,350
	514,400
	514,450
	514,500
	514,550
	514,600
	514,650
	514,700
	514,750
	514,800
	514,850
	514,900
	514,950
	515,000
	515,050
	515,100
	515,150
	515,200
	515,250
	515,300
	515,350
	515,400
	515,450
	515,500
	515,550
	515,600
	515,650
	515,700
	515,750
	515,800
	515,850
	515,900
	515,950
	516,000

◇ The Number 516,001 to the Number 518,000

Start | | Notes

	516,050
	516,100
	516,150
	516,200
	516,250
	516,300
	516,350
	516,400
	516,450
	516,500
	516,550
	516,600
	516,650
	516,700
	516,750
	516,800
	516,850
	516,900
	516,950
	517,000
	517,050
	517,100
	517,150
	517,200
	517,250
	517,300
	517,350
	517,400
	517,450
	517,500
	517,550
	517,600
	517,650
	517,700
	517,750
	517,800
	517,850
	517,900
	517,950
	518,000

The Number 518,001 to the Number 520,000

Start	Notes
	518,050
	518,100
	518,150
	518,200
	518,250
	518,300
	518,350
	518,400
	518,450
	518,500
	518,550
	518,600
	518,650
	518,700
	518,750
	518,800
	518,850
	518,900
	518,950
	519,000
	519,050
	519,100
	519,150
	519,200
	519,250
	519,300
	519,350
	519,400
	519,450
	519,500
	519,550
	519,600
	519,650
	519,700
	519,750
	519,800
	519,850
	519,900
	519,950
	520,000

The Number 520,001 to the Number 522,000

Start

	Notes
520,050	
520,100	
520,150	
520,200	
520,250	
520,300	
520,350	
520,400	
520,450	
520,500	
520,550	
520,600	
520,650	
520,700	
520,750	
520,800	
520,850	
520,900	
520,950	
521,000	
521,050	
521,100	
521,150	
521,200	
521,250	
521,300	
521,350	
521,400	
521,450	
521,500	
521,550	
521,600	
521,650	
521,700	
521,750	
521,800	
521,850	
521,900	
521,950	
522,000	

The Number 522,001 to the Number 524,000

Start		Notes
522,050
522,100
522,150
522,200
522,250
522,300
522,350
522,400
522,450
522,500
522,550
522,600
522,650
522,700
522,750
522,800
522,850
522,900
522,950
523,000
523,050
523,100
523,150
523,200
523,250
523,300
523,350
523,400
523,450
523,500
523,550
523,600
523,650
523,700
523,750
523,800
523,850
523,900
523,950
524,000

The Number 524,001 to the Number 526,000

Start
Notes

	524,050
	524,100
	524,150
	524,200
	524,250
	524,300
	524,350
	524,400
	524,450
	524,500
	524,550
	524,600
	524,650
	524,700
	524,750
	524,800
	524,850
	524,900
	524,950
	525,000
	525,050
	525,100
	525,150
	525,200
	525,250
	525,300
	525,350
	525,400
	525,450
	525,500
	525,550
	525,600
	525,650
	525,700
	525,750
	525,800
	525,850
	525,900
	525,950
	526,000

The Number 526,001 to the Number 528,000

Start		Notes

Number	Notes
526,050	
526,100	
526,150	
526,200	
526,250	
526,300	
526,350	
526,400	
526,450	
526,500	
526,550	
526,600	
526,650	
526,700	
526,750	
526,800	
526,850	
526,900	
526,950	
527,000	
527,050	
527,100	
527,150	
527,200	
527,250	
527,300	
527,350	
527,400	
527,450	
527,500	
527,550	
527,600	
527,650	
527,700	
527,750	
527,800	
527,850	
527,900	
527,950	
528,000	

The Number 528,001 to the Number 530,000

Start

Notes

	528,050
	528,100
	528,150
	528,200
	528,250
	528,300
	528,350
	528,400
	528,450
	528,500
	528,550
	528,600
	528,650
	528,700
	528,750
	528,800
	528,850
	528,900
	528,950
	529,000
	529,050
	529,100
	529,150
	529,200
	529,250
	529,300
	529,350
	529,400
	529,450
	529,500
	529,550
	529,600
	529,650
	529,700
	529,750
	529,800
	529,850
	529,900
	529,950
	530,000

The Number 530,001 to the Number 532,000

Start		Notes
	530,050	
	530,100	
	530,150	
	530,200	
	530,250	
	530,300	
	530,350	
	530,400	
	530,450	
	530,500	
	530,550	
	530,600	
	530,650	
	530,700	
	530,750	
	530,800	
	530,850	
	530,900	
	530,950	
	531,000	
	531,050	
	531,100	
	531,150	
	531,200	
	531,250	
	531,300	
	531,350	
	531,400	
	531,450	
	531,500	
	531,550	
	531,600	
	531,650	
	531,700	
	531,750	
	531,800	
	531,850	
	531,900	
	531,950	
	532,000	

The Number 532,001 to the Number 534,000

Start   Notes

	532,050
	532,100
	532,150
	532,200
	532,250
	532,300
	532,350
	532,400
	532,450
	532,500
	532,550
	532,600
	532,650
	532,700
	532,750
	532,800
	532,850
	532,900
	532,950
	533,000
	533,050
	533,100
	533,150
	533,200
	533,250
	533,300
	533,350
	533,400
	533,450
	533,500
	533,550
	533,600
	533,650
	533,700
	533,750
	533,800
	533,850
	533,900
	533,950
	534,000

The Number 534,001 to the Number 536,000

Start		Notes
	534,050	
	534,100	
	534,150	
	534,200	
	534,250	
	534,300	
	534,350	
	534,400	
	534,450	
	534,500	
	534,550	
	534,600	
	534,650	
	534,700	
	534,750	
	534,800	
	534,850	
	534,900	
	534,950	
	535,000	
	535,050	
	535,100	
	535,150	
	535,200	
	535,250	
	535,300	
	535,350	
	535,400	
	535,450	
	535,500	
	535,550	
	535,600	
	535,650	
	535,700	
	535,750	
	535,800	
	535,850	
	535,900	
	535,950	
	536,000	

The Number 536,001 to the Number 538,000

Start   Notes

	536,050
	536,100
	536,150
	536,200
	536,250
	536,300
	536,350
	536,400
	536,450
	536,500
	536,550
	536,600
	536,650
	536,700
	536,750
	536,800
	536,850
	536,900
	536,950
	537,000
	537,050
	537,100
	537,150
	537,200
	537,250
	537,300
	537,350
	537,400
	537,450
	537,500
	537,550
	537,600
	537,650
	537,700
	537,750
	537,800
	537,850
	537,900
	537,950
	538,000

The Number 538,001 to the Number 540,000

	Notes
538,050	
538,100	
538,150	
538,200	
538,250	
538,300	
538,350	
538,400	
538,450	
538,500	
538,550	
538,600	
538,650	
538,700	
538,750	
538,800	
538,850	
538,900	
538,950	
539,000	
539,050	
539,100	
539,150	
539,200	
539,250	
539,300	
539,350	
539,400	
539,450	
539,500	
539,550	
539,600	
539,650	
539,700	
539,750	
539,800	
539,850	
539,900	
539,950	
540,000	

The Number 540,001 to the Number 542,000

Start   Notes

	540,050
	540,100
	540,150
	540,200
	540,250
	540,300
	540,350
	540,400
	540,450
	540,500
	540,550
	540,600
	540,650
	540,700
	540,750
	540,800
	540,850
	540,900
	540,950
	541,000
	541,050
	541,100
	541,150
	541,200
	541,250
	541,300
	541,350
	541,400
	541,450
	541,500
	541,550
	541,600
	541,650
	541,700
	541,750
	541,800
	541,850
	541,900
	541,950
	542,000

◇ The Number 542,001 to the Number 544,000

Start   **Notes**

	542,050
	542,100
	542,150
	542,200
	542,250
	542,300
	542,350
	542,400
	542,450
	542,500
	542,550
	542,600
	542,650
	542,700
	542,750
	542,800
	542,850
	542,900
	542,950
	543,000
	543,050
	543,100
	543,150
	543,200
	543,250
	543,300
	543,350
	543,400
	543,450
	543,500
	543,550
	543,600
	543,650
	543,700
	543,750
	543,800
	543,850
	543,900
	543,950
	544,000

The Number 544,001 to the Number 546,000

Start   Notes

	544,050
	544,100
	544,150
	544,200
	544,250
	544,300
	544,350
	544,400
	544,450
	544,500
	544,550
	544,600
	544,650
	544,700
	544,750
	544,800
	544,850
	544,900
	544,950
	545,000
	545,050
	545,100
	545,150
	545,200
	545,250
	545,300
	545,350
	545,400
	545,450
	545,500
	545,550
	545,600
	545,650
	545,700
	545,750
	545,800
	545,850
	545,900
	545,950
	546,000

The Number 546,001 to the Number 548,000

Start	Notes
	546,050
	546,100
	546,150
	546,200
	546,250
	546,300
	546,350
	546,400
	546,450
	546,500
	546,550
	546,600
	546,650
	546,700
	546,750
	546,800
	546,850
	546,900
	546,950
	547,000
	547,050
	547,100
	547,150
	547,200
	547,250
	547,300
	547,350
	547,400
	547,450
	547,500
	547,550
	547,600
	547,650
	547,700
	547,750
	547,800
	547,850
	547,900
	547,950
	548,000

The Number 548,001 to the Number 550,000

Start | | Notes

	548,050
	548,100
	548,150
	548,200
	548,250
	548,300
	548,350
	548,400
	548,450
	548,500
	548,550
	548,600
	548,650
	548,700
	548,750
	548,800
	548,850
	548,900
	548,950
	549,000
	549,050
	549,100
	549,150
	549,200
	549,250
	549,300
	549,350
	549,400
	549,450
	549,500
	549,550
	549,600
	549,650
	549,700
	549,750
	549,800
	549,850
	549,900
	549,950
	550,000

The Number 550,001 to the Number 552,000

Start	Notes
	550,050
	550,100
	550,150
	550,200
	550,250
	550,300
	550,350
	550,400
	550,450
	550,500
	550,550
	550,600
	550,650
	550,700
	550,750
	550,800
	550,850
	550,900
	550,950
	551,000
	551,050
	551,100
	551,150
	551,200
	551,250
	551,300
	551,350
	551,400
	551,450
	551,500
	551,550
	551,600
	551,650
	551,700
	551,750
	551,800
	551,850
	551,900
	551,950
	552,000

The Number 552,001 to the Number 554,000

Start

Notes

	552,050
	552,100
	552,150
	552,200
	552,250
	552,300
	552,350
	552,400
	552,450
	552,500
	552,550
	552,600
	552,650
	552,700
	552,750
	552,800
	552,850
	552,900
	552,950
	553,000
	553,050
	553,100
	553,150
	553,200
	553,250
	553,300
	553,350
	553,400
	553,450
	553,500
	553,550
	553,600
	553,650
	553,700
	553,750
	553,800
	553,850
	553,900
	553,950
	554,000

The Number 554,001 to the Number 556,000

Start		Notes
	554,050	
	554,100	
	554,150	
	554,200	
	554,250	
	554,300	
	554,350	
	554,400	
	554,450	
	554,500	
	554,550	
	554,600	
	554,650	
	554,700	
	554,750	
	554,800	
	554,850	
	554,900	
	554,950	
	555,000	
	555,050	
	555,100	
	555,150	
	555,200	
	555,250	
	555,300	
	555,350	
	555,400	
	555,450	
	555,500	
	555,550	
	555,600	
	555,650	
	555,700	
	555,750	
	555,800	
	555,850	
	555,900	
	555,950	
	556,000	

The Number 556,001 to the Number 558,000

Start

Notes

556,050	
556,100	
556,150	
556,200	
556,250	
556,300	
556,350	
556,400	
556,450	
556,500	
556,550	
556,600	
556,650	
556,700	
556,750	
556,800	
556,850	
556,900	
556,950	
557,000	
557,050	
557,100	
557,150	
557,200	
557,250	
557,300	
557,350	
557,400	
557,450	
557,500	
557,550	
557,600	
557,650	
557,700	
557,750	
557,800	
557,850	
557,900	
557,950	
558,000	

The Number 558,001 to the Number 560,000

Start Notes

	558,050
	558,100
	558,150
	558,200
	558,250
	558,300
	558,350
	558,400
	558,450
	558,500
	558,550
	558,600
	558,650
	558,700
	558,750
	558,800
	558,850
	558,900
	558,950
	559,000
	559,050
	559,100
	559,150
	559,200
	559,250
	559,300
	559,350
	559,400
	559,450
	559,500
	559,550
	559,600
	559,650
	559,700
	559,750
	559,800
	559,850
	559,900
	559,950
	560,000

The Number 560,001 to the Number 562,000

Start		Notes

	Numbers	Notes
	560,050	
	560,100	
	560,150	
	560,200	
	560,250	
	560,300	
	560,350	
	560,400	
	560,450	
	560,500	
	560,550	
	560,600	
	560,650	
	560,700	
	560,750	
	560,800	
	560,850	
	560,900	
	560,950	
	561,000	
	561,050	
	561,100	
	561,150	
	561,200	
	561,250	
	561,300	
	561,350	
	561,400	
	561,450	
	561,500	
	561,550	
	561,600	
	561,650	
	561,700	
	561,750	
	561,800	
	561,850	
	561,900	
	561,950	
	562,000	

The Number 562,001 to the Number 564,000

Start		Notes
	562,050	
	562,100	
	562,150	
	562,200	
	562,250	
	562,300	
	562,350	
	562,400	
	562,450	
	562,500	
	562,550	
	562,600	
	562,650	
	562,700	
	562,750	
	562,800	
	562,850	
	562,900	
	562,950	
	563,000	
	563,050	
	563,100	
	563,150	
	563,200	
	563,250	
	563,300	
	563,350	
	563,400	
	563,450	
	563,500	
	563,550	
	563,600	
	563,650	
	563,700	
	563,750	
	563,800	
	563,850	
	563,900	
	563,950	
	564,000	

The Number 564,001 to the Number 566,000

Start   Notes

564,050
564,100
564,150
564,200
564,250
564,300
564,350
564,400
564,450
564,500
564,550
564,600
564,650
564,700
564,750
564,800
564,850
564,900
564,950
565,000
565,050
565,100
565,150
565,200
565,250
565,300
565,350
565,400
565,450
565,500
565,550
565,600
565,650
565,700
565,750
565,800
565,850
565,900
565,950
566,000

The Number 566,001 to the Number 568,000

Start		Notes
	566,050	
	566,100	
	566,150	
	566,200	
	566,250	
	566,300	
	566,350	
	566,400	
	566,450	
	566,500	
	566,550	
	566,600	
	566,650	
	566,700	
	566,750	
	566,800	
	566,850	
	566,900	
	566,950	
	567,000	
	567,050	
	567,100	
	567,150	
	567,200	
	567,250	
	567,300	
	567,350	
	567,400	
	567,450	
	567,500	
	567,550	
	567,600	
	567,650	
	567,700	
	567,750	
	567,800	
	567,850	
	567,900	
	567,950	
	568,000	

The Number 568,001 to the Number 570,000

Start		Notes
568,050	
568,100	
568,150	
568,200	
568,250	
568,300	
568,350	
568,400	
568,450	
568,500	
568,550	
568,600	
568,650	
568,700	
568,750	
568,800	
568,850	
568,900	
568,950	
569,000	
569,050	
569,100	
569,150	
569,200	
569,250	
569,300	
569,350	
569,400	
569,450	
569,500	
569,550	
569,600	
569,650	
569,700	
569,750	
569,800	
569,850	
569,900	
569,950	
570,000	

The Number 570,001 to the Number 572,000

Start		Notes
	570,050	
	570,100	
	570,150	
	570,200	
	570,250	
	570,300	
	570,350	
	570,400	
	570,450	
	570,500	
	570,550	
	570,600	
	570,650	
	570,700	
	570,750	
	570,800	
	570,850	
	570,900	
	570,950	
	571,000	
	571,050	
	571,100	
	571,150	
	571,200	
	571,250	
	571,300	
	571,350	
	571,400	
	571,450	
	571,500	
	571,550	
	571,600	
	571,650	
	571,700	
	571,750	
	571,800	
	571,850	
	571,900	
	571,950	
	572,000	

The Number 572,001 to the Number 574,000

Start | | Notes

	572,050
	572,100
	572,150
	572,200
	572,250
	572,300
	572,350
	572,400
	572,450
	572,500
	572,550
	572,600
	572,650
	572,700
	572,750
	572,800
	572,850
	572,900
	572,950
	573,000
	573,050
	573,100
	573,150
	573,200
	573,250
	573,300
	573,350
	573,400
	573,450
	573,500
	573,550
	573,600
	573,650
	573,700
	573,750
	573,800
	573,850
	573,900
	573,950
	574,000

The Number 574,001 to the Number 576,000

Start

◊◊◊◊◊ ◊◊◊◊◊ ◊◊◊◊◊ ◊◊◊◊◊ ◊◊◊◊◊ ◊◊◊◊◊ ◊◊◊◊◊ ◊◊◊◊◊ ◊◊◊◊◊ ◊◊◊◊◊	574,050
◊◊◊◊◊ ◊◊◊◊◊ ◊◊◊◊◊ ◊◊◊◊◊ ◊◊◊◊◊ ◊◊◊◊◊ ◊◊◊◊◊ ◊◊◊◊◊ ◊◊◊◊◊ ◊◊◊◊◊	574,100
◊◊◊◊◊ ◊◊◊◊◊ ◊◊◊◊◊ ◊◊◊◊◊ ◊◊◊◊◊ ◊◊◊◊◊ ◊◊◊◊◊ ◊◊◊◊◊ ◊◊◊◊◊ ◊◊◊◊◊	574,150
◊◊◊◊◊ ◊◊◊◊◊ ◊◊◊◊◊ ◊◊◊◊◊ ◊◊◊◊◊ ◊◊◊◊◊ ◊◊◊◊◊ ◊◊◊◊◊ ◊◊◊◊◊ ◊◊◊◊◊	574,200
◊◊◊◊◊ ◊◊◊◊◊ ◊◊◊◊◊ ◊◊◊◊◊ ◊◊◊◊◊ ◊◊◊◊◊ ◊◊◊◊◊ ◊◊◊◊◊ ◊◊◊◊◊ ◊◊◊◊◊	574,250
◊◊◊◊◊ ◊◊◊◊◊ ◊◊◊◊◊ ◊◊◊◊◊ ◊◊◊◊◊ ◊◊◊◊◊ ◊◊◊◊◊ ◊◊◊◊◊ ◊◊◊◊◊ ◊◊◊◊◊	574,300
◊◊◊◊◊ ◊◊◊◊◊ ◊◊◊◊◊ ◊◊◊◊◊ ◊◊◊◊◊ ◊◊◊◊◊ ◊◊◊◊◊ ◊◊◊◊◊ ◊◊◊◊◊ ◊◊◊◊◊	574,350
◊◊◊◊◊ ◊◊◊◊◊ ◊◊◊◊◊ ◊◊◊◊◊ ◊◊◊◊◊ ◊◊◊◊◊ ◊◊◊◊◊ ◊◊◊◊◊ ◊◊◊◊◊ ◊◊◊◊◊	574,400
◊◊◊◊◊ ◊◊◊◊◊ ◊◊◊◊◊ ◊◊◊◊◊ ◊◊◊◊◊ ◊◊◊◊◊ ◊◊◊◊◊ ◊◊◊◊◊ ◊◊◊◊◊ ◊◊◊◊◊	574,450
◊◊◊◊◊ ◊◊◊◊◊ ◊◊◊◊◊ ◊◊◊◊◊ ◊◊◊◊◊ ◊◊◊◊◊ ◊◊◊◊◊ ◊◊◊◊◊ ◊◊◊◊◊ ◊◊◊◊◊	574,500
◊◊◊◊◊ ◊◊◊◊◊ ◊◊◊◊◊ ◊◊◊◊◊ ◊◊◊◊◊ ◊◊◊◊◊ ◊◊◊◊◊ ◊◊◊◊◊ ◊◊◊◊◊ ◊◊◊◊◊	574,550
◊◊◊◊◊ ◊◊◊◊◊ ◊◊◊◊◊ ◊◊◊◊◊ ◊◊◊◊◊ ◊◊◊◊◊ ◊◊◊◊◊ ◊◊◊◊◊ ◊◊◊◊◊ ◊◊◊◊◊	574,600
◊◊◊◊◊ ◊◊◊◊◊ ◊◊◊◊◊ ◊◊◊◊◊ ◊◊◊◊◊ ◊◊◊◊◊ ◊◊◊◊◊ ◊◊◊◊◊ ◊◊◊◊◊ ◊◊◊◊◊	574,650
◊◊◊◊◊ ◊◊◊◊◊ ◊◊◊◊◊ ◊◊◊◊◊ ◊◊◊◊◊ ◊◊◊◊◊ ◊◊◊◊◊ ◊◊◊◊◊ ◊◊◊◊◊ ◊◊◊◊◊	574,700
◊◊◊◊◊ ◊◊◊◊◊ ◊◊◊◊◊ ◊◊◊◊◊ ◊◊◊◊◊ ◊◊◊◊◊ ◊◊◊◊◊ ◊◊◊◊◊ ◊◊◊◊◊ ◊◊◊◊◊	574,750
◊◊◊◊◊ ◊◊◊◊◊ ◊◊◊◊◊ ◊◊◊◊◊ ◊◊◊◊◊ ◊◊◊◊◊ ◊◊◊◊◊ ◊◊◊◊◊ ◊◊◊◊◊ ◊◊◊◊◊	574,800
◊◊◊◊◊ ◊◊◊◊◊ ◊◊◊◊◊ ◊◊◊◊◊ ◊◊◊◊◊ ◊◊◊◊◊ ◊◊◊◊◊ ◊◊◊◊◊ ◊◊◊◊◊ ◊◊◊◊◊	574,850
◊◊◊◊◊ ◊◊◊◊◊ ◊◊◊◊◊ ◊◊◊◊◊ ◊◊◊◊◊ ◊◊◊◊◊ ◊◊◊◊◊ ◊◊◊◊◊ ◊◊◊◊◊ ◊◊◊◊◊	574,900
◊◊◊◊◊ ◊◊◊◊◊ ◊◊◊◊◊ ◊◊◊◊◊ ◊◊◊◊◊ ◊◊◊◊◊ ◊◊◊◊◊ ◊◊◊◊◊ ◊◊◊◊◊ ◊◊◊◊◊	574,950
◊◊◊◊◊ ◊◊◊◊◊ ◊◊◊◊◊ ◊◊◊◊◊ ◊◊◊◊◊ ◊◊◊◊◊ ◊◊◊◊◊ ◊◊◊◊◊ ◊◊◊◊◊ ◊◊◊◊◊	**575,000**
◊◊◊◊◊ ◊◊◊◊◊ ◊◊◊◊◊ ◊◊◊◊◊ ◊◊◊◊◊ ◊◊◊◊◊ ◊◊◊◊◊ ◊◊◊◊◊ ◊◊◊◊◊ ◊◊◊◊◊	575,050
◊◊◊◊◊ ◊◊◊◊◊ ◊◊◊◊◊ ◊◊◊◊◊ ◊◊◊◊◊ ◊◊◊◊◊ ◊◊◊◊◊ ◊◊◊◊◊ ◊◊◊◊◊ ◊◊◊◊◊	575,100
◊◊◊◊◊ ◊◊◊◊◊ ◊◊◊◊◊ ◊◊◊◊◊ ◊◊◊◊◊ ◊◊◊◊◊ ◊◊◊◊◊ ◊◊◊◊◊ ◊◊◊◊◊ ◊◊◊◊◊	575,150
◊◊◊◊◊ ◊◊◊◊◊ ◊◊◊◊◊ ◊◊◊◊◊ ◊◊◊◊◊ ◊◊◊◊◊ ◊◊◊◊◊ ◊◊◊◊◊ ◊◊◊◊◊ ◊◊◊◊◊	575,200
◊◊◊◊◊ ◊◊◊◊◊ ◊◊◊◊◊ ◊◊◊◊◊ ◊◊◊◊◊ ◊◊◊◊◊ ◊◊◊◊◊ ◊◊◊◊◊ ◊◊◊◊◊ ◊◊◊◊◊	575,250
◊◊◊◊◊ ◊◊◊◊◊ ◊◊◊◊◊ ◊◊◊◊◊ ◊◊◊◊◊ ◊◊◊◊◊ ◊◊◊◊◊ ◊◊◊◊◊ ◊◊◊◊◊ ◊◊◊◊◊	575,300
◊◊◊◊◊ ◊◊◊◊◊ ◊◊◊◊◊ ◊◊◊◊◊ ◊◊◊◊◊ ◊◊◊◊◊ ◊◊◊◊◊ ◊◊◊◊◊ ◊◊◊◊◊ ◊◊◊◊◊	575,350
◊◊◊◊◊ ◊◊◊◊◊ ◊◊◊◊◊ ◊◊◊◊◊ ◊◊◊◊◊ ◊◊◊◊◊ ◊◊◊◊◊ ◊◊◊◊◊ ◊◊◊◊◊ ◊◊◊◊◊	575,400
◊◊◊◊◊ ◊◊◊◊◊ ◊◊◊◊◊ ◊◊◊◊◊ ◊◊◊◊◊ ◊◊◊◊◊ ◊◊◊◊◊ ◊◊◊◊◊ ◊◊◊◊◊ ◊◊◊◊◊	575,450
◊◊◊◊◊ ◊◊◊◊◊ ◊◊◊◊◊ ◊◊◊◊◊ ◊◊◊◊◊ ◊◊◊◊◊ ◊◊◊◊◊ ◊◊◊◊◊ ◊◊◊◊◊ ◊◊◊◊◊	575,500
◊◊◊◊◊ ◊◊◊◊◊ ◊◊◊◊◊ ◊◊◊◊◊ ◊◊◊◊◊ ◊◊◊◊◊ ◊◊◊◊◊ ◊◊◊◊◊ ◊◊◊◊◊ ◊◊◊◊◊	575,550
◊◊◊◊◊ ◊◊◊◊◊ ◊◊◊◊◊ ◊◊◊◊◊ ◊◊◊◊◊ ◊◊◊◊◊ ◊◊◊◊◊ ◊◊◊◊◊ ◊◊◊◊◊ ◊◊◊◊◊	575,600
◊◊◊◊◊ ◊◊◊◊◊ ◊◊◊◊◊ ◊◊◊◊◊ ◊◊◊◊◊ ◊◊◊◊◊ ◊◊◊◊◊ ◊◊◊◊◊ ◊◊◊◊◊ ◊◊◊◊◊	575,650
◊◊◊◊◊ ◊◊◊◊◊ ◊◊◊◊◊ ◊◊◊◊◊ ◊◊◊◊◊ ◊◊◊◊◊ ◊◊◊◊◊ ◊◊◊◊◊ ◊◊◊◊◊ ◊◊◊◊◊	575,700
◊◊◊◊◊ ◊◊◊◊◊ ◊◊◊◊◊ ◊◊◊◊◊ ◊◊◊◊◊ ◊◊◊◊◊ ◊◊◊◊◊ ◊◊◊◊◊ ◊◊◊◊◊ ◊◊◊◊◊	575,750
◊◊◊◊◊ ◊◊◊◊◊ ◊◊◊◊◊ ◊◊◊◊◊ ◊◊◊◊◊ ◊◊◊◊◊ ◊◊◊◊◊ ◊◊◊◊◊ ◊◊◊◊◊ ◊◊◊◊◊	575,800
◊◊◊◊◊ ◊◊◊◊◊ ◊◊◊◊◊ ◊◊◊◊◊ ◊◊◊◊◊ ◊◊◊◊◊ ◊◊◊◊◊ ◊◊◊◊◊ ◊◊◊◊◊ ◊◊◊◊◊	575,850
◊◊◊◊◊ ◊◊◊◊◊ ◊◊◊◊◊ ◊◊◊◊◊ ◊◊◊◊◊ ◊◊◊◊◊ ◊◊◊◊◊ ◊◊◊◊◊ ◊◊◊◊◊ ◊◊◊◊◊	575,900
◊◊◊◊◊ ◊◊◊◊◊ ◊◊◊◊◊ ◊◊◊◊◊ ◊◊◊◊◊ ◊◊◊◊◊ ◊◊◊◊◊ ◊◊◊◊◊ ◊◊◊◊◊ ◊◊◊◊◊	575,950
◊◊◊◊◊ ◊◊◊◊◊ ◊◊◊◊◊ ◊◊◊◊◊ ◊◊◊◊◊ ◊◊◊◊◊ ◊◊◊◊◊ ◊◊◊◊◊ ◊◊◊◊◊ ◊◊◊◊◊	**576,000**

◊ The Number 576,001 to the Number 578,000

Start   **Notes**

	576,050
	576,100
	576,150
	576,200
	576,250
	576,300
	576,350
	576,400
	576,450
	576,500
	576,550
	576,600
	576,650
	576,700
	576,750
	576,800
	576,850
	576,900
	576,950
	577,000
	577,050
	577,100
	577,150
	577,200
	577,250
	577,300
	577,350
	577,400
	577,450
	577,500
	577,550
	577,600
	577,650
	577,700
	577,750
	577,800
	577,850
	577,900
	577,950
	578,000

The Number 578,001 to the Number 580,000

Start	Notes
578,050	
578,100	
578,150	
578,200	
578,250	
578,300	
578,350	
578,400	
578,450	
578,500	
578,550	
578,600	
578,650	
578,700	
578,750	
578,800	
578,850	
578,900	
578,950	
579,000	
579,050	
579,100	
579,150	
579,200	
579,250	
579,300	
579,350	
579,400	
579,450	
579,500	
579,550	
579,600	
579,650	
579,700	
579,750	
579,800	
579,850	
579,900	
579,950	
580,000	

The Number 580,001 to the Number 582,000

Start

Notes

	580,050
	580,100
	580,150
	580,200
	580,250
	580,300
	580,350
	580,400
	580,450
	580,500
	580,550
	580,600
	580,650
	580,700
	580,750
	580,800
	580,850
	580,900
	580,950
	581,000
	581,050
	581,100
	581,150
	581,200
	581,250
	581,300
	581,350
	581,400
	581,450
	581,500
	581,550
	581,600
	581,650
	581,700
	581,750
	581,800
	581,850
	581,900
	581,950
	582,000

The Number 582,001 to the Number 584,000

Start

Notes

	582,050
	582,100
	582,150
	582,200
	582,250
	582,300
	582,350
	582,400
	582,450
	582,500
	582,550
	582,600
	582,650
	582,700
	582,750
	582,800
	582,850
	582,900
	582,950
	583,000
	583,050
	583,100
	583,150
	583,200
	583,250
	583,300
	583,350
	583,400
	583,450
	583,500
	583,550
	583,600
	583,650
	583,700
	583,750
	583,800
	583,850
	583,900
	583,950
	584,000

The Number 584,001 to the Number 586,000

Start

Notes

	584,050
	584,100
	584,150
	584,200
	584,250
	584,300
	584,350
	584,400
	584,450
	584,500
	584,550
	584,600
	584,650
	584,700
	584,750
	584,800
	584,850
	584,900
	584,950
	585,000
	585,050
	585,100
	585,150
	585,200
	585,250
	585,300
	585,350
	585,400
	585,450
	585,500
	585,550
	585,600
	585,650
	585,700
	585,750
	585,800
	585,850
	585,900
	585,950
	586,000

◊ The Number 586,001 to the Number 588,000

Start   Notes

	586,050
	586,100
	586,150
	586,200
	586,250
	586,300
	586,350
	586,400
	586,450
	586,500
	586,550
	586,600
	586,650
	586,700
	586,750
	586,800
	586,850
	586,900
	586,950
	587,000
	587,050
	587,100
	587,150
	587,200
	587,250
	587,300
	587,350
	587,400
	587,450
	587,500
	587,550
	587,600
	587,650
	587,700
	587,750
	587,800
	587,850
	587,900
	587,950
	588,000

The Number 588,001 to the Number 590,000

Start		Notes

	588,050
	588,100
	588,150
	588,200
	588,250
	588,300
	588,350
	588,400
	588,450
	588,500
	588,550
	588,600
	588,650
	588,700
	588,750
	588,800
	588,850
	588,900
	588,950
	589,000
	589,050
	589,100
	589,150
	589,200
	589,250
	589,300
	589,350
	589,400
	589,450
	589,500
	589,550
	589,600
	589,650
	589,700
	589,750
	589,800
	589,850
	589,900
	589,950
	590,000

The Number 590,001 to the Number 592,000

Start		Notes
	590,050
	590,100
	590,150
	590,200
	590,250
	590,300
	590,350
	590,400
	590,450
	590,500
	590,550
	590,600
	590,650
	590,700
	590,750
	590,800
	590,850
	590,900
	590,950
	591,000
	591,050
	591,100
	591,150
	591,200
	591,250
	591,300
	591,350
	591,400
	591,450
	591,500
	591,550
	591,600
	591,650
	591,700
	591,750
	591,800
	591,850
	591,900
	591,950
	592,000

The Number 592,001 to the Number 594,000

Start

Notes

	592,050
	592,100
	592,150
	592,200
	592,250
	592,300
	592,350
	592,400
	592,450
	592,500
	592,550
	592,600
	592,650
	592,700
	592,750
	592,800
	592,850
	592,900
	592,950
	593,000
	593,050
	593,100
	593,150
	593,200
	593,250
	593,300
	593,350
	593,400
	593,450
	593,500
	593,550
	593,600
	593,650
	593,700
	593,750
	593,800
	593,850
	593,900
	593,950
	594,000

The Number 594,001 to the Number 596,000

Start		Notes
	594,050	
	594,100	
	594,150	
	594,200	
	594,250	
	594,300	
	594,350	
	594,400	
	594,450	
	594,500	
	594,550	
	594,600	
	594,650	
	594,700	
	594,750	
	594,800	
	594,850	
	594,900	
	594,950	
	595,000	
	595,050	
	595,100	
	595,150	
	595,200	
	595,250	
	595,300	
	595,350	
	595,400	
	595,450	
	595,500	
	595,550	
	595,600	
	595,650	
	595,700	
	595,750	
	595,800	
	595,850	
	595,900	
	595,950	
	596,000	

The Number 596,001 to the Number 598,000

Start   Notes

	596,050
	596,100
	596,150
	596,200
	596,250
	596,300
	596,350
	596,400
	596,450
	596,500
	596,550
	596,600
	596,650
	596,700
	596,750
	596,800
	596,850
	596,900
	596,950
	597,000
	597,050
	597,100
	597,150
	597,200
	597,250
	597,300
	597,350
	597,400
	597,450
	597,500
	597,550
	597,600
	597,650
	597,700
	597,750
	597,800
	597,850
	597,900
	597,950
	598,000

The Number 598,001 to the Number 600,000

Start   Notes

	598,050
	598,100
	598,150
	598,200
	598,250
	598,300
	598,350
	598,400
	598,450
	598,500
	598,550
	598,600
	598,650
	598,700
	598,750
	598,800
	598,850
	598,900
	598,950
	599,000
	599,050
	599,100
	599,150
	599,200
	599,250
	599,300
	599,350
	599,400
	599,450
	599,500
	599,550
	599,600
	599,650
	599,700
	599,750
	599,800
	599,850
	599,900
	599,950
	600,000

The Number 600,001 to the Number 602,000

Start　　　　　　　　　　　　　　　　　　　　　　　　　　　　　**Notes**

	600,050
	600,100
	600,150
	600,200
	600,250
	600,300
	600,350
	600,400
	600,450
	600,500
	600,550
	600,600
	600,650
	600,700
	600,750
	600,800
	600,850
	600,900
	600,950
	601,000
	601,050
	601,100
	601,150
	601,200
	601,250
	601,300
	601,350
	601,400
	601,450
	601,500
	601,550
	601,600
	601,650
	601,700
	601,750
	601,800
	601,850
	601,900
	601,950
	602,000

The Number 602,001 to the Number 604,000

Start		Notes
	602,050	
	602,100	
	602,150	
	602,200	
	602,250	
	602,300	
	602,350	
	602,400	
	602,450	
	602,500	
	602,550	
	602,600	
	602,650	
	602,700	
	602,750	
	602,800	
	602,850	
	602,900	
	602,950	
	603,000	
	603,050	
	603,100	
	603,150	
	603,200	
	603,250	
	603,300	
	603,350	
	603,400	
	603,450	
	603,500	
	603,550	
	603,600	
	603,650	
	603,700	
	603,750	
	603,800	
	603,850	
	603,900	
	603,950	
	604,000	

◇ The Number 604,001 to the Number 606,000

Start		Notes

Diamonds	Number	Notes
◇◇◇◇◇ ◇◇◇◇◇ ◇◇◇◇◇ ◇◇◇◇◇ ◇◇◇◇◇ ◇◇◇◇◇ ◇◇◇◇◇ ◇◇◇◇◇ ◇◇◇◇◇ ◇◇◇◇◇	604,050	
◇◇◇◇◇ ◇◇◇◇◇ ◇◇◇◇◇ ◇◇◇◇◇ ◇◇◇◇◇ ◇◇◇◇◇ ◇◇◇◇◇ ◇◇◇◇◇ ◇◇◇◇◇ ◇◇◇◇◇	604,100	
◇◇◇◇◇ ◇◇◇◇◇ ◇◇◇◇◇ ◇◇◇◇◇ ◇◇◇◇◇ ◇◇◇◇◇ ◇◇◇◇◇ ◇◇◇◇◇ ◇◇◇◇◇ ◇◇◇◇◇	604,150	
◇◇◇◇◇ ◇◇◇◇◇ ◇◇◇◇◇ ◇◇◇◇◇ ◇◇◇◇◇ ◇◇◇◇◇ ◇◇◇◇◇ ◇◇◇◇◇ ◇◇◇◇◇ ◇◇◇◇◇	604,200	
◇◇◇◇◇ ◇◇◇◇◇ ◇◇◇◇◇ ◇◇◇◇◇ ◇◇◇◇◇ ◇◇◇◇◇ ◇◇◇◇◇ ◇◇◇◇◇ ◇◇◇◇◇ ◇◇◇◇◇	604,250	
◇◇◇◇◇ ◇◇◇◇◇ ◇◇◇◇◇ ◇◇◇◇◇ ◇◇◇◇◇ ◇◇◇◇◇ ◇◇◇◇◇ ◇◇◇◇◇ ◇◇◇◇◇ ◇◇◇◇◇	604,300	
◇◇◇◇◇ ◇◇◇◇◇ ◇◇◇◇◇ ◇◇◇◇◇ ◇◇◇◇◇ ◇◇◇◇◇ ◇◇◇◇◇ ◇◇◇◇◇ ◇◇◇◇◇ ◇◇◇◇◇	604,350	
◇◇◇◇◇ ◇◇◇◇◇ ◇◇◇◇◇ ◇◇◇◇◇ ◇◇◇◇◇ ◇◇◇◇◇ ◇◇◇◇◇ ◇◇◇◇◇ ◇◇◇◇◇ ◇◇◇◇◇	604,400	
◇◇◇◇◇ ◇◇◇◇◇ ◇◇◇◇◇ ◇◇◇◇◇ ◇◇◇◇◇ ◇◇◇◇◇ ◇◇◇◇◇ ◇◇◇◇◇ ◇◇◇◇◇ ◇◇◇◇◇	604,450	
◇◇◇◇◇ ◇◇◇◇◇ ◇◇◇◇◇ ◇◇◇◇◇ ◇◇◇◇◇ ◇◇◇◇◇ ◇◇◇◇◇ ◇◇◇◇◇ ◇◇◇◇◇ ◇◇◇◇◇	604,500	
◇◇◇◇◇ ◇◇◇◇◇ ◇◇◇◇◇ ◇◇◇◇◇ ◇◇◇◇◇ ◇◇◇◇◇ ◇◇◇◇◇ ◇◇◇◇◇ ◇◇◇◇◇ ◇◇◇◇◇	604,550	
◇◇◇◇◇ ◇◇◇◇◇ ◇◇◇◇◇ ◇◇◇◇◇ ◇◇◇◇◇ ◇◇◇◇◇ ◇◇◇◇◇ ◇◇◇◇◇ ◇◇◇◇◇ ◇◇◇◇◇	604,600	
◇◇◇◇◇ ◇◇◇◇◇ ◇◇◇◇◇ ◇◇◇◇◇ ◇◇◇◇◇ ◇◇◇◇◇ ◇◇◇◇◇ ◇◇◇◇◇ ◇◇◇◇◇ ◇◇◇◇◇	604,650	
◇◇◇◇◇ ◇◇◇◇◇ ◇◇◇◇◇ ◇◇◇◇◇ ◇◇◇◇◇ ◇◇◇◇◇ ◇◇◇◇◇ ◇◇◇◇◇ ◇◇◇◇◇ ◇◇◇◇◇	604,700	
◇◇◇◇◇ ◇◇◇◇◇ ◇◇◇◇◇ ◇◇◇◇◇ ◇◇◇◇◇ ◇◇◇◇◇ ◇◇◇◇◇ ◇◇◇◇◇ ◇◇◇◇◇ ◇◇◇◇◇	604,750	
◇◇◇◇◇ ◇◇◇◇◇ ◇◇◇◇◇ ◇◇◇◇◇ ◇◇◇◇◇ ◇◇◇◇◇ ◇◇◇◇◇ ◇◇◇◇◇ ◇◇◇◇◇ ◇◇◇◇◇	604,800	
◇◇◇◇◇ ◇◇◇◇◇ ◇◇◇◇◇ ◇◇◇◇◇ ◇◇◇◇◇ ◇◇◇◇◇ ◇◇◇◇◇ ◇◇◇◇◇ ◇◇◇◇◇ ◇◇◇◇◇	604,850	
◇◇◇◇◇ ◇◇◇◇◇ ◇◇◇◇◇ ◇◇◇◇◇ ◇◇◇◇◇ ◇◇◇◇◇ ◇◇◇◇◇ ◇◇◇◇◇ ◇◇◇◇◇ ◇◇◇◇◇	604,900	
◇◇◇◇◇ ◇◇◇◇◇ ◇◇◇◇◇ ◇◇◇◇◇ ◇◇◇◇◇ ◇◇◇◇◇ ◇◇◇◇◇ ◇◇◇◇◇ ◇◇◇◇◇ ◇◇◇◇◇	604,950	
◇◇◇◇◇ ◇◇◇◇◇ ◇◇◇◇◇ ◇◇◇◇◇ ◇◇◇◇◇ ◇◇◇◇◇ ◇◇◇◇◇ ◇◇◇◇◇ ◇◇◇◇◇ ◇◇◇◇◇	**605,000**	
◇◇◇◇◇ ◇◇◇◇◇ ◇◇◇◇◇ ◇◇◇◇◇ ◇◇◇◇◇ ◇◇◇◇◇ ◇◇◇◇◇ ◇◇◇◇◇ ◇◇◇◇◇ ◇◇◇◇◇	605,050	
◇◇◇◇◇ ◇◇◇◇◇ ◇◇◇◇◇ ◇◇◇◇◇ ◇◇◇◇◇ ◇◇◇◇◇ ◇◇◇◇◇ ◇◇◇◇◇ ◇◇◇◇◇ ◇◇◇◇◇	605,100	
◇◇◇◇◇ ◇◇◇◇◇ ◇◇◇◇◇ ◇◇◇◇◇ ◇◇◇◇◇ ◇◇◇◇◇ ◇◇◇◇◇ ◇◇◇◇◇ ◇◇◇◇◇ ◇◇◇◇◇	605,150	
◇◇◇◇◇ ◇◇◇◇◇ ◇◇◇◇◇ ◇◇◇◇◇ ◇◇◇◇◇ ◇◇◇◇◇ ◇◇◇◇◇ ◇◇◇◇◇ ◇◇◇◇◇ ◇◇◇◇◇	605,200	
◇◇◇◇◇ ◇◇◇◇◇ ◇◇◇◇◇ ◇◇◇◇◇ ◇◇◇◇◇ ◇◇◇◇◇ ◇◇◇◇◇ ◇◇◇◇◇ ◇◇◇◇◇ ◇◇◇◇◇	605,250	
◇◇◇◇◇ ◇◇◇◇◇ ◇◇◇◇◇ ◇◇◇◇◇ ◇◇◇◇◇ ◇◇◇◇◇ ◇◇◇◇◇ ◇◇◇◇◇ ◇◇◇◇◇ ◇◇◇◇◇	605,300	
◇◇◇◇◇ ◇◇◇◇◇ ◇◇◇◇◇ ◇◇◇◇◇ ◇◇◇◇◇ ◇◇◇◇◇ ◇◇◇◇◇ ◇◇◇◇◇ ◇◇◇◇◇ ◇◇◇◇◇	605,350	
◇◇◇◇◇ ◇◇◇◇◇ ◇◇◇◇◇ ◇◇◇◇◇ ◇◇◇◇◇ ◇◇◇◇◇ ◇◇◇◇◇ ◇◇◇◇◇ ◇◇◇◇◇ ◇◇◇◇◇	605,400	
◇◇◇◇◇ ◇◇◇◇◇ ◇◇◇◇◇ ◇◇◇◇◇ ◇◇◇◇◇ ◇◇◇◇◇ ◇◇◇◇◇ ◇◇◇◇◇ ◇◇◇◇◇ ◇◇◇◇◇	605,450	
◇◇◇◇◇ ◇◇◇◇◇ ◇◇◇◇◇ ◇◇◇◇◇ ◇◇◇◇◇ ◇◇◇◇◇ ◇◇◇◇◇ ◇◇◇◇◇ ◇◇◇◇◇ ◇◇◇◇◇	605,500	
◇◇◇◇◇ ◇◇◇◇◇ ◇◇◇◇◇ ◇◇◇◇◇ ◇◇◇◇◇ ◇◇◇◇◇ ◇◇◇◇◇ ◇◇◇◇◇ ◇◇◇◇◇ ◇◇◇◇◇	605,550	
◇◇◇◇◇ ◇◇◇◇◇ ◇◇◇◇◇ ◇◇◇◇◇ ◇◇◇◇◇ ◇◇◇◇◇ ◇◇◇◇◇ ◇◇◇◇◇ ◇◇◇◇◇ ◇◇◇◇◇	605,600	
◇◇◇◇◇ ◇◇◇◇◇ ◇◇◇◇◇ ◇◇◇◇◇ ◇◇◇◇◇ ◇◇◇◇◇ ◇◇◇◇◇ ◇◇◇◇◇ ◇◇◇◇◇ ◇◇◇◇◇	605,650	
◇◇◇◇◇ ◇◇◇◇◇ ◇◇◇◇◇ ◇◇◇◇◇ ◇◇◇◇◇ ◇◇◇◇◇ ◇◇◇◇◇ ◇◇◇◇◇ ◇◇◇◇◇ ◇◇◇◇◇	605,700	
◇◇◇◇◇ ◇◇◇◇◇ ◇◇◇◇◇ ◇◇◇◇◇ ◇◇◇◇◇ ◇◇◇◇◇ ◇◇◇◇◇ ◇◇◇◇◇ ◇◇◇◇◇ ◇◇◇◇◇	605,750	
◇◇◇◇◇ ◇◇◇◇◇ ◇◇◇◇◇ ◇◇◇◇◇ ◇◇◇◇◇ ◇◇◇◇◇ ◇◇◇◇◇ ◇◇◇◇◇ ◇◇◇◇◇ ◇◇◇◇◇	605,800	
◇◇◇◇◇ ◇◇◇◇◇ ◇◇◇◇◇ ◇◇◇◇◇ ◇◇◇◇◇ ◇◇◇◇◇ ◇◇◇◇◇ ◇◇◇◇◇ ◇◇◇◇◇ ◇◇◇◇◇	605,850	
◇◇◇◇◇ ◇◇◇◇◇ ◇◇◇◇◇ ◇◇◇◇◇ ◇◇◇◇◇ ◇◇◇◇◇ ◇◇◇◇◇ ◇◇◇◇◇ ◇◇◇◇◇ ◇◇◇◇◇	605,900	
◇◇◇◇◇ ◇◇◇◇◇ ◇◇◇◇◇ ◇◇◇◇◇ ◇◇◇◇◇ ◇◇◇◇◇ ◇◇◇◇◇ ◇◇◇◇◇ ◇◇◇◇◇ ◇◇◇◇◇	605,950	
◇◇◇◇◇ ◇◇◇◇◇ ◇◇◇◇◇ ◇◇◇◇◇ ◇◇◇◇◇ ◇◇◇◇◇ ◇◇◇◇◇ ◇◇◇◇◇ ◇◇◇◇◇ ◇◇◇◇◇	**606,000**	

The Number 606,001 to the Number 608,000

Start		Notes

	606,050
	606,100
	606,150
	606,200
	606,250
	606,300
	606,350
	606,400
	606,450
	606,500
	606,550
	606,600
	606,650
	606,700
	606,750
	606,800
	606,850
	606,900
	606,950
	607,000
	607,050
	607,100
	607,150
	607,200
	607,250
	607,300
	607,350
	607,400
	607,450
	607,500
	607,550
	607,600
	607,650
	607,700
	607,750
	607,800
	607,850
	607,900
	607,950
	608,000

The Number 608,001 to the Number 610,000

Start   Notes

	608,050
	608,100
	608,150
	608,200
	608,250
	608,300
	608,350
	608,400
	608,450
	608,500
	608,550
	608,600
	608,650
	608,700
	608,750
	608,800
	608,850
	608,900
	608,950
	609,000
	609,050
	609,100
	609,150
	609,200
	609,250
	609,300
	609,350
	609,400
	609,450
	609,500
	609,550
	609,600
	609,650
	609,700
	609,750
	609,800
	609,850
	609,900
	609,950
	610,000

The Number 610,001 to the Number 612,000

Start   Notes

	610,050
	610,100
	610,150
	610,200
	610,250
	610,300
	610,350
	610,400
	610,450
	610,500
	610,550
	610,600
	610,650
	610,700
	610,750
	610,800
	610,850
	610,900
	610,950
	611,000
	611,050
	611,100
	611,150
	611,200
	611,250
	611,300
	611,350
	611,400
	611,450
	611,500
	611,550
	611,600
	611,650
	611,700
	611,750
	611,800
	611,850
	611,900
	611,950
	612,000

The Number 612,001 to the Number 614,000

Start

	Notes
	612,050
	612,100
	612,150
	612,200
	612,250
	612,300
	612,350
	612,400
	612,450
	612,500
	612,550
	612,600
	612,650
	612,700
	612,750
	612,800
	612,850
	612,900
	612,950
	613,000
	613,050
	613,100
	613,150
	613,200
	613,250
	613,300
	613,350
	613,400
	613,450
	613,500
	613,550
	613,600
	613,650
	613,700
	613,750
	613,800
	613,850
	613,900
	613,950
	614,000

The Number 614,001 to the Number 616,000

Start   Notes

Number
614,050
614,100
614,150
614,200
614,250
614,300
614,350
614,400
614,450
614,500
614,550
614,600
614,650
614,700
614,750
614,800
614,850
614,900
614,950
615,000
615,050
615,100
615,150
615,200
615,250
615,300
615,350
615,400
615,450
615,500
615,550
615,600
615,650
615,700
615,750
615,800
615,850
615,900
615,950
616,000

The Number 616,001 to the Number 618,000

Start		Notes
◊◊◊◊◊ ◊◊◊◊◊ ◊◊◊◊◊ ◊◊◊◊◊ ◊◊◊◊◊ ◊◊◊◊◊ ◊◊◊◊◊ ◊◊◊◊◊ ◊◊◊◊◊ ◊◊◊◊◊	616,050
◊◊◊◊◊ ◊◊◊◊◊ ◊◊◊◊◊ ◊◊◊◊◊ ◊◊◊◊◊ ◊◊◊◊◊ ◊◊◊◊◊ ◊◊◊◊◊ ◊◊◊◊◊ ◊◊◊◊◊	616,100
◊◊◊◊◊ ◊◊◊◊◊ ◊◊◊◊◊ ◊◊◊◊◊ ◊◊◊◊◊ ◊◊◊◊◊ ◊◊◊◊◊ ◊◊◊◊◊ ◊◊◊◊◊ ◊◊◊◊◊	616,150
◊◊◊◊◊ ◊◊◊◊◊ ◊◊◊◊◊ ◊◊◊◊◊ ◊◊◊◊◊ ◊◊◊◊◊ ◊◊◊◊◊ ◊◊◊◊◊ ◊◊◊◊◊ ◊◊◊◊◊	616,200
◊◊◊◊◊ ◊◊◊◊◊ ◊◊◊◊◊ ◊◊◊◊◊ ◊◊◊◊◊ ◊◊◊◊◊ ◊◊◊◊◊ ◊◊◊◊◊ ◊◊◊◊◊ ◊◊◊◊◊	616,250
◊◊◊◊◊ ◊◊◊◊◊ ◊◊◊◊◊ ◊◊◊◊◊ ◊◊◊◊◊ ◊◊◊◊◊ ◊◊◊◊◊ ◊◊◊◊◊ ◊◊◊◊◊ ◊◊◊◊◊	616,300
◊◊◊◊◊ ◊◊◊◊◊ ◊◊◊◊◊ ◊◊◊◊◊ ◊◊◊◊◊ ◊◊◊◊◊ ◊◊◊◊◊ ◊◊◊◊◊ ◊◊◊◊◊ ◊◊◊◊◊	616,350
◊◊◊◊◊ ◊◊◊◊◊ ◊◊◊◊◊ ◊◊◊◊◊ ◊◊◊◊◊ ◊◊◊◊◊ ◊◊◊◊◊ ◊◊◊◊◊ ◊◊◊◊◊ ◊◊◊◊◊	616,400
◊◊◊◊◊ ◊◊◊◊◊ ◊◊◊◊◊ ◊◊◊◊◊ ◊◊◊◊◊ ◊◊◊◊◊ ◊◊◊◊◊ ◊◊◊◊◊ ◊◊◊◊◊ ◊◊◊◊◊	616,450
◊◊◊◊◊ ◊◊◊◊◊ ◊◊◊◊◊ ◊◊◊◊◊ ◊◊◊◊◊ ◊◊◊◊◊ ◊◊◊◊◊ ◊◊◊◊◊ ◊◊◊◊◊ ◊◊◊◊◊	616,500
◊◊◊◊◊ ◊◊◊◊◊ ◊◊◊◊◊ ◊◊◊◊◊ ◊◊◊◊◊ ◊◊◊◊◊ ◊◊◊◊◊ ◊◊◊◊◊ ◊◊◊◊◊ ◊◊◊◊◊	616,550
◊◊◊◊◊ ◊◊◊◊◊ ◊◊◊◊◊ ◊◊◊◊◊ ◊◊◊◊◊ ◊◊◊◊◊ ◊◊◊◊◊ ◊◊◊◊◊ ◊◊◊◊◊ ◊◊◊◊◊	616,600
◊◊◊◊◊ ◊◊◊◊◊ ◊◊◊◊◊ ◊◊◊◊◊ ◊◊◊◊◊ ◊◊◊◊◊ ◊◊◊◊◊ ◊◊◊◊◊ ◊◊◊◊◊ ◊◊◊◊◊	616,650
◊◊◊◊◊ ◊◊◊◊◊ ◊◊◊◊◊ ◊◊◊◊◊ ◊◊◊◊◊ ◊◊◊◊◊ ◊◊◊◊◊ ◊◊◊◊◊ ◊◊◊◊◊ ◊◊◊◊◊	616,700
◊◊◊◊◊ ◊◊◊◊◊ ◊◊◊◊◊ ◊◊◊◊◊ ◊◊◊◊◊ ◊◊◊◊◊ ◊◊◊◊◊ ◊◊◊◊◊ ◊◊◊◊◊ ◊◊◊◊◊	616,750
◊◊◊◊◊ ◊◊◊◊◊ ◊◊◊◊◊ ◊◊◊◊◊ ◊◊◊◊◊ ◊◊◊◊◊ ◊◊◊◊◊ ◊◊◊◊◊ ◊◊◊◊◊ ◊◊◊◊◊	616,800
◊◊◊◊◊ ◊◊◊◊◊ ◊◊◊◊◊ ◊◊◊◊◊ ◊◊◊◊◊ ◊◊◊◊◊ ◊◊◊◊◊ ◊◊◊◊◊ ◊◊◊◊◊ ◊◊◊◊◊	616,850
◊◊◊◊◊ ◊◊◊◊◊ ◊◊◊◊◊ ◊◊◊◊◊ ◊◊◊◊◊ ◊◊◊◊◊ ◊◊◊◊◊ ◊◊◊◊◊ ◊◊◊◊◊ ◊◊◊◊◊	616,900
◊◊◊◊◊ ◊◊◊◊◊ ◊◊◊◊◊ ◊◊◊◊◊ ◊◊◊◊◊ ◊◊◊◊◊ ◊◊◊◊◊ ◊◊◊◊◊ ◊◊◊◊◊ ◊◊◊◊◊	616,950
◊◊◊◊◊ ◊◊◊◊◊ ◊◊◊◊◊ ◊◊◊◊◊ ◊◊◊◊◊ ◊◊◊◊◊ ◊◊◊◊◊ ◊◊◊◊◊ ◊◊◊◊◊ ◊◊◊◊◊	**617,000**
◊◊◊◊◊ ◊◊◊◊◊ ◊◊◊◊◊ ◊◊◊◊◊ ◊◊◊◊◊ ◊◊◊◊◊ ◊◊◊◊◊ ◊◊◊◊◊ ◊◊◊◊◊ ◊◊◊◊◊	617,050
◊◊◊◊◊ ◊◊◊◊◊ ◊◊◊◊◊ ◊◊◊◊◊ ◊◊◊◊◊ ◊◊◊◊◊ ◊◊◊◊◊ ◊◊◊◊◊ ◊◊◊◊◊ ◊◊◊◊◊	617,100
◊◊◊◊◊ ◊◊◊◊◊ ◊◊◊◊◊ ◊◊◊◊◊ ◊◊◊◊◊ ◊◊◊◊◊ ◊◊◊◊◊ ◊◊◊◊◊ ◊◊◊◊◊ ◊◊◊◊◊	617,150
◊◊◊◊◊ ◊◊◊◊◊ ◊◊◊◊◊ ◊◊◊◊◊ ◊◊◊◊◊ ◊◊◊◊◊ ◊◊◊◊◊ ◊◊◊◊◊ ◊◊◊◊◊ ◊◊◊◊◊	617,200
◊◊◊◊◊ ◊◊◊◊◊ ◊◊◊◊◊ ◊◊◊◊◊ ◊◊◊◊◊ ◊◊◊◊◊ ◊◊◊◊◊ ◊◊◊◊◊ ◊◊◊◊◊ ◊◊◊◊◊	617,250
◊◊◊◊◊ ◊◊◊◊◊ ◊◊◊◊◊ ◊◊◊◊◊ ◊◊◊◊◊ ◊◊◊◊◊ ◊◊◊◊◊ ◊◊◊◊◊ ◊◊◊◊◊ ◊◊◊◊◊	617,300
◊◊◊◊◊ ◊◊◊◊◊ ◊◊◊◊◊ ◊◊◊◊◊ ◊◊◊◊◊ ◊◊◊◊◊ ◊◊◊◊◊ ◊◊◊◊◊ ◊◊◊◊◊ ◊◊◊◊◊	617,350
◊◊◊◊◊ ◊◊◊◊◊ ◊◊◊◊◊ ◊◊◊◊◊ ◊◊◊◊◊ ◊◊◊◊◊ ◊◊◊◊◊ ◊◊◊◊◊ ◊◊◊◊◊ ◊◊◊◊◊	617,400
◊◊◊◊◊ ◊◊◊◊◊ ◊◊◊◊◊ ◊◊◊◊◊ ◊◊◊◊◊ ◊◊◊◊◊ ◊◊◊◊◊ ◊◊◊◊◊ ◊◊◊◊◊ ◊◊◊◊◊	617,450
◊◊◊◊◊ ◊◊◊◊◊ ◊◊◊◊◊ ◊◊◊◊◊ ◊◊◊◊◊ ◊◊◊◊◊ ◊◊◊◊◊ ◊◊◊◊◊ ◊◊◊◊◊ ◊◊◊◊◊	617,500
◊◊◊◊◊ ◊◊◊◊◊ ◊◊◊◊◊ ◊◊◊◊◊ ◊◊◊◊◊ ◊◊◊◊◊ ◊◊◊◊◊ ◊◊◊◊◊ ◊◊◊◊◊ ◊◊◊◊◊	617,550
◊◊◊◊◊ ◊◊◊◊◊ ◊◊◊◊◊ ◊◊◊◊◊ ◊◊◊◊◊ ◊◊◊◊◊ ◊◊◊◊◊ ◊◊◊◊◊ ◊◊◊◊◊ ◊◊◊◊◊	617,600
◊◊◊◊◊ ◊◊◊◊◊ ◊◊◊◊◊ ◊◊◊◊◊ ◊◊◊◊◊ ◊◊◊◊◊ ◊◊◊◊◊ ◊◊◊◊◊ ◊◊◊◊◊ ◊◊◊◊◊	617,650
◊◊◊◊◊ ◊◊◊◊◊ ◊◊◊◊◊ ◊◊◊◊◊ ◊◊◊◊◊ ◊◊◊◊◊ ◊◊◊◊◊ ◊◊◊◊◊ ◊◊◊◊◊ ◊◊◊◊◊	617,700
◊◊◊◊◊ ◊◊◊◊◊ ◊◊◊◊◊ ◊◊◊◊◊ ◊◊◊◊◊ ◊◊◊◊◊ ◊◊◊◊◊ ◊◊◊◊◊ ◊◊◊◊◊ ◊◊◊◊◊	617,750
◊◊◊◊◊ ◊◊◊◊◊ ◊◊◊◊◊ ◊◊◊◊◊ ◊◊◊◊◊ ◊◊◊◊◊ ◊◊◊◊◊ ◊◊◊◊◊ ◊◊◊◊◊ ◊◊◊◊◊	617,800
◊◊◊◊◊ ◊◊◊◊◊ ◊◊◊◊◊ ◊◊◊◊◊ ◊◊◊◊◊ ◊◊◊◊◊ ◊◊◊◊◊ ◊◊◊◊◊ ◊◊◊◊◊ ◊◊◊◊◊	617,850
◊◊◊◊◊ ◊◊◊◊◊ ◊◊◊◊◊ ◊◊◊◊◊ ◊◊◊◊◊ ◊◊◊◊◊ ◊◊◊◊◊ ◊◊◊◊◊ ◊◊◊◊◊ ◊◊◊◊◊	617,900
◊◊◊◊◊ ◊◊◊◊◊ ◊◊◊◊◊ ◊◊◊◊◊ ◊◊◊◊◊ ◊◊◊◊◊ ◊◊◊◊◊ ◊◊◊◊◊ ◊◊◊◊◊ ◊◊◊◊◊	617,950
◊◊◊◊◊ ◊◊◊◊◊ ◊◊◊◊◊ ◊◊◊◊◊ ◊◊◊◊◊ ◊◊◊◊◊ ◊◊◊◊◊ ◊◊◊◊◊ ◊◊◊◊◊ ◊◊◊◊◊	**618,000**

The Number 618,001 to the Number 620,000

Start		Notes

		Notes
	618,050	
	618,100	
	618,150	
	618,200	
	618,250	
	618,300	
	618,350	
	618,400	
	618,450	
	618,500	
	618,550	
	618,600	
	618,650	
	618,700	
	618,750	
	618,800	
	618,850	
	618,900	
	618,950	
	619,000	
	619,050	
	619,100	
	619,150	
	619,200	
	619,250	
	619,300	
	619,350	
	619,400	
	619,450	
	619,500	
	619,550	
	619,600	
	619,650	
	619,700	
	619,750	
	619,800	
	619,850	
	619,900	
	619,950	
	620,000	

The Number 620,001 to the Number 622,000

Start   Notes

	620,050
	620,100
	620,150
	620,200
	620,250
	620,300
	620,350
	620,400
	620,450
	620,500
	620,550
	620,600
	620,650
	620,700
	620,750
	620,800
	620,850
	620,900
	620,950
	621,000
	621,050
	621,100
	621,150
	621,200
	621,250
	621,300
	621,350
	621,400
	621,450
	621,500
	621,550
	621,600
	621,650
	621,700
	621,750
	621,800
	621,850
	621,900
	621,950
	622,000

The Number 622,001 to the Number 624,000

Start		Notes

		Notes
	622,050	
	622,100	
	622,150	
	622,200	
	622,250	
	622,300	
	622,350	
	622,400	
	622,450	
	622,500	
	622,550	
	622,600	
	622,650	
	622,700	
	622,750	
	622,800	
	622,850	
	622,900	
	622,950	
	623,000	
	623,050	
	623,100	
	623,150	
	623,200	
	623,250	
	623,300	
	623,350	
	623,400	
	623,450	
	623,500	
	623,550	
	623,600	
	623,650	
	623,700	
	623,750	
	623,800	
	623,850	
	623,900	
	623,950	
	624,000	

The Number 624,001 to the Number 626,000

Start | | Notes

	624,050
	624,100
	624,150
	624,200
	624,250
	624,300
	624,350
	624,400
	624,450
	624,500
	624,550
	624,600
	624,650
	624,700
	624,750
	624,800
	624,850
	624,900
	624,950
	625,000
	625,050
	625,100
	625,150
	625,200
	625,250
	625,300
	625,350
	625,400
	625,450
	625,500
	625,550
	625,600
	625,650
	625,700
	625,750
	625,800
	625,850
	625,900
	625,950
	626,000

346

The Number 626,001 to the Number 628,000

Start	Notes
626,050	
626,100	
626,150	
626,200	
626,250	
626,300	
626,350	
626,400	
626,450	
626,500	
626,550	
626,600	
626,650	
626,700	
626,750	
626,800	
626,850	
626,900	
626,950	
627,000	
627,050	
627,100	
627,150	
627,200	
627,250	
627,300	
627,350	
627,400	
627,450	
627,500	
627,550	
627,600	
627,650	
627,700	
627,750	
627,800	
627,850	
627,900	
627,950	
628,000	

The Number 628,001 to the Number 630,000

Start | | Notes

	Number	Notes
◇◇◇◇◇ ◇◇◇◇◇ ◇◇◇◇◇ ◇◇◇◇◇ ◇◇◇◇◇ ◇◇◇◇◇ ◇◇◇◇◇ ◇◇◇◇◇ ◇◇◇◇◇ ◇◇◇◇◇	628,050	
◇◇◇◇◇ ◇◇◇◇◇ ◇◇◇◇◇ ◇◇◇◇◇ ◇◇◇◇◇ ◇◇◇◇◇ ◇◇◇◇◇ ◇◇◇◇◇ ◇◇◇◇◇ ◇◇◇◇◇	628,100	
◇◇◇◇◇ ◇◇◇◇◇ ◇◇◇◇◇ ◇◇◇◇◇ ◇◇◇◇◇ ◇◇◇◇◇ ◇◇◇◇◇ ◇◇◇◇◇ ◇◇◇◇◇ ◇◇◇◇◇	628,150	
◇◇◇◇◇ ◇◇◇◇◇ ◇◇◇◇◇ ◇◇◇◇◇ ◇◇◇◇◇ ◇◇◇◇◇ ◇◇◇◇◇ ◇◇◇◇◇ ◇◇◇◇◇ ◇◇◇◇◇	628,200	
◇◇◇◇◇ ◇◇◇◇◇ ◇◇◇◇◇ ◇◇◇◇◇ ◇◇◇◇◇ ◇◇◇◇◇ ◇◇◇◇◇ ◇◇◇◇◇ ◇◇◇◇◇ ◇◇◇◇◇	628,250	
◇◇◇◇◇ ◇◇◇◇◇ ◇◇◇◇◇ ◇◇◇◇◇ ◇◇◇◇◇ ◇◇◇◇◇ ◇◇◇◇◇ ◇◇◇◇◇ ◇◇◇◇◇ ◇◇◇◇◇	628,300	
◇◇◇◇◇ ◇◇◇◇◇ ◇◇◇◇◇ ◇◇◇◇◇ ◇◇◇◇◇ ◇◇◇◇◇ ◇◇◇◇◇ ◇◇◇◇◇ ◇◇◇◇◇ ◇◇◇◇◇	628,350	
◇◇◇◇◇ ◇◇◇◇◇ ◇◇◇◇◇ ◇◇◇◇◇ ◇◇◇◇◇ ◇◇◇◇◇ ◇◇◇◇◇ ◇◇◇◇◇ ◇◇◇◇◇ ◇◇◇◇◇	628,400	
◇◇◇◇◇ ◇◇◇◇◇ ◇◇◇◇◇ ◇◇◇◇◇ ◇◇◇◇◇ ◇◇◇◇◇ ◇◇◇◇◇ ◇◇◇◇◇ ◇◇◇◇◇ ◇◇◇◇◇	628,450	
◇◇◇◇◇ ◇◇◇◇◇ ◇◇◇◇◇ ◇◇◇◇◇ ◇◇◇◇◇ ◇◇◇◇◇ ◇◇◇◇◇ ◇◇◇◇◇ ◇◇◇◇◇ ◇◇◇◇◇	628,500	
◇◇◇◇◇ ◇◇◇◇◇ ◇◇◇◇◇ ◇◇◇◇◇ ◇◇◇◇◇ ◇◇◇◇◇ ◇◇◇◇◇ ◇◇◇◇◇ ◇◇◇◇◇ ◇◇◇◇◇	628,550	
◇◇◇◇◇ ◇◇◇◇◇ ◇◇◇◇◇ ◇◇◇◇◇ ◇◇◇◇◇ ◇◇◇◇◇ ◇◇◇◇◇ ◇◇◇◇◇ ◇◇◇◇◇ ◇◇◇◇◇	628,600	
◇◇◇◇◇ ◇◇◇◇◇ ◇◇◇◇◇ ◇◇◇◇◇ ◇◇◇◇◇ ◇◇◇◇◇ ◇◇◇◇◇ ◇◇◇◇◇ ◇◇◇◇◇ ◇◇◇◇◇	628,650	
◇◇◇◇◇ ◇◇◇◇◇ ◇◇◇◇◇ ◇◇◇◇◇ ◇◇◇◇◇ ◇◇◇◇◇ ◇◇◇◇◇ ◇◇◇◇◇ ◇◇◇◇◇ ◇◇◇◇◇	628,700	
◇◇◇◇◇ ◇◇◇◇◇ ◇◇◇◇◇ ◇◇◇◇◇ ◇◇◇◇◇ ◇◇◇◇◇ ◇◇◇◇◇ ◇◇◇◇◇ ◇◇◇◇◇ ◇◇◇◇◇	628,750	
◇◇◇◇◇ ◇◇◇◇◇ ◇◇◇◇◇ ◇◇◇◇◇ ◇◇◇◇◇ ◇◇◇◇◇ ◇◇◇◇◇ ◇◇◇◇◇ ◇◇◇◇◇ ◇◇◇◇◇	628,800	
◇◇◇◇◇ ◇◇◇◇◇ ◇◇◇◇◇ ◇◇◇◇◇ ◇◇◇◇◇ ◇◇◇◇◇ ◇◇◇◇◇ ◇◇◇◇◇ ◇◇◇◇◇ ◇◇◇◇◇	628,850	
◇◇◇◇◇ ◇◇◇◇◇ ◇◇◇◇◇ ◇◇◇◇◇ ◇◇◇◇◇ ◇◇◇◇◇ ◇◇◇◇◇ ◇◇◇◇◇ ◇◇◇◇◇ ◇◇◇◇◇	628,900	
◇◇◇◇◇ ◇◇◇◇◇ ◇◇◇◇◇ ◇◇◇◇◇ ◇◇◇◇◇ ◇◇◇◇◇ ◇◇◇◇◇ ◇◇◇◇◇ ◇◇◇◇◇ ◇◇◇◇◇	628,950	
◇◇◇◇◇ ◇◇◇◇◇ ◇◇◇◇◇ ◇◇◇◇◇ ◇◇◇◇◇ ◇◇◇◇◇ ◇◇◇◇◇ ◇◇◇◇◇ ◇◇◇◇◇ ◇◇◇◇◇	**629,000**	
◇◇◇◇◇ ◇◇◇◇◇ ◇◇◇◇◇ ◇◇◇◇◇ ◇◇◇◇◇ ◇◇◇◇◇ ◇◇◇◇◇ ◇◇◇◇◇ ◇◇◇◇◇ ◇◇◇◇◇	629,050	
◇◇◇◇◇ ◇◇◇◇◇ ◇◇◇◇◇ ◇◇◇◇◇ ◇◇◇◇◇ ◇◇◇◇◇ ◇◇◇◇◇ ◇◇◇◇◇ ◇◇◇◇◇ ◇◇◇◇◇	629,100	
◇◇◇◇◇ ◇◇◇◇◇ ◇◇◇◇◇ ◇◇◇◇◇ ◇◇◇◇◇ ◇◇◇◇◇ ◇◇◇◇◇ ◇◇◇◇◇ ◇◇◇◇◇ ◇◇◇◇◇	629,150	
◇◇◇◇◇ ◇◇◇◇◇ ◇◇◇◇◇ ◇◇◇◇◇ ◇◇◇◇◇ ◇◇◇◇◇ ◇◇◇◇◇ ◇◇◇◇◇ ◇◇◇◇◇ ◇◇◇◇◇	629,200	
◇◇◇◇◇ ◇◇◇◇◇ ◇◇◇◇◇ ◇◇◇◇◇ ◇◇◇◇◇ ◇◇◇◇◇ ◇◇◇◇◇ ◇◇◇◇◇ ◇◇◇◇◇ ◇◇◇◇◇	629,250	
◇◇◇◇◇ ◇◇◇◇◇ ◇◇◇◇◇ ◇◇◇◇◇ ◇◇◇◇◇ ◇◇◇◇◇ ◇◇◇◇◇ ◇◇◇◇◇ ◇◇◇◇◇ ◇◇◇◇◇	629,300	
◇◇◇◇◇ ◇◇◇◇◇ ◇◇◇◇◇ ◇◇◇◇◇ ◇◇◇◇◇ ◇◇◇◇◇ ◇◇◇◇◇ ◇◇◇◇◇ ◇◇◇◇◇ ◇◇◇◇◇	629,350	
◇◇◇◇◇ ◇◇◇◇◇ ◇◇◇◇◇ ◇◇◇◇◇ ◇◇◇◇◇ ◇◇◇◇◇ ◇◇◇◇◇ ◇◇◇◇◇ ◇◇◇◇◇ ◇◇◇◇◇	629,400	
◇◇◇◇◇ ◇◇◇◇◇ ◇◇◇◇◇ ◇◇◇◇◇ ◇◇◇◇◇ ◇◇◇◇◇ ◇◇◇◇◇ ◇◇◇◇◇ ◇◇◇◇◇ ◇◇◇◇◇	629,450	
◇◇◇◇◇ ◇◇◇◇◇ ◇◇◇◇◇ ◇◇◇◇◇ ◇◇◇◇◇ ◇◇◇◇◇ ◇◇◇◇◇ ◇◇◇◇◇ ◇◇◇◇◇ ◇◇◇◇◇	629,500	
◇◇◇◇◇ ◇◇◇◇◇ ◇◇◇◇◇ ◇◇◇◇◇ ◇◇◇◇◇ ◇◇◇◇◇ ◇◇◇◇◇ ◇◇◇◇◇ ◇◇◇◇◇ ◇◇◇◇◇	629,550	
◇◇◇◇◇ ◇◇◇◇◇ ◇◇◇◇◇ ◇◇◇◇◇ ◇◇◇◇◇ ◇◇◇◇◇ ◇◇◇◇◇ ◇◇◇◇◇ ◇◇◇◇◇ ◇◇◇◇◇	629,600	
◇◇◇◇◇ ◇◇◇◇◇ ◇◇◇◇◇ ◇◇◇◇◇ ◇◇◇◇◇ ◇◇◇◇◇ ◇◇◇◇◇ ◇◇◇◇◇ ◇◇◇◇◇ ◇◇◇◇◇	629,650	
◇◇◇◇◇ ◇◇◇◇◇ ◇◇◇◇◇ ◇◇◇◇◇ ◇◇◇◇◇ ◇◇◇◇◇ ◇◇◇◇◇ ◇◇◇◇◇ ◇◇◇◇◇ ◇◇◇◇◇	629,700	
◇◇◇◇◇ ◇◇◇◇◇ ◇◇◇◇◇ ◇◇◇◇◇ ◇◇◇◇◇ ◇◇◇◇◇ ◇◇◇◇◇ ◇◇◇◇◇ ◇◇◇◇◇ ◇◇◇◇◇	629,750	
◇◇◇◇◇ ◇◇◇◇◇ ◇◇◇◇◇ ◇◇◇◇◇ ◇◇◇◇◇ ◇◇◇◇◇ ◇◇◇◇◇ ◇◇◇◇◇ ◇◇◇◇◇ ◇◇◇◇◇	629,800	
◇◇◇◇◇ ◇◇◇◇◇ ◇◇◇◇◇ ◇◇◇◇◇ ◇◇◇◇◇ ◇◇◇◇◇ ◇◇◇◇◇ ◇◇◇◇◇ ◇◇◇◇◇ ◇◇◇◇◇	629,850	
◇◇◇◇◇ ◇◇◇◇◇ ◇◇◇◇◇ ◇◇◇◇◇ ◇◇◇◇◇ ◇◇◇◇◇ ◇◇◇◇◇ ◇◇◇◇◇ ◇◇◇◇◇ ◇◇◇◇◇	629,900	
◇◇◇◇◇ ◇◇◇◇◇ ◇◇◇◇◇ ◇◇◇◇◇ ◇◇◇◇◇ ◇◇◇◇◇ ◇◇◇◇◇ ◇◇◇◇◇ ◇◇◇◇◇ ◇◇◇◇◇	629,950	
◇◇◇◇◇ ◇◇◇◇◇ ◇◇◇◇◇ ◇◇◇◇◇ ◇◇◇◇◇ ◇◇◇◇◇ ◇◇◇◇◇ ◇◇◇◇◇ ◇◇◇◇◇ ◇◇◇◇◇	**630,000**	

The Number 630,001 to the Number 632,000

Start		Notes
	630,050	
	630,100	
	630,150	
	630,200	
	630,250	
	630,300	
	630,350	
	630,400	
	630,450	
	630,500	
	630,550	
	630,600	
	630,650	
	630,700	
	630,750	
	630,800	
	630,850	
	630,900	
	630,950	
	631,000	
	631,050	
	631,100	
	631,150	
	631,200	
	631,250	
	631,300	
	631,350	
	631,400	
	631,450	
	631,500	
	631,550	
	631,600	
	631,650	
	631,700	
	631,750	
	631,800	
	631,850	
	631,900	
	631,950	
	632,000	

The Number 632,001 to the Number 634,000

Start		Notes

	632,050
	632,100
	632,150
	632,200
	632,250
	632,300
	632,350
	632,400
	632,450
	632,500
	632,550
	632,600
	632,650
	632,700
	632,750
	632,800
	632,850
	632,900
	632,950
	633,000
	633,050
	633,100
	633,150
	633,200
	633,250
	633,300
	633,350
	633,400
	633,450
	633,500
	633,550
	633,600
	633,650
	633,700
	633,750
	633,800
	633,850
	633,900
	633,950
	634,000

The Number 634,001 to the Number 636,000

Start		Notes

634,050
634,100
634,150
634,200
634,250
634,300
634,350
634,400
634,450
634,500
634,550
634,600
634,650
634,700
634,750
634,800
634,850
634,900
634,950
635,000
635,050
635,100
635,150
635,200
635,250
635,300
635,350
635,400
635,450
635,500
635,550
635,600
635,650
635,700
635,750
635,800
635,850
635,900
635,950
636,000

The Number 636,001 to the Number 638,000

Start

Notes

	Notes
636,050	
636,100	
636,150	
636,200	
636,250	
636,300	
636,350	
636,400	
636,450	
636,500	
636,550	
636,600	
636,650	
636,700	
636,750	
636,800	
636,850	
636,900	
636,950	
637,000	
637,050	
637,100	
637,150	
637,200	
637,250	
637,300	
637,350	
637,400	
637,450	
637,500	
637,550	
637,600	
637,650	
637,700	
637,750	
637,800	
637,850	
637,900	
637,950	
638,000	

The Number 638,001 to the Number 640,000

Start	Notes
638,050	
638,100	
638,150	
638,200	
638,250	
638,300	
638,350	
638,400	
638,450	
638,500	
638,550	
638,600	
638,650	
638,700	
638,750	
638,800	
638,850	
638,900	
638,950	
639,000	
639,050	
639,100	
639,150	
639,200	
639,250	
639,300	
639,350	
639,400	
639,450	
639,500	
639,550	
639,600	
639,650	
639,700	
639,750	
639,800	
639,850	
639,900	
639,950	
640,000	

The Number 640,001 to the Number 642,000

Start		Notes
◊◊◊◊◊ ...	640,050
	640,100
	640,150
	640,200
	640,250
	640,300
	640,350
	640,400
	640,450
	640,500
	640,550
	640,600
	640,650
	640,700
	640,750
	640,800
	640,850
	640,900
	640,950
	641,000
	641,050
	641,100
	641,150
	641,200
	641,250
	641,300
	641,350
	641,400
	641,450
	641,500
	641,550
	641,600
	641,650
	641,700
	641,750
	641,800
	641,850
	641,900
	641,950
	642,000

The Number 642,001 to the Number 644,000

Start		Notes
642,050
642,100
642,150
642,200
642,250
642,300
642,350
642,400
642,450
642,500
642,550
642,600
642,650
642,700
642,750
642,800
642,850
642,900
642,950
643,000
643,050
643,100
643,150
643,200
643,250
643,300
643,350
643,400
643,450
643,500
643,550
643,600
643,650
643,700
643,750
643,800
643,850
643,900
643,950
644,000

The Number 644,001 to the Number 646,000

Start

	Number	Notes
◇◇◇◇◇ ◇◇◇◇◇ ◇◇◇◇◇ ◇◇◇◇◇ ◇◇◇◇◇ ◇◇◇◇◇ ◇◇◇◇◇ ◇◇◇◇◇ ◇◇◇◇◇ ◇◇◇◇◇	644,050	
◇◇◇◇◇ ◇◇◇◇◇ ◇◇◇◇◇ ◇◇◇◇◇ ◇◇◇◇◇ ◇◇◇◇◇ ◇◇◇◇◇ ◇◇◇◇◇ ◇◇◇◇◇ ◇◇◇◇◇	644,100	
◇◇◇◇◇ ◇◇◇◇◇ ◇◇◇◇◇ ◇◇◇◇◇ ◇◇◇◇◇ ◇◇◇◇◇ ◇◇◇◇◇ ◇◇◇◇◇ ◇◇◇◇◇ ◇◇◇◇◇	644,150	
◇◇◇◇◇ ◇◇◇◇◇ ◇◇◇◇◇ ◇◇◇◇◇ ◇◇◇◇◇ ◇◇◇◇◇ ◇◇◇◇◇ ◇◇◇◇◇ ◇◇◇◇◇ ◇◇◇◇◇	644,200	
◇◇◇◇◇ ◇◇◇◇◇ ◇◇◇◇◇ ◇◇◇◇◇ ◇◇◇◇◇ ◇◇◇◇◇ ◇◇◇◇◇ ◇◇◇◇◇ ◇◇◇◇◇ ◇◇◇◇◇	644,250	
◇◇◇◇◇ ◇◇◇◇◇ ◇◇◇◇◇ ◇◇◇◇◇ ◇◇◇◇◇ ◇◇◇◇◇ ◇◇◇◇◇ ◇◇◇◇◇ ◇◇◇◇◇ ◇◇◇◇◇	644,300	
◇◇◇◇◇ ◇◇◇◇◇ ◇◇◇◇◇ ◇◇◇◇◇ ◇◇◇◇◇ ◇◇◇◇◇ ◇◇◇◇◇ ◇◇◇◇◇ ◇◇◇◇◇ ◇◇◇◇◇	644,350	
◇◇◇◇◇ ◇◇◇◇◇ ◇◇◇◇◇ ◇◇◇◇◇ ◇◇◇◇◇ ◇◇◇◇◇ ◇◇◇◇◇ ◇◇◇◇◇ ◇◇◇◇◇ ◇◇◇◇◇	644,400	
◇◇◇◇◇ ◇◇◇◇◇ ◇◇◇◇◇ ◇◇◇◇◇ ◇◇◇◇◇ ◇◇◇◇◇ ◇◇◇◇◇ ◇◇◇◇◇ ◇◇◇◇◇ ◇◇◇◇◇	644,450	
◇◇◇◇◇ ◇◇◇◇◇ ◇◇◇◇◇ ◇◇◇◇◇ ◇◇◇◇◇ ◇◇◇◇◇ ◇◇◇◇◇ ◇◇◇◇◇ ◇◇◇◇◇ ◇◇◇◇◇	644,500	
◇◇◇◇◇ ◇◇◇◇◇ ◇◇◇◇◇ ◇◇◇◇◇ ◇◇◇◇◇ ◇◇◇◇◇ ◇◇◇◇◇ ◇◇◇◇◇ ◇◇◇◇◇ ◇◇◇◇◇	644,550	
◇◇◇◇◇ ◇◇◇◇◇ ◇◇◇◇◇ ◇◇◇◇◇ ◇◇◇◇◇ ◇◇◇◇◇ ◇◇◇◇◇ ◇◇◇◇◇ ◇◇◇◇◇ ◇◇◇◇◇	644,600	
◇◇◇◇◇ ◇◇◇◇◇ ◇◇◇◇◇ ◇◇◇◇◇ ◇◇◇◇◇ ◇◇◇◇◇ ◇◇◇◇◇ ◇◇◇◇◇ ◇◇◇◇◇ ◇◇◇◇◇	644,650	
◇◇◇◇◇ ◇◇◇◇◇ ◇◇◇◇◇ ◇◇◇◇◇ ◇◇◇◇◇ ◇◇◇◇◇ ◇◇◇◇◇ ◇◇◇◇◇ ◇◇◇◇◇ ◇◇◇◇◇	644,700	
◇◇◇◇◇ ◇◇◇◇◇ ◇◇◇◇◇ ◇◇◇◇◇ ◇◇◇◇◇ ◇◇◇◇◇ ◇◇◇◇◇ ◇◇◇◇◇ ◇◇◇◇◇ ◇◇◇◇◇	644,750	
◇◇◇◇◇ ◇◇◇◇◇ ◇◇◇◇◇ ◇◇◇◇◇ ◇◇◇◇◇ ◇◇◇◇◇ ◇◇◇◇◇ ◇◇◇◇◇ ◇◇◇◇◇ ◇◇◇◇◇	644,800	
◇◇◇◇◇ ◇◇◇◇◇ ◇◇◇◇◇ ◇◇◇◇◇ ◇◇◇◇◇ ◇◇◇◇◇ ◇◇◇◇◇ ◇◇◇◇◇ ◇◇◇◇◇ ◇◇◇◇◇	644,850	
◇◇◇◇◇ ◇◇◇◇◇ ◇◇◇◇◇ ◇◇◇◇◇ ◇◇◇◇◇ ◇◇◇◇◇ ◇◇◇◇◇ ◇◇◇◇◇ ◇◇◇◇◇ ◇◇◇◇◇	644,900	
◇◇◇◇◇ ◇◇◇◇◇ ◇◇◇◇◇ ◇◇◇◇◇ ◇◇◇◇◇ ◇◇◇◇◇ ◇◇◇◇◇ ◇◇◇◇◇ ◇◇◇◇◇ ◇◇◇◇◇	644,950	
◇◇◇◇◇ ◇◇◇◇◇ ◇◇◇◇◇ ◇◇◇◇◇ ◇◇◇◇◇ ◇◇◇◇◇ ◇◇◇◇◇ ◇◇◇◇◇ ◇◇◇◇◇ ◇◇◇◇◇	**645,000**	
◇◇◇◇◇ ◇◇◇◇◇ ◇◇◇◇◇ ◇◇◇◇◇ ◇◇◇◇◇ ◇◇◇◇◇ ◇◇◇◇◇ ◇◇◇◇◇ ◇◇◇◇◇ ◇◇◇◇◇	645,050	
◇◇◇◇◇ ◇◇◇◇◇ ◇◇◇◇◇ ◇◇◇◇◇ ◇◇◇◇◇ ◇◇◇◇◇ ◇◇◇◇◇ ◇◇◇◇◇ ◇◇◇◇◇ ◇◇◇◇◇	645,100	
◇◇◇◇◇ ◇◇◇◇◇ ◇◇◇◇◇ ◇◇◇◇◇ ◇◇◇◇◇ ◇◇◇◇◇ ◇◇◇◇◇ ◇◇◇◇◇ ◇◇◇◇◇ ◇◇◇◇◇	645,150	
◇◇◇◇◇ ◇◇◇◇◇ ◇◇◇◇◇ ◇◇◇◇◇ ◇◇◇◇◇ ◇◇◇◇◇ ◇◇◇◇◇ ◇◇◇◇◇ ◇◇◇◇◇ ◇◇◇◇◇	645,200	
◇◇◇◇◇ ◇◇◇◇◇ ◇◇◇◇◇ ◇◇◇◇◇ ◇◇◇◇◇ ◇◇◇◇◇ ◇◇◇◇◇ ◇◇◇◇◇ ◇◇◇◇◇ ◇◇◇◇◇	645,250	
◇◇◇◇◇ ◇◇◇◇◇ ◇◇◇◇◇ ◇◇◇◇◇ ◇◇◇◇◇ ◇◇◇◇◇ ◇◇◇◇◇ ◇◇◇◇◇ ◇◇◇◇◇ ◇◇◇◇◇	645,300	
◇◇◇◇◇ ◇◇◇◇◇ ◇◇◇◇◇ ◇◇◇◇◇ ◇◇◇◇◇ ◇◇◇◇◇ ◇◇◇◇◇ ◇◇◇◇◇ ◇◇◇◇◇ ◇◇◇◇◇	645,350	
◇◇◇◇◇ ◇◇◇◇◇ ◇◇◇◇◇ ◇◇◇◇◇ ◇◇◇◇◇ ◇◇◇◇◇ ◇◇◇◇◇ ◇◇◇◇◇ ◇◇◇◇◇ ◇◇◇◇◇	645,400	
◇◇◇◇◇ ◇◇◇◇◇ ◇◇◇◇◇ ◇◇◇◇◇ ◇◇◇◇◇ ◇◇◇◇◇ ◇◇◇◇◇ ◇◇◇◇◇ ◇◇◇◇◇ ◇◇◇◇◇	645,450	
◇◇◇◇◇ ◇◇◇◇◇ ◇◇◇◇◇ ◇◇◇◇◇ ◇◇◇◇◇ ◇◇◇◇◇ ◇◇◇◇◇ ◇◇◇◇◇ ◇◇◇◇◇ ◇◇◇◇◇	645,500	
◇◇◇◇◇ ◇◇◇◇◇ ◇◇◇◇◇ ◇◇◇◇◇ ◇◇◇◇◇ ◇◇◇◇◇ ◇◇◇◇◇ ◇◇◇◇◇ ◇◇◇◇◇ ◇◇◇◇◇	645,550	
◇◇◇◇◇ ◇◇◇◇◇ ◇◇◇◇◇ ◇◇◇◇◇ ◇◇◇◇◇ ◇◇◇◇◇ ◇◇◇◇◇ ◇◇◇◇◇ ◇◇◇◇◇ ◇◇◇◇◇	645,600	
◇◇◇◇◇ ◇◇◇◇◇ ◇◇◇◇◇ ◇◇◇◇◇ ◇◇◇◇◇ ◇◇◇◇◇ ◇◇◇◇◇ ◇◇◇◇◇ ◇◇◇◇◇ ◇◇◇◇◇	645,650	
◇◇◇◇◇ ◇◇◇◇◇ ◇◇◇◇◇ ◇◇◇◇◇ ◇◇◇◇◇ ◇◇◇◇◇ ◇◇◇◇◇ ◇◇◇◇◇ ◇◇◇◇◇ ◇◇◇◇◇	645,700	
◇◇◇◇◇ ◇◇◇◇◇ ◇◇◇◇◇ ◇◇◇◇◇ ◇◇◇◇◇ ◇◇◇◇◇ ◇◇◇◇◇ ◇◇◇◇◇ ◇◇◇◇◇ ◇◇◇◇◇	645,750	
◇◇◇◇◇ ◇◇◇◇◇ ◇◇◇◇◇ ◇◇◇◇◇ ◇◇◇◇◇ ◇◇◇◇◇ ◇◇◇◇◇ ◇◇◇◇◇ ◇◇◇◇◇ ◇◇◇◇◇	645,800	
◇◇◇◇◇ ◇◇◇◇◇ ◇◇◇◇◇ ◇◇◇◇◇ ◇◇◇◇◇ ◇◇◇◇◇ ◇◇◇◇◇ ◇◇◇◇◇ ◇◇◇◇◇ ◇◇◇◇◇	645,850	
◇◇◇◇◇ ◇◇◇◇◇ ◇◇◇◇◇ ◇◇◇◇◇ ◇◇◇◇◇ ◇◇◇◇◇ ◇◇◇◇◇ ◇◇◇◇◇ ◇◇◇◇◇ ◇◇◇◇◇	645,900	
◇◇◇◇◇ ◇◇◇◇◇ ◇◇◇◇◇ ◇◇◇◇◇ ◇◇◇◇◇ ◇◇◇◇◇ ◇◇◇◇◇ ◇◇◇◇◇ ◇◇◇◇◇ ◇◇◇◇◇	645,950	
◇◇◇◇◇ ◇◇◇◇◇ ◇◇◇◇◇ ◇◇◇◇◇ ◇◇◇◇◇ ◇◇◇◇◇ ◇◇◇◇◇ ◇◇◇◇◇ ◇◇◇◇◇ ◇◇◇◇◇	**646,000**	

The Number 646,001 to the Number 648,000

Start   Notes

	646,050
	646,100
	646,150
	646,200
	646,250
	646,300
	646,350
	646,400
	646,450
	646,500
	646,550
	646,600
	646,650
	646,700
	646,750
	646,800
	646,850
	646,900
	646,950
	647,000
	647,050
	647,100
	647,150
	647,200
	647,250
	647,300
	647,350
	647,400
	647,450
	647,500
	647,550
	647,600
	647,650
	647,700
	647,750
	647,800
	647,850
	647,900
	647,950
	648,000

◇ The Number 648,001 to the Number 650,000

Start   Notes

	648,050
	648,100
	648,150
	648,200
	648,250
	648,300
	648,350
	648,400
	648,450
	648,500
	648,550
	648,600
	648,650
	648,700
	648,750
	648,800
	648,850
	648,900
	648,950
	649,000
	649,050
	649,100
	649,150
	649,200
	649,250
	649,300
	649,350
	649,400
	649,450
	649,500
	649,550
	649,600
	649,650
	649,700
	649,750
	649,800
	649,850
	649,900
	649,950
	650,000

The Number 650,001 to the Number 652,000

Start

Notes

◇◇◇◇◇ ◇◇◇◇◇ ◇◇◇◇◇ ◇◇◇◇◇ ◇◇◇◇◇ ◇◇◇◇◇ ◇◇◇◇◇ ◇◇◇◇◇ ◇◇◇◇◇ ◇◇◇◇◇	650,050
◇◇◇◇◇ ◇◇◇◇◇ ◇◇◇◇◇ ◇◇◇◇◇ ◇◇◇◇◇ ◇◇◇◇◇ ◇◇◇◇◇ ◇◇◇◇◇ ◇◇◇◇◇ ◇◇◇◇◇	650,100
◇◇◇◇◇ ◇◇◇◇◇ ◇◇◇◇◇ ◇◇◇◇◇ ◇◇◇◇◇ ◇◇◇◇◇ ◇◇◇◇◇ ◇◇◇◇◇ ◇◇◇◇◇ ◇◇◇◇◇	650,150
◇◇◇◇◇ ◇◇◇◇◇ ◇◇◇◇◇ ◇◇◇◇◇ ◇◇◇◇◇ ◇◇◇◇◇ ◇◇◇◇◇ ◇◇◇◇◇ ◇◇◇◇◇ ◇◇◇◇◇	650,200
◇◇◇◇◇ ◇◇◇◇◇ ◇◇◇◇◇ ◇◇◇◇◇ ◇◇◇◇◇ ◇◇◇◇◇ ◇◇◇◇◇ ◇◇◇◇◇ ◇◇◇◇◇ ◇◇◇◇◇	650,250
◇◇◇◇◇ ◇◇◇◇◇ ◇◇◇◇◇ ◇◇◇◇◇ ◇◇◇◇◇ ◇◇◇◇◇ ◇◇◇◇◇ ◇◇◇◇◇ ◇◇◇◇◇ ◇◇◇◇◇	650,300
◇◇◇◇◇ ◇◇◇◇◇ ◇◇◇◇◇ ◇◇◇◇◇ ◇◇◇◇◇ ◇◇◇◇◇ ◇◇◇◇◇ ◇◇◇◇◇ ◇◇◇◇◇ ◇◇◇◇◇	650,350
◇◇◇◇◇ ◇◇◇◇◇ ◇◇◇◇◇ ◇◇◇◇◇ ◇◇◇◇◇ ◇◇◇◇◇ ◇◇◇◇◇ ◇◇◇◇◇ ◇◇◇◇◇ ◇◇◇◇◇	650,400
◇◇◇◇◇ ◇◇◇◇◇ ◇◇◇◇◇ ◇◇◇◇◇ ◇◇◇◇◇ ◇◇◇◇◇ ◇◇◇◇◇ ◇◇◇◇◇ ◇◇◇◇◇ ◇◇◇◇◇	650,450
◇◇◇◇◇ ◇◇◇◇◇ ◇◇◇◇◇ ◇◇◇◇◇ ◇◇◇◇◇ ◇◇◇◇◇ ◇◇◇◇◇ ◇◇◇◇◇ ◇◇◇◇◇ ◇◇◇◇◇	650,500
◇◇◇◇◇ ◇◇◇◇◇ ◇◇◇◇◇ ◇◇◇◇◇ ◇◇◇◇◇ ◇◇◇◇◇ ◇◇◇◇◇ ◇◇◇◇◇ ◇◇◇◇◇ ◇◇◇◇◇	650,550
◇◇◇◇◇ ◇◇◇◇◇ ◇◇◇◇◇ ◇◇◇◇◇ ◇◇◇◇◇ ◇◇◇◇◇ ◇◇◇◇◇ ◇◇◇◇◇ ◇◇◇◇◇ ◇◇◇◇◇	650,600
◇◇◇◇◇ ◇◇◇◇◇ ◇◇◇◇◇ ◇◇◇◇◇ ◇◇◇◇◇ ◇◇◇◇◇ ◇◇◇◇◇ ◇◇◇◇◇ ◇◇◇◇◇ ◇◇◇◇◇	650,650
◇◇◇◇◇ ◇◇◇◇◇ ◇◇◇◇◇ ◇◇◇◇◇ ◇◇◇◇◇ ◇◇◇◇◇ ◇◇◇◇◇ ◇◇◇◇◇ ◇◇◇◇◇ ◇◇◇◇◇	650,700
◇◇◇◇◇ ◇◇◇◇◇ ◇◇◇◇◇ ◇◇◇◇◇ ◇◇◇◇◇ ◇◇◇◇◇ ◇◇◇◇◇ ◇◇◇◇◇ ◇◇◇◇◇ ◇◇◇◇◇	650,750
◇◇◇◇◇ ◇◇◇◇◇ ◇◇◇◇◇ ◇◇◇◇◇ ◇◇◇◇◇ ◇◇◇◇◇ ◇◇◇◇◇ ◇◇◇◇◇ ◇◇◇◇◇ ◇◇◇◇◇	650,800
◇◇◇◇◇ ◇◇◇◇◇ ◇◇◇◇◇ ◇◇◇◇◇ ◇◇◇◇◇ ◇◇◇◇◇ ◇◇◇◇◇ ◇◇◇◇◇ ◇◇◇◇◇ ◇◇◇◇◇	650,850
◇◇◇◇◇ ◇◇◇◇◇ ◇◇◇◇◇ ◇◇◇◇◇ ◇◇◇◇◇ ◇◇◇◇◇ ◇◇◇◇◇ ◇◇◇◇◇ ◇◇◇◇◇ ◇◇◇◇◇	650,900
◇◇◇◇◇ ◇◇◇◇◇ ◇◇◇◇◇ ◇◇◇◇◇ ◇◇◇◇◇ ◇◇◇◇◇ ◇◇◇◇◇ ◇◇◇◇◇ ◇◇◇◇◇ ◇◇◇◇◇	650,950
◇◇◇◇◇ ◇◇◇◇◇ ◇◇◇◇◇ ◇◇◇◇◇ ◇◇◇◇◇ ◇◇◇◇◇ ◇◇◇◇◇ ◇◇◇◇◇ ◇◇◇◇◇ ◇◇◇◇◇	**651,000**
◇◇◇◇◇ ◇◇◇◇◇ ◇◇◇◇◇ ◇◇◇◇◇ ◇◇◇◇◇ ◇◇◇◇◇ ◇◇◇◇◇ ◇◇◇◇◇ ◇◇◇◇◇ ◇◇◇◇◇	651,050
◇◇◇◇◇ ◇◇◇◇◇ ◇◇◇◇◇ ◇◇◇◇◇ ◇◇◇◇◇ ◇◇◇◇◇ ◇◇◇◇◇ ◇◇◇◇◇ ◇◇◇◇◇ ◇◇◇◇◇	651,100
◇◇◇◇◇ ◇◇◇◇◇ ◇◇◇◇◇ ◇◇◇◇◇ ◇◇◇◇◇ ◇◇◇◇◇ ◇◇◇◇◇ ◇◇◇◇◇ ◇◇◇◇◇ ◇◇◇◇◇	651,150
◇◇◇◇◇ ◇◇◇◇◇ ◇◇◇◇◇ ◇◇◇◇◇ ◇◇◇◇◇ ◇◇◇◇◇ ◇◇◇◇◇ ◇◇◇◇◇ ◇◇◇◇◇ ◇◇◇◇◇	651,200
◇◇◇◇◇ ◇◇◇◇◇ ◇◇◇◇◇ ◇◇◇◇◇ ◇◇◇◇◇ ◇◇◇◇◇ ◇◇◇◇◇ ◇◇◇◇◇ ◇◇◇◇◇ ◇◇◇◇◇	651,250
◇◇◇◇◇ ◇◇◇◇◇ ◇◇◇◇◇ ◇◇◇◇◇ ◇◇◇◇◇ ◇◇◇◇◇ ◇◇◇◇◇ ◇◇◇◇◇ ◇◇◇◇◇ ◇◇◇◇◇	651,300
◇◇◇◇◇ ◇◇◇◇◇ ◇◇◇◇◇ ◇◇◇◇◇ ◇◇◇◇◇ ◇◇◇◇◇ ◇◇◇◇◇ ◇◇◇◇◇ ◇◇◇◇◇ ◇◇◇◇◇	651,350
◇◇◇◇◇ ◇◇◇◇◇ ◇◇◇◇◇ ◇◇◇◇◇ ◇◇◇◇◇ ◇◇◇◇◇ ◇◇◇◇◇ ◇◇◇◇◇ ◇◇◇◇◇ ◇◇◇◇◇	651,400
◇◇◇◇◇ ◇◇◇◇◇ ◇◇◇◇◇ ◇◇◇◇◇ ◇◇◇◇◇ ◇◇◇◇◇ ◇◇◇◇◇ ◇◇◇◇◇ ◇◇◇◇◇ ◇◇◇◇◇	651,450
◇◇◇◇◇ ◇◇◇◇◇ ◇◇◇◇◇ ◇◇◇◇◇ ◇◇◇◇◇ ◇◇◇◇◇ ◇◇◇◇◇ ◇◇◇◇◇ ◇◇◇◇◇ ◇◇◇◇◇	651,500
◇◇◇◇◇ ◇◇◇◇◇ ◇◇◇◇◇ ◇◇◇◇◇ ◇◇◇◇◇ ◇◇◇◇◇ ◇◇◇◇◇ ◇◇◇◇◇ ◇◇◇◇◇ ◇◇◇◇◇	651,550
◇◇◇◇◇ ◇◇◇◇◇ ◇◇◇◇◇ ◇◇◇◇◇ ◇◇◇◇◇ ◇◇◇◇◇ ◇◇◇◇◇ ◇◇◇◇◇ ◇◇◇◇◇ ◇◇◇◇◇	651,600
◇◇◇◇◇ ◇◇◇◇◇ ◇◇◇◇◇ ◇◇◇◇◇ ◇◇◇◇◇ ◇◇◇◇◇ ◇◇◇◇◇ ◇◇◇◇◇ ◇◇◇◇◇ ◇◇◇◇◇	651,650
◇◇◇◇◇ ◇◇◇◇◇ ◇◇◇◇◇ ◇◇◇◇◇ ◇◇◇◇◇ ◇◇◇◇◇ ◇◇◇◇◇ ◇◇◇◇◇ ◇◇◇◇◇ ◇◇◇◇◇	651,700
◇◇◇◇◇ ◇◇◇◇◇ ◇◇◇◇◇ ◇◇◇◇◇ ◇◇◇◇◇ ◇◇◇◇◇ ◇◇◇◇◇ ◇◇◇◇◇ ◇◇◇◇◇ ◇◇◇◇◇	651,750
◇◇◇◇◇ ◇◇◇◇◇ ◇◇◇◇◇ ◇◇◇◇◇ ◇◇◇◇◇ ◇◇◇◇◇ ◇◇◇◇◇ ◇◇◇◇◇ ◇◇◇◇◇ ◇◇◇◇◇	651,800
◇◇◇◇◇ ◇◇◇◇◇ ◇◇◇◇◇ ◇◇◇◇◇ ◇◇◇◇◇ ◇◇◇◇◇ ◇◇◇◇◇ ◇◇◇◇◇ ◇◇◇◇◇ ◇◇◇◇◇	651,850
◇◇◇◇◇ ◇◇◇◇◇ ◇◇◇◇◇ ◇◇◇◇◇ ◇◇◇◇◇ ◇◇◇◇◇ ◇◇◇◇◇ ◇◇◇◇◇ ◇◇◇◇◇ ◇◇◇◇◇	651,900
◇◇◇◇◇ ◇◇◇◇◇ ◇◇◇◇◇ ◇◇◇◇◇ ◇◇◇◇◇ ◇◇◇◇◇ ◇◇◇◇◇ ◇◇◇◇◇ ◇◇◇◇◇ ◇◇◇◇◇	651,950
◇◇◇◇◇ ◇◇◇◇◇ ◇◇◇◇◇ ◇◇◇◇◇ ◇◇◇◇◇ ◇◇◇◇◇ ◇◇◇◇◇ ◇◇◇◇◇ ◇◇◇◇◇ ◇◇◇◇◇	**652,000**

The Number 652,001 to the Number 654,000

Start		Notes

652,050	
652,100	
652,150	
652,200	
652,250	
652,300	
652,350	
652,400	
652,450	
652,500	
652,550	
652,600	
652,650	
652,700	
652,750	
652,800	
652,850	
652,900	
652,950	
653,000	
653,050	
653,100	
653,150	
653,200	
653,250	
653,300	
653,350	
653,400	
653,450	
653,500	
653,550	
653,600	
653,650	
653,700	
653,750	
653,800	
653,850	
653,900	
653,950	
654,000	

The Number 654,001 to the Number 656,000

Start | | Notes

654,050	
654,100	
654,150	
654,200	
654,250	
654,300	
654,350	
654,400	
654,450	
654,500	
654,550	
654,600	
654,650	
654,700	
654,750	
654,800	
654,850	
654,900	
654,950	
655,000	
655,050	
655,100	
655,150	
655,200	
655,250	
655,300	
655,350	
655,400	
655,450	
655,500	
655,550	
655,600	
655,650	
655,700	
655,750	
655,800	
655,850	
655,900	
655,950	
656,000	

The Number 656,001 to the Number 658,000

Start | | Notes

656,050	
656,100	
656,150	
656,200	
656,250	
656,300	
656,350	
656,400	
656,450	
656,500	
656,550	
656,600	
656,650	
656,700	
656,750	
656,800	
656,850	
656,900	
656,950	
657,000	
657,050	
657,100	
657,150	
657,200	
657,250	
657,300	
657,350	
657,400	
657,450	
657,500	
657,550	
657,600	
657,650	
657,700	
657,750	
657,800	
657,850	
657,900	
657,950	
658,000	

The Number 658,001 to the Number 660,000

Start		Notes
	658,050	
	658,100	
	658,150	
	658,200	
	658,250	
	658,300	
	658,350	
	658,400	
	658,450	
	658,500	
	658,550	
	658,600	
	658,650	
	658,700	
	658,750	
	658,800	
	658,850	
	658,900	
	658,950	
	659,000	
	659,050	
	659,100	
	659,150	
	659,200	
	659,250	
	659,300	
	659,350	
	659,400	
	659,450	
	659,500	
	659,550	
	659,600	
	659,650	
	659,700	
	659,750	
	659,800	
	659,850	
	659,900	
	659,950	
	660,000	

The Number 660,001 to the Number 662,000

Start

Notes

660,050
660,100
660,150
660,200
660,250
660,300
660,350
660,400
660,450
660,500
660,550
660,600
660,650
660,700
660,750
660,800
660,850
660,900
660,950
661,000
661,050
661,100
661,150
661,200
661,250
661,300
661,350
661,400
661,450
661,500
661,550
661,600
661,650
661,700
661,750
661,800
661,850
661,900
661,950
662,000

The Number 662,001 to the Number 664,000

Start		Notes
◊◊◊◊◊ ◊◊◊◊◊ ◊◊◊◊◊ ◊◊◊◊◊ ◊◊◊◊◊ ◊◊◊◊◊ ◊◊◊◊◊ ◊◊◊◊◊ ◊◊◊◊◊ ◊◊◊◊◊	662,050	
◊◊◊◊◊ ◊◊◊◊◊ ◊◊◊◊◊ ◊◊◊◊◊ ◊◊◊◊◊ ◊◊◊◊◊ ◊◊◊◊◊ ◊◊◊◊◊ ◊◊◊◊◊ ◊◊◊◊◊	662,100	
◊◊◊◊◊ ◊◊◊◊◊ ◊◊◊◊◊ ◊◊◊◊◊ ◊◊◊◊◊ ◊◊◊◊◊ ◊◊◊◊◊ ◊◊◊◊◊ ◊◊◊◊◊ ◊◊◊◊◊	662,150	
◊◊◊◊◊ ◊◊◊◊◊ ◊◊◊◊◊ ◊◊◊◊◊ ◊◊◊◊◊ ◊◊◊◊◊ ◊◊◊◊◊ ◊◊◊◊◊ ◊◊◊◊◊ ◊◊◊◊◊	662,200	
◊◊◊◊◊ ◊◊◊◊◊ ◊◊◊◊◊ ◊◊◊◊◊ ◊◊◊◊◊ ◊◊◊◊◊ ◊◊◊◊◊ ◊◊◊◊◊ ◊◊◊◊◊ ◊◊◊◊◊	662,250	
◊◊◊◊◊ ◊◊◊◊◊ ◊◊◊◊◊ ◊◊◊◊◊ ◊◊◊◊◊ ◊◊◊◊◊ ◊◊◊◊◊ ◊◊◊◊◊ ◊◊◊◊◊ ◊◊◊◊◊	662,300	
◊◊◊◊◊ ◊◊◊◊◊ ◊◊◊◊◊ ◊◊◊◊◊ ◊◊◊◊◊ ◊◊◊◊◊ ◊◊◊◊◊ ◊◊◊◊◊ ◊◊◊◊◊ ◊◊◊◊◊	662,350	
◊◊◊◊◊ ◊◊◊◊◊ ◊◊◊◊◊ ◊◊◊◊◊ ◊◊◊◊◊ ◊◊◊◊◊ ◊◊◊◊◊ ◊◊◊◊◊ ◊◊◊◊◊ ◊◊◊◊◊	662,400	
◊◊◊◊◊ ◊◊◊◊◊ ◊◊◊◊◊ ◊◊◊◊◊ ◊◊◊◊◊ ◊◊◊◊◊ ◊◊◊◊◊ ◊◊◊◊◊ ◊◊◊◊◊ ◊◊◊◊◊	662,450	
◊◊◊◊◊ ◊◊◊◊◊ ◊◊◊◊◊ ◊◊◊◊◊ ◊◊◊◊◊ ◊◊◊◊◊ ◊◊◊◊◊ ◊◊◊◊◊ ◊◊◊◊◊ ◊◊◊◊◊	662,500	
◊◊◊◊◊ ◊◊◊◊◊ ◊◊◊◊◊ ◊◊◊◊◊ ◊◊◊◊◊ ◊◊◊◊◊ ◊◊◊◊◊ ◊◊◊◊◊ ◊◊◊◊◊ ◊◊◊◊◊	662,550	
◊◊◊◊◊ ◊◊◊◊◊ ◊◊◊◊◊ ◊◊◊◊◊ ◊◊◊◊◊ ◊◊◊◊◊ ◊◊◊◊◊ ◊◊◊◊◊ ◊◊◊◊◊ ◊◊◊◊◊	662,600	
◊◊◊◊◊ ◊◊◊◊◊ ◊◊◊◊◊ ◊◊◊◊◊ ◊◊◊◊◊ ◊◊◊◊◊ ◊◊◊◊◊ ◊◊◊◊◊ ◊◊◊◊◊ ◊◊◊◊◊	662,650	
◊◊◊◊◊ ◊◊◊◊◊ ◊◊◊◊◊ ◊◊◊◊◊ ◊◊◊◊◊ ◊◊◊◊◊ ◊◊◊◊◊ ◊◊◊◊◊ ◊◊◊◊◊ ◊◊◊◊◊	662,700	
◊◊◊◊◊ ◊◊◊◊◊ ◊◊◊◊◊ ◊◊◊◊◊ ◊◊◊◊◊ ◊◊◊◊◊ ◊◊◊◊◊ ◊◊◊◊◊ ◊◊◊◊◊ ◊◊◊◊◊	662,750	
◊◊◊◊◊ ◊◊◊◊◊ ◊◊◊◊◊ ◊◊◊◊◊ ◊◊◊◊◊ ◊◊◊◊◊ ◊◊◊◊◊ ◊◊◊◊◊ ◊◊◊◊◊ ◊◊◊◊◊	662,800	
◊◊◊◊◊ ◊◊◊◊◊ ◊◊◊◊◊ ◊◊◊◊◊ ◊◊◊◊◊ ◊◊◊◊◊ ◊◊◊◊◊ ◊◊◊◊◊ ◊◊◊◊◊ ◊◊◊◊◊	662,850	
◊◊◊◊◊ ◊◊◊◊◊ ◊◊◊◊◊ ◊◊◊◊◊ ◊◊◊◊◊ ◊◊◊◊◊ ◊◊◊◊◊ ◊◊◊◊◊ ◊◊◊◊◊ ◊◊◊◊◊	662,900	
◊◊◊◊◊ ◊◊◊◊◊ ◊◊◊◊◊ ◊◊◊◊◊ ◊◊◊◊◊ ◊◊◊◊◊ ◊◊◊◊◊ ◊◊◊◊◊ ◊◊◊◊◊ ◊◊◊◊◊	662,950	
◊◊◊◊◊ ◊◊◊◊◊ ◊◊◊◊◊ ◊◊◊◊◊ ◊◊◊◊◊ ◊◊◊◊◊ ◊◊◊◊◊ ◊◊◊◊◊ ◊◊◊◊◊ ◊◊◊◊◊	**663,000**	
◊◊◊◊◊ ◊◊◊◊◊ ◊◊◊◊◊ ◊◊◊◊◊ ◊◊◊◊◊ ◊◊◊◊◊ ◊◊◊◊◊ ◊◊◊◊◊ ◊◊◊◊◊ ◊◊◊◊◊	663,050	
◊◊◊◊◊ ◊◊◊◊◊ ◊◊◊◊◊ ◊◊◊◊◊ ◊◊◊◊◊ ◊◊◊◊◊ ◊◊◊◊◊ ◊◊◊◊◊ ◊◊◊◊◊ ◊◊◊◊◊	663,100	
◊◊◊◊◊ ◊◊◊◊◊ ◊◊◊◊◊ ◊◊◊◊◊ ◊◊◊◊◊ ◊◊◊◊◊ ◊◊◊◊◊ ◊◊◊◊◊ ◊◊◊◊◊ ◊◊◊◊◊	663,150	
◊◊◊◊◊ ◊◊◊◊◊ ◊◊◊◊◊ ◊◊◊◊◊ ◊◊◊◊◊ ◊◊◊◊◊ ◊◊◊◊◊ ◊◊◊◊◊ ◊◊◊◊◊ ◊◊◊◊◊	663,200	
◊◊◊◊◊ ◊◊◊◊◊ ◊◊◊◊◊ ◊◊◊◊◊ ◊◊◊◊◊ ◊◊◊◊◊ ◊◊◊◊◊ ◊◊◊◊◊ ◊◊◊◊◊ ◊◊◊◊◊	663,250	
◊◊◊◊◊ ◊◊◊◊◊ ◊◊◊◊◊ ◊◊◊◊◊ ◊◊◊◊◊ ◊◊◊◊◊ ◊◊◊◊◊ ◊◊◊◊◊ ◊◊◊◊◊ ◊◊◊◊◊	663,300	
◊◊◊◊◊ ◊◊◊◊◊ ◊◊◊◊◊ ◊◊◊◊◊ ◊◊◊◊◊ ◊◊◊◊◊ ◊◊◊◊◊ ◊◊◊◊◊ ◊◊◊◊◊ ◊◊◊◊◊	663,350	
◊◊◊◊◊ ◊◊◊◊◊ ◊◊◊◊◊ ◊◊◊◊◊ ◊◊◊◊◊ ◊◊◊◊◊ ◊◊◊◊◊ ◊◊◊◊◊ ◊◊◊◊◊ ◊◊◊◊◊	663,400	
◊◊◊◊◊ ◊◊◊◊◊ ◊◊◊◊◊ ◊◊◊◊◊ ◊◊◊◊◊ ◊◊◊◊◊ ◊◊◊◊◊ ◊◊◊◊◊ ◊◊◊◊◊ ◊◊◊◊◊	663,450	
◊◊◊◊◊ ◊◊◊◊◊ ◊◊◊◊◊ ◊◊◊◊◊ ◊◊◊◊◊ ◊◊◊◊◊ ◊◊◊◊◊ ◊◊◊◊◊ ◊◊◊◊◊ ◊◊◊◊◊	663,500	
◊◊◊◊◊ ◊◊◊◊◊ ◊◊◊◊◊ ◊◊◊◊◊ ◊◊◊◊◊ ◊◊◊◊◊ ◊◊◊◊◊ ◊◊◊◊◊ ◊◊◊◊◊ ◊◊◊◊◊	663,550	
◊◊◊◊◊ ◊◊◊◊◊ ◊◊◊◊◊ ◊◊◊◊◊ ◊◊◊◊◊ ◊◊◊◊◊ ◊◊◊◊◊ ◊◊◊◊◊ ◊◊◊◊◊ ◊◊◊◊◊	663,600	
◊◊◊◊◊ ◊◊◊◊◊ ◊◊◊◊◊ ◊◊◊◊◊ ◊◊◊◊◊ ◊◊◊◊◊ ◊◊◊◊◊ ◊◊◊◊◊ ◊◊◊◊◊ ◊◊◊◊◊	663,650	
◊◊◊◊◊ ◊◊◊◊◊ ◊◊◊◊◊ ◊◊◊◊◊ ◊◊◊◊◊ ◊◊◊◊◊ ◊◊◊◊◊ ◊◊◊◊◊ ◊◊◊◊◊ ◊◊◊◊◊	663,700	
◊◊◊◊◊ ◊◊◊◊◊ ◊◊◊◊◊ ◊◊◊◊◊ ◊◊◊◊◊ ◊◊◊◊◊ ◊◊◊◊◊ ◊◊◊◊◊ ◊◊◊◊◊ ◊◊◊◊◊	663,750	
◊◊◊◊◊ ◊◊◊◊◊ ◊◊◊◊◊ ◊◊◊◊◊ ◊◊◊◊◊ ◊◊◊◊◊ ◊◊◊◊◊ ◊◊◊◊◊ ◊◊◊◊◊ ◊◊◊◊◊	663,800	
◊◊◊◊◊ ◊◊◊◊◊ ◊◊◊◊◊ ◊◊◊◊◊ ◊◊◊◊◊ ◊◊◊◊◊ ◊◊◊◊◊ ◊◊◊◊◊ ◊◊◊◊◊ ◊◊◊◊◊	663,850	
◊◊◊◊◊ ◊◊◊◊◊ ◊◊◊◊◊ ◊◊◊◊◊ ◊◊◊◊◊ ◊◊◊◊◊ ◊◊◊◊◊ ◊◊◊◊◊ ◊◊◊◊◊ ◊◊◊◊◊	663,900	
◊◊◊◊◊ ◊◊◊◊◊ ◊◊◊◊◊ ◊◊◊◊◊ ◊◊◊◊◊ ◊◊◊◊◊ ◊◊◊◊◊ ◊◊◊◊◊ ◊◊◊◊◊ ◊◊◊◊◊	663,950	
◊◊◊◊◊ ◊◊◊◊◊ ◊◊◊◊◊ ◊◊◊◊◊ ◊◊◊◊◊ ◊◊◊◊◊ ◊◊◊◊◊ ◊◊◊◊◊ ◊◊◊◊◊ ◊◊◊◊◊	**664,000**	

The Number 664,001 to the Number 666,000

Start		Notes
664,050
664,100
664,150
664,200
664,250
664,300
664,350
664,400
664,450
664,500
664,550
664,600
664,650
664,700
664,750
664,800
664,850
664,900
664,950
665,000
665,050
665,100
665,150
665,200
665,250
665,300
665,350
665,400
665,450
665,500
665,550
665,600
665,650
665,700
665,750
665,800
665,850
665,900
665,950
666,000

The Number 666,001 to the Number 668,000

Start	Notes
	666,050
	666,100
	666,150
	666,200
	666,250
	666,300
	666,350
	666,400
	666,450
	666,500
	666,550
	666,600
	666,650
	666,700
	666,750
	666,800
	666,850
	666,900
	666,950
	667,000
	667,050
	667,100
	667,150
	667,200
	667,250
	667,300
	667,350
	667,400
	667,450
	667,500
	667,550
	667,600
	667,650
	667,700
	667,750
	667,800
	667,850
	667,900
	667,950
	668,000

The Number 668,001 to the Number 670,000

Start   Notes

	668,050
	668,100
	668,150
	668,200
	668,250
	668,300
	668,350
	668,400
	668,450
	668,500
	668,550
	668,600
	668,650
	668,700
	668,750
	668,800
	668,850
	668,900
	668,950
	669,000
	669,050
	669,100
	669,150
	669,200
	669,250
	669,300
	669,350
	669,400
	669,450
	669,500
	669,550
	669,600
	669,650
	669,700
	669,750
	669,800
	669,850
	669,900
	669,950
	670,000

The Number 670,001 to the Number 672,000

Start		Notes
	670,050	
	670,100	
	670,150	
	670,200	
	670,250	
	670,300	
	670,350	
	670,400	
	670,450	
	670,500	
	670,550	
	670,600	
	670,650	
	670,700	
	670,750	
	670,800	
	670,850	
	670,900	
	670,950	
	671,000	
	671,050	
	671,100	
	671,150	
	671,200	
	671,250	
	671,300	
	671,350	
	671,400	
	671,450	
	671,500	
	671,550	
	671,600	
	671,650	
	671,700	
	671,750	
	671,800	
	671,850	
	671,900	
	671,950	
	672,000	

The Number 672,001 to the Number 674,000

Start		Notes

| 672,050 |
| 672,100 |
| 672,150 |
| 672,200 |
| 672,250 |
| 672,300 |
| 672,350 |
| 672,400 |
| 672,450 |
| 672,500 |
| 672,550 |
| 672,600 |
| 672,650 |
| 672,700 |
| 672,750 |
| 672,800 |
| 672,850 |
| 672,900 |
| 672,950 |
| **673,000** |
| 673,050 |
| 673,100 |
| 673,150 |
| 673,200 |
| 673,250 |
| 673,300 |
| 673,350 |
| 673,400 |
| 673,450 |
| 673,500 |
| 673,550 |
| 673,600 |
| 673,650 |
| 673,700 |
| 673,750 |
| 673,800 |
| 673,850 |
| 673,900 |
| 673,950 |
| **674,000** |

The Number 674,001 to the Number 676,000

Start		Notes
674,050
674,100
674,150
674,200
674,250
674,300
674,350
674,400
674,450
674,500
674,550
674,600
674,650
674,700
674,750
674,800
674,850
674,900
674,950
675,000
675,050
675,100
675,150
675,200
675,250
675,300
675,350
675,400
675,450
675,500
675,550
675,600
675,650
675,700
675,750
675,800
675,850
675,900
675,950
676,000

The Number 676,001 to the Number 678,000

Start

Notes

	676,050
	676,100
	676,150
	676,200
	676,250
	676,300
	676,350
	676,400
	676,450
	676,500
	676,550
	676,600
	676,650
	676,700
	676,750
	676,800
	676,850
	676,900
	676,950
	677,000
	677,050
	677,100
	677,150
	677,200
	677,250
	677,300
	677,350
	677,400
	677,450
	677,500
	677,550
	677,600
	677,650
	677,700
	677,750
	677,800
	677,850
	677,900
	677,950
	678,000

The Number 678,001 to the Number 680,000

Start

Notes

	678,050
	678,100
	678,150
	678,200
	678,250
	678,300
	678,350
	678,400
	678,450
	678,500
	678,550
	678,600
	678,650
	678,700
	678,750
	678,800
	678,850
	678,900
	678,950
	679,000
	679,050
	679,100
	679,150
	679,200
	679,250
	679,300
	679,350
	679,400
	679,450
	679,500
	679,550
	679,600
	679,650
	679,700
	679,750
	679,800
	679,850
	679,900
	679,950
	680,000

The Number 680,001 to the Number 682,000

Start   **Notes**

680,050
680,100
680,150
680,200
680,250
680,300
680,350
680,400
680,450
680,500
680,550
680,600
680,650
680,700
680,750
680,800
680,850
680,900
680,950
681,000
681,050
681,100
681,150
681,200
681,250
681,300
681,350
681,400
681,450
681,500
681,550
681,600
681,650
681,700
681,750
681,800
681,850
681,900
681,950
682,000

The Number 682,001 to the Number 684,000

Start		Notes
	682,050	
	682,100	
	682,150	
	682,200	
	682,250	
	682,300	
	682,350	
	682,400	
	682,450	
	682,500	
	682,550	
	682,600	
	682,650	
	682,700	
	682,750	
	682,800	
	682,850	
	682,900	
	682,950	
	683,000	
	683,050	
	683,100	
	683,150	
	683,200	
	683,250	
	683,300	
	683,350	
	683,400	
	683,450	
	683,500	
	683,550	
	683,600	
	683,650	
	683,700	
	683,750	
	683,800	
	683,850	
	683,900	
	683,950	
	684,000	

◊ The Number 684,001 to the Number 686,000

Start

Notes

	684,050
	684,100
	684,150
	684,200
	684,250
	684,300
	684,350
	684,400
	684,450
	684,500
	684,550
	684,600
	684,650
	684,700
	684,750
	684,800
	684,850
	684,900
	684,950
	685,000
	685,050
	685,100
	685,150
	685,200
	685,250
	685,300
	685,350
	685,400
	685,450
	685,500
	685,550
	685,600
	685,650
	685,700
	685,750
	685,800
	685,850
	685,900
	685,950
	686,000

The Number 686,001 to the Number 688,000

Start   Notes

	686,050
	686,100
	686,150
	686,200
	686,250
	686,300
	686,350
	686,400
	686,450
	686,500
	686,550
	686,600
	686,650
	686,700
	686,750
	686,800
	686,850
	686,900
	686,950
	687,000
	687,050
	687,100
	687,150
	687,200
	687,250
	687,300
	687,350
	687,400
	687,450
	687,500
	687,550
	687,600
	687,650
	687,700
	687,750
	687,800
	687,850
	687,900
	687,950
	688,000

The Number 688,001 to the Number 690,000

Start   Notes

	688,050
	688,100
	688,150
	688,200
	688,250
	688,300
	688,350
	688,400
	688,450
	688,500
	688,550
	688,600
	688,650
	688,700
	688,750
	688,800
	688,850
	688,900
	688,950
	689,000
	689,050
	689,100
	689,150
	689,200
	689,250
	689,300
	689,350
	689,400
	689,450
	689,500
	689,550
	689,600
	689,650
	689,700
	689,750
	689,800
	689,850
	689,900
	689,950
	690,000

The Number 690,001 to the Number 692,000

Start		Notes
	690,050	
	690,100	
	690,150	
	690,200	
	690,250	
	690,300	
	690,350	
	690,400	
	690,450	
	690,500	
	690,550	
	690,600	
	690,650	
	690,700	
	690,750	
	690,800	
	690,850	
	690,900	
	690,950	
	691,000	
	691,050	
	691,100	
	691,150	
	691,200	
	691,250	
	691,300	
	691,350	
	691,400	
	691,450	
	691,500	
	691,550	
	691,600	
	691,650	
	691,700	
	691,750	
	691,800	
	691,850	
	691,900	
	691,950	
	692,000	

The Number 692,001 to the Number 694,000

Start		Notes
	692,050
	692,100
	692,150
	692,200
	692,250
	692,300
	692,350
	692,400
	692,450
	692,500
	692,550
	692,600
	692,650
	692,700
	692,750
	692,800
	692,850
	692,900
	692,950
	693,000
	693,050
	693,100
	693,150
	693,200
	693,250
	693,300
	693,350
	693,400
	693,450
	693,500
	693,550
	693,600
	693,650
	693,700
	693,750
	693,800
	693,850
	693,900
	693,950
	694,000

The Number 694,001 to the Number 696,000

Start   Notes

694,050	
694,100	
694,150	
694,200	
694,250	
694,300	
694,350	
694,400	
694,450	
694,500	
694,550	
694,600	
694,650	
694,700	
694,750	
694,800	
694,850	
694,900	
694,950	
695,000	
695,050	
695,100	
695,150	
695,200	
695,250	
695,300	
695,350	
695,400	
695,450	
695,500	
695,550	
695,600	
695,650	
695,700	
695,750	
695,800	
695,850	
695,900	
695,950	
696,000	

The Number 696,001 to the Number 698,000

Start		Notes
	696,050	
	696,100	
	696,150	
	696,200	
	696,250	
	696,300	
	696,350	
	696,400	
	696,450	
	696,500	
	696,550	
	696,600	
	696,650	
	696,700	
	696,750	
	696,800	
	696,850	
	696,900	
	696,950	
	697,000	
	697,050	
	697,100	
	697,150	
	697,200	
	697,250	
	697,300	
	697,350	
	697,400	
	697,450	
	697,500	
	697,550	
	697,600	
	697,650	
	697,700	
	697,750	
	697,800	
	697,850	
	697,900	
	697,950	
	698,000	

The Number 698,001 to the Number 700,000

Start		Notes
	698,050	
	698,100	
	698,150	
	698,200	
	698,250	
	698,300	
	698,350	
	698,400	
	698,450	
	698,500	
	698,550	
	698,600	
	698,650	
	698,700	
	698,750	
	698,800	
	698,850	
	698,900	
	698,950	
	699,000	
	699,050	
	699,100	
	699,150	
	699,200	
	699,250	
	699,300	
	699,350	
	699,400	
	699,450	
	699,500	
	699,550	
	699,600	
	699,650	
	699,700	
	699,750	
	699,800	
	699,850	
	699,900	
	699,950	
	700,000	

The Number 700,001 to the Number 702,000

Start		Notes

Number	Notes
700,050	
700,100	
700,150	
700,200	
700,250	
700,300	
700,350	
700,400	
700,450	
700,500	
700,550	
700,600	
700,650	
700,700	
700,750	
700,800	
700,850	
700,900	
700,950	
701,000	
701,050	
701,100	
701,150	
701,200	
701,250	
701,300	
701,350	
701,400	
701,450	
701,500	
701,550	
701,600	
701,650	
701,700	
701,750	
701,800	
701,850	
701,900	
701,950	
702,000	

The Number 702,001 to the Number 704,000

Start
Notes

	702,050
	702,100
	702,150
	702,200
	702,250
	702,300
	702,350
	702,400
	702,450
	702,500
	702,550
	702,600
	702,650
	702,700
	702,750
	702,800
	702,850
	702,900
	702,950
	703,000
	703,050
	703,100
	703,150
	703,200
	703,250
	703,300
	703,350
	703,400
	703,450
	703,500
	703,550
	703,600
	703,650
	703,700
	703,750
	703,800
	703,850
	703,900
	703,950
	704,000

The Number 704,001 to the Number 706,000

Start   Notes

	704,050
	704,100
	704,150
	704,200
	704,250
	704,300
	704,350
	704,400
	704,450
	704,500
	704,550
	704,600
	704,650
	704,700
	704,750
	704,800
	704,850
	704,900
	704,950
	705,000
	705,050
	705,100
	705,150
	705,200
	705,250
	705,300
	705,350
	705,400
	705,450
	705,500
	705,550
	705,600
	705,650
	705,700
	705,750
	705,800
	705,850
	705,900
	705,950
	706,000

The Number 706,001 to the Number 708,000

Start · Notes

	706,050
	706,100
	706,150
	706,200
	706,250
	706,300
	706,350
	706,400
	706,450
	706,500
	706,550
	706,600
	706,650
	706,700
	706,750
	706,800
	706,850
	706,900
	706,950
	707,000
	707,050
	707,100
	707,150
	707,200
	707,250
	707,300
	707,350
	707,400
	707,450
	707,500
	707,550
	707,600
	707,650
	707,700
	707,750
	707,800
	707,850
	707,900
	707,950
	708,000

The Number 708,001 to the Number 710,000

Start   Notes

	708,050
	708,100
	708,150
	708,200
	708,250
	708,300
	708,350
	708,400
	708,450
	708,500
	708,550
	708,600
	708,650
	708,700
	708,750
	708,800
	708,850
	708,900
	708,950
	709,000
	709,050
	709,100
	709,150
	709,200
	709,250
	709,300
	709,350
	709,400
	709,450
	709,500
	709,550
	709,600
	709,650
	709,700
	709,750
	709,800
	709,850
	709,900
	709,950
	710,000

The Number 710,001 to the Number 712,000

Start   Notes

	Number	Notes
	710,050	
	710,100	
	710,150	
	710,200	
	710,250	
	710,300	
	710,350	
	710,400	
	710,450	
	710,500	
	710,550	
	710,600	
	710,650	
	710,700	
	710,750	
	710,800	
	710,850	
	710,900	
	710,950	
	711,000	
	711,050	
	711,100	
	711,150	
	711,200	
	711,250	
	711,300	
	711,350	
	711,400	
	711,450	
	711,500	
	711,550	
	711,600	
	711,650	
	711,700	
	711,750	
	711,800	
	711,850	
	711,900	
	711,950	
	712,000	

The Number 712,001 to the Number 714,000

Start   Notes

	712,050
	712,100
	712,150
	712,200
	712,250
	712,300
	712,350
	712,400
	712,450
	712,500
	712,550
	712,600
	712,650
	712,700
	712,750
	712,800
	712,850
	712,900
	712,950
	713,000
	713,050
	713,100
	713,150
	713,200
	713,250
	713,300
	713,350
	713,400
	713,450
	713,500
	713,550
	713,600
	713,650
	713,700
	713,750
	713,800
	713,850
	713,900
	713,950
	714,000

The Number 714,001 to the Number 716,000

Start	Notes
	714,050
	714,100
	714,150
	714,200
	714,250
	714,300
	714,350
	714,400
	714,450
	714,500
	714,550
	714,600
	714,650
	714,700
	714,750
	714,800
	714,850
	714,900
	714,950
	715,000
	715,050
	715,100
	715,150
	715,200
	715,250
	715,300
	715,350
	715,400
	715,450
	715,500
	715,550
	715,600
	715,650
	715,700
	715,750
	715,800
	715,850
	715,900
	715,950
	716,000

The Number 716,001 to the Number 718,000

Start   Notes

	716,050
	716,100
	716,150
	716,200
	716,250
	716,300
	716,350
	716,400
	716,450
	716,500
	716,550
	716,600
	716,650
	716,700
	716,750
	716,800
	716,850
	716,900
	716,950
	717,000
	717,050
	717,100
	717,150
	717,200
	717,250
	717,300
	717,350
	717,400
	717,450
	717,500
	717,550
	717,600
	717,650
	717,700
	717,750
	717,800
	717,850
	717,900
	717,950
	718,000

The Number 718,001 to the Number 720,000

Start

Notes

	718,050
	718,100
	718,150
	718,200
	718,250
	718,300
	718,350
	718,400
	718,450
	718,500
	718,550
	718,600
	718,650
	718,700
	718,750
	718,800
	718,850
	718,900
	718,950
	719,000
	719,050
	719,100
	719,150
	719,200
	719,250
	719,300
	719,350
	719,400
	719,450
	719,500
	719,550
	719,600
	719,650
	719,700
	719,750
	719,800
	719,850
	719,900
	719,950
	720,000

The Number 720,001 to the Number 722,000

Start		Notes
◊◊◊◊◊ ◊◊◊◊◊ ◊◊◊◊◊ ◊◊◊◊◊ ◊◊◊◊◊ ◊◊◊◊◊ ◊◊◊◊◊ ◊◊◊◊◊ ◊◊◊◊◊ ◊◊◊◊◊	720,050
◊◊◊◊◊ ◊◊◊◊◊ ◊◊◊◊◊ ◊◊◊◊◊ ◊◊◊◊◊ ◊◊◊◊◊ ◊◊◊◊◊ ◊◊◊◊◊ ◊◊◊◊◊ ◊◊◊◊◊	720,100
◊◊◊◊◊ ◊◊◊◊◊ ◊◊◊◊◊ ◊◊◊◊◊ ◊◊◊◊◊ ◊◊◊◊◊ ◊◊◊◊◊ ◊◊◊◊◊ ◊◊◊◊◊ ◊◊◊◊◊	720,150
◊◊◊◊◊ ◊◊◊◊◊ ◊◊◊◊◊ ◊◊◊◊◊ ◊◊◊◊◊ ◊◊◊◊◊ ◊◊◊◊◊ ◊◊◊◊◊ ◊◊◊◊◊ ◊◊◊◊◊	720,200
◊◊◊◊◊ ◊◊◊◊◊ ◊◊◊◊◊ ◊◊◊◊◊ ◊◊◊◊◊ ◊◊◊◊◊ ◊◊◊◊◊ ◊◊◊◊◊ ◊◊◊◊◊ ◊◊◊◊◊	720,250
◊◊◊◊◊ ◊◊◊◊◊ ◊◊◊◊◊ ◊◊◊◊◊ ◊◊◊◊◊ ◊◊◊◊◊ ◊◊◊◊◊ ◊◊◊◊◊ ◊◊◊◊◊ ◊◊◊◊◊	720,300
◊◊◊◊◊ ◊◊◊◊◊ ◊◊◊◊◊ ◊◊◊◊◊ ◊◊◊◊◊ ◊◊◊◊◊ ◊◊◊◊◊ ◊◊◊◊◊ ◊◊◊◊◊ ◊◊◊◊◊	720,350
◊◊◊◊◊ ◊◊◊◊◊ ◊◊◊◊◊ ◊◊◊◊◊ ◊◊◊◊◊ ◊◊◊◊◊ ◊◊◊◊◊ ◊◊◊◊◊ ◊◊◊◊◊ ◊◊◊◊◊	720,400
◊◊◊◊◊ ◊◊◊◊◊ ◊◊◊◊◊ ◊◊◊◊◊ ◊◊◊◊◊ ◊◊◊◊◊ ◊◊◊◊◊ ◊◊◊◊◊ ◊◊◊◊◊ ◊◊◊◊◊	720,450
◊◊◊◊◊ ◊◊◊◊◊ ◊◊◊◊◊ ◊◊◊◊◊ ◊◊◊◊◊ ◊◊◊◊◊ ◊◊◊◊◊ ◊◊◊◊◊ ◊◊◊◊◊ ◊◊◊◊◊	720,500
◊◊◊◊◊ ◊◊◊◊◊ ◊◊◊◊◊ ◊◊◊◊◊ ◊◊◊◊◊ ◊◊◊◊◊ ◊◊◊◊◊ ◊◊◊◊◊ ◊◊◊◊◊ ◊◊◊◊◊	720,550
◊◊◊◊◊ ◊◊◊◊◊ ◊◊◊◊◊ ◊◊◊◊◊ ◊◊◊◊◊ ◊◊◊◊◊ ◊◊◊◊◊ ◊◊◊◊◊ ◊◊◊◊◊ ◊◊◊◊◊	720,600
◊◊◊◊◊ ◊◊◊◊◊ ◊◊◊◊◊ ◊◊◊◊◊ ◊◊◊◊◊ ◊◊◊◊◊ ◊◊◊◊◊ ◊◊◊◊◊ ◊◊◊◊◊ ◊◊◊◊◊	720,650
◊◊◊◊◊ ◊◊◊◊◊ ◊◊◊◊◊ ◊◊◊◊◊ ◊◊◊◊◊ ◊◊◊◊◊ ◊◊◊◊◊ ◊◊◊◊◊ ◊◊◊◊◊ ◊◊◊◊◊	720,700
◊◊◊◊◊ ◊◊◊◊◊ ◊◊◊◊◊ ◊◊◊◊◊ ◊◊◊◊◊ ◊◊◊◊◊ ◊◊◊◊◊ ◊◊◊◊◊ ◊◊◊◊◊ ◊◊◊◊◊	720,750
◊◊◊◊◊ ◊◊◊◊◊ ◊◊◊◊◊ ◊◊◊◊◊ ◊◊◊◊◊ ◊◊◊◊◊ ◊◊◊◊◊ ◊◊◊◊◊ ◊◊◊◊◊ ◊◊◊◊◊	720,800
◊◊◊◊◊ ◊◊◊◊◊ ◊◊◊◊◊ ◊◊◊◊◊ ◊◊◊◊◊ ◊◊◊◊◊ ◊◊◊◊◊ ◊◊◊◊◊ ◊◊◊◊◊ ◊◊◊◊◊	720,850
◊◊◊◊◊ ◊◊◊◊◊ ◊◊◊◊◊ ◊◊◊◊◊ ◊◊◊◊◊ ◊◊◊◊◊ ◊◊◊◊◊ ◊◊◊◊◊ ◊◊◊◊◊ ◊◊◊◊◊	720,900
◊◊◊◊◊ ◊◊◊◊◊ ◊◊◊◊◊ ◊◊◊◊◊ ◊◊◊◊◊ ◊◊◊◊◊ ◊◊◊◊◊ ◊◊◊◊◊ ◊◊◊◊◊ ◊◊◊◊◊	720,950
◊◊◊◊◊ ◊◊◊◊◊ ◊◊◊◊◊ ◊◊◊◊◊ ◊◊◊◊◊ ◊◊◊◊◊ ◊◊◊◊◊ ◊◊◊◊◊ ◊◊◊◊◊ ◊◊◊◊◊	**721,000**
◊◊◊◊◊ ◊◊◊◊◊ ◊◊◊◊◊ ◊◊◊◊◊ ◊◊◊◊◊ ◊◊◊◊◊ ◊◊◊◊◊ ◊◊◊◊◊ ◊◊◊◊◊ ◊◊◊◊◊	721,050
◊◊◊◊◊ ◊◊◊◊◊ ◊◊◊◊◊ ◊◊◊◊◊ ◊◊◊◊◊ ◊◊◊◊◊ ◊◊◊◊◊ ◊◊◊◊◊ ◊◊◊◊◊ ◊◊◊◊◊	721,100
◊◊◊◊◊ ◊◊◊◊◊ ◊◊◊◊◊ ◊◊◊◊◊ ◊◊◊◊◊ ◊◊◊◊◊ ◊◊◊◊◊ ◊◊◊◊◊ ◊◊◊◊◊ ◊◊◊◊◊	721,150
◊◊◊◊◊ ◊◊◊◊◊ ◊◊◊◊◊ ◊◊◊◊◊ ◊◊◊◊◊ ◊◊◊◊◊ ◊◊◊◊◊ ◊◊◊◊◊ ◊◊◊◊◊ ◊◊◊◊◊	721,200
◊◊◊◊◊ ◊◊◊◊◊ ◊◊◊◊◊ ◊◊◊◊◊ ◊◊◊◊◊ ◊◊◊◊◊ ◊◊◊◊◊ ◊◊◊◊◊ ◊◊◊◊◊ ◊◊◊◊◊	721,250
◊◊◊◊◊ ◊◊◊◊◊ ◊◊◊◊◊ ◊◊◊◊◊ ◊◊◊◊◊ ◊◊◊◊◊ ◊◊◊◊◊ ◊◊◊◊◊ ◊◊◊◊◊ ◊◊◊◊◊	721,300
◊◊◊◊◊ ◊◊◊◊◊ ◊◊◊◊◊ ◊◊◊◊◊ ◊◊◊◊◊ ◊◊◊◊◊ ◊◊◊◊◊ ◊◊◊◊◊ ◊◊◊◊◊ ◊◊◊◊◊	721,350
◊◊◊◊◊ ◊◊◊◊◊ ◊◊◊◊◊ ◊◊◊◊◊ ◊◊◊◊◊ ◊◊◊◊◊ ◊◊◊◊◊ ◊◊◊◊◊ ◊◊◊◊◊ ◊◊◊◊◊	721,400
◊◊◊◊◊ ◊◊◊◊◊ ◊◊◊◊◊ ◊◊◊◊◊ ◊◊◊◊◊ ◊◊◊◊◊ ◊◊◊◊◊ ◊◊◊◊◊ ◊◊◊◊◊ ◊◊◊◊◊	721,450
◊◊◊◊◊ ◊◊◊◊◊ ◊◊◊◊◊ ◊◊◊◊◊ ◊◊◊◊◊ ◊◊◊◊◊ ◊◊◊◊◊ ◊◊◊◊◊ ◊◊◊◊◊ ◊◊◊◊◊	721,500
◊◊◊◊◊ ◊◊◊◊◊ ◊◊◊◊◊ ◊◊◊◊◊ ◊◊◊◊◊ ◊◊◊◊◊ ◊◊◊◊◊ ◊◊◊◊◊ ◊◊◊◊◊ ◊◊◊◊◊	721,550
◊◊◊◊◊ ◊◊◊◊◊ ◊◊◊◊◊ ◊◊◊◊◊ ◊◊◊◊◊ ◊◊◊◊◊ ◊◊◊◊◊ ◊◊◊◊◊ ◊◊◊◊◊ ◊◊◊◊◊	721,600
◊◊◊◊◊ ◊◊◊◊◊ ◊◊◊◊◊ ◊◊◊◊◊ ◊◊◊◊◊ ◊◊◊◊◊ ◊◊◊◊◊ ◊◊◊◊◊ ◊◊◊◊◊ ◊◊◊◊◊	721,650
◊◊◊◊◊ ◊◊◊◊◊ ◊◊◊◊◊ ◊◊◊◊◊ ◊◊◊◊◊ ◊◊◊◊◊ ◊◊◊◊◊ ◊◊◊◊◊ ◊◊◊◊◊ ◊◊◊◊◊	721,700
◊◊◊◊◊ ◊◊◊◊◊ ◊◊◊◊◊ ◊◊◊◊◊ ◊◊◊◊◊ ◊◊◊◊◊ ◊◊◊◊◊ ◊◊◊◊◊ ◊◊◊◊◊ ◊◊◊◊◊	721,750
◊◊◊◊◊ ◊◊◊◊◊ ◊◊◊◊◊ ◊◊◊◊◊ ◊◊◊◊◊ ◊◊◊◊◊ ◊◊◊◊◊ ◊◊◊◊◊ ◊◊◊◊◊ ◊◊◊◊◊	721,800
◊◊◊◊◊ ◊◊◊◊◊ ◊◊◊◊◊ ◊◊◊◊◊ ◊◊◊◊◊ ◊◊◊◊◊ ◊◊◊◊◊ ◊◊◊◊◊ ◊◊◊◊◊ ◊◊◊◊◊	721,850
◊◊◊◊◊ ◊◊◊◊◊ ◊◊◊◊◊ ◊◊◊◊◊ ◊◊◊◊◊ ◊◊◊◊◊ ◊◊◊◊◊ ◊◊◊◊◊ ◊◊◊◊◊ ◊◊◊◊◊	721,900
◊◊◊◊◊ ◊◊◊◊◊ ◊◊◊◊◊ ◊◊◊◊◊ ◊◊◊◊◊ ◊◊◊◊◊ ◊◊◊◊◊ ◊◊◊◊◊ ◊◊◊◊◊ ◊◊◊◊◊	721,950
◊◊◊◊◊ ◊◊◊◊◊ ◊◊◊◊◊ ◊◊◊◊◊ ◊◊◊◊◊ ◊◊◊◊◊ ◊◊◊◊◊ ◊◊◊◊◊ ◊◊◊◊◊ ◊◊◊◊◊	**722,000**

The Number 722,001 to the Number 724,000

Start | | Notes

	722,050
	722,100
	722,150
	722,200
	722,250
	722,300
	722,350
	722,400
	722,450
	722,500
	722,550
	722,600
	722,650
	722,700
	722,750
	722,800
	722,850
	722,900
	722,950
	723,000
	723,050
	723,100
	723,150
	723,200
	723,250
	723,300
	723,350
	723,400
	723,450
	723,500
	723,550
	723,600
	723,650
	723,700
	723,750
	723,800
	723,850
	723,900
	723,950
	724,000

The Number 724,001 to the Number 726,000

Start

Notes

	724,050
	724,100
	724,150
	724,200
	724,250
	724,300
	724,350
	724,400
	724,450
	724,500
	724,550
	724,600
	724,650
	724,700
	724,750
	724,800
	724,850
	724,900
	724,950
	725,000
	725,050
	725,100
	725,150
	725,200
	725,250
	725,300
	725,350
	725,400
	725,450
	725,500
	725,550
	725,600
	725,650
	725,700
	725,750
	725,800
	725,850
	725,900
	725,950
	726,000

◊ The Number 726,001 to the Number 728,000

Start | | Notes

	Number	Notes
◊◊◊◊◊ ◊◊◊◊◊ ◊◊◊◊◊ ◊◊◊◊◊ ◊◊◊◊◊ ◊◊◊◊◊ ◊◊◊◊◊ ◊◊◊◊◊ ◊◊◊◊◊ ◊◊◊◊◊	726,050	
◊◊◊◊◊ ◊◊◊◊◊ ◊◊◊◊◊ ◊◊◊◊◊ ◊◊◊◊◊ ◊◊◊◊◊ ◊◊◊◊◊ ◊◊◊◊◊ ◊◊◊◊◊ ◊◊◊◊◊	726,100	
◊◊◊◊◊ ◊◊◊◊◊ ◊◊◊◊◊ ◊◊◊◊◊ ◊◊◊◊◊ ◊◊◊◊◊ ◊◊◊◊◊ ◊◊◊◊◊ ◊◊◊◊◊ ◊◊◊◊◊	726,150	
◊◊◊◊◊ ◊◊◊◊◊ ◊◊◊◊◊ ◊◊◊◊◊ ◊◊◊◊◊ ◊◊◊◊◊ ◊◊◊◊◊ ◊◊◊◊◊ ◊◊◊◊◊ ◊◊◊◊◊	726,200	
◊◊◊◊◊ ◊◊◊◊◊ ◊◊◊◊◊ ◊◊◊◊◊ ◊◊◊◊◊ ◊◊◊◊◊ ◊◊◊◊◊ ◊◊◊◊◊ ◊◊◊◊◊ ◊◊◊◊◊	726,250	
◊◊◊◊◊ ◊◊◊◊◊ ◊◊◊◊◊ ◊◊◊◊◊ ◊◊◊◊◊ ◊◊◊◊◊ ◊◊◊◊◊ ◊◊◊◊◊ ◊◊◊◊◊ ◊◊◊◊◊	726,300	
◊◊◊◊◊ ◊◊◊◊◊ ◊◊◊◊◊ ◊◊◊◊◊ ◊◊◊◊◊ ◊◊◊◊◊ ◊◊◊◊◊ ◊◊◊◊◊ ◊◊◊◊◊ ◊◊◊◊◊	726,350	
◊◊◊◊◊ ◊◊◊◊◊ ◊◊◊◊◊ ◊◊◊◊◊ ◊◊◊◊◊ ◊◊◊◊◊ ◊◊◊◊◊ ◊◊◊◊◊ ◊◊◊◊◊ ◊◊◊◊◊	726,400	
◊◊◊◊◊ ◊◊◊◊◊ ◊◊◊◊◊ ◊◊◊◊◊ ◊◊◊◊◊ ◊◊◊◊◊ ◊◊◊◊◊ ◊◊◊◊◊ ◊◊◊◊◊ ◊◊◊◊◊	726,450	
◊◊◊◊◊ ◊◊◊◊◊ ◊◊◊◊◊ ◊◊◊◊◊ ◊◊◊◊◊ ◊◊◊◊◊ ◊◊◊◊◊ ◊◊◊◊◊ ◊◊◊◊◊ ◊◊◊◊◊	726,500	
◊◊◊◊◊ ◊◊◊◊◊ ◊◊◊◊◊ ◊◊◊◊◊ ◊◊◊◊◊ ◊◊◊◊◊ ◊◊◊◊◊ ◊◊◊◊◊ ◊◊◊◊◊ ◊◊◊◊◊	726,550	
◊◊◊◊◊ ◊◊◊◊◊ ◊◊◊◊◊ ◊◊◊◊◊ ◊◊◊◊◊ ◊◊◊◊◊ ◊◊◊◊◊ ◊◊◊◊◊ ◊◊◊◊◊ ◊◊◊◊◊	726,600	
◊◊◊◊◊ ◊◊◊◊◊ ◊◊◊◊◊ ◊◊◊◊◊ ◊◊◊◊◊ ◊◊◊◊◊ ◊◊◊◊◊ ◊◊◊◊◊ ◊◊◊◊◊ ◊◊◊◊◊	726,650	
◊◊◊◊◊ ◊◊◊◊◊ ◊◊◊◊◊ ◊◊◊◊◊ ◊◊◊◊◊ ◊◊◊◊◊ ◊◊◊◊◊ ◊◊◊◊◊ ◊◊◊◊◊ ◊◊◊◊◊	726,700	
◊◊◊◊◊ ◊◊◊◊◊ ◊◊◊◊◊ ◊◊◊◊◊ ◊◊◊◊◊ ◊◊◊◊◊ ◊◊◊◊◊ ◊◊◊◊◊ ◊◊◊◊◊ ◊◊◊◊◊	726,750	
◊◊◊◊◊ ◊◊◊◊◊ ◊◊◊◊◊ ◊◊◊◊◊ ◊◊◊◊◊ ◊◊◊◊◊ ◊◊◊◊◊ ◊◊◊◊◊ ◊◊◊◊◊ ◊◊◊◊◊	726,800	
◊◊◊◊◊ ◊◊◊◊◊ ◊◊◊◊◊ ◊◊◊◊◊ ◊◊◊◊◊ ◊◊◊◊◊ ◊◊◊◊◊ ◊◊◊◊◊ ◊◊◊◊◊ ◊◊◊◊◊	726,850	
◊◊◊◊◊ ◊◊◊◊◊ ◊◊◊◊◊ ◊◊◊◊◊ ◊◊◊◊◊ ◊◊◊◊◊ ◊◊◊◊◊ ◊◊◊◊◊ ◊◊◊◊◊ ◊◊◊◊◊	726,900	
◊◊◊◊◊ ◊◊◊◊◊ ◊◊◊◊◊ ◊◊◊◊◊ ◊◊◊◊◊ ◊◊◊◊◊ ◊◊◊◊◊ ◊◊◊◊◊ ◊◊◊◊◊ ◊◊◊◊◊	726,950	
◊◊◊◊◊ ◊◊◊◊◊ ◊◊◊◊◊ ◊◊◊◊◊ ◊◊◊◊◊ ◊◊◊◊◊ ◊◊◊◊◊ ◊◊◊◊◊ ◊◊◊◊◊ ◊◊◊◊◊	**727,000**	
◊◊◊◊◊ ◊◊◊◊◊ ◊◊◊◊◊ ◊◊◊◊◊ ◊◊◊◊◊ ◊◊◊◊◊ ◊◊◊◊◊ ◊◊◊◊◊ ◊◊◊◊◊ ◊◊◊◊◊	727,050	
◊◊◊◊◊ ◊◊◊◊◊ ◊◊◊◊◊ ◊◊◊◊◊ ◊◊◊◊◊ ◊◊◊◊◊ ◊◊◊◊◊ ◊◊◊◊◊ ◊◊◊◊◊ ◊◊◊◊◊	727,100	
◊◊◊◊◊ ◊◊◊◊◊ ◊◊◊◊◊ ◊◊◊◊◊ ◊◊◊◊◊ ◊◊◊◊◊ ◊◊◊◊◊ ◊◊◊◊◊ ◊◊◊◊◊ ◊◊◊◊◊	727,150	
◊◊◊◊◊ ◊◊◊◊◊ ◊◊◊◊◊ ◊◊◊◊◊ ◊◊◊◊◊ ◊◊◊◊◊ ◊◊◊◊◊ ◊◊◊◊◊ ◊◊◊◊◊ ◊◊◊◊◊	727,200	
◊◊◊◊◊ ◊◊◊◊◊ ◊◊◊◊◊ ◊◊◊◊◊ ◊◊◊◊◊ ◊◊◊◊◊ ◊◊◊◊◊ ◊◊◊◊◊ ◊◊◊◊◊ ◊◊◊◊◊	727,250	
◊◊◊◊◊ ◊◊◊◊◊ ◊◊◊◊◊ ◊◊◊◊◊ ◊◊◊◊◊ ◊◊◊◊◊ ◊◊◊◊◊ ◊◊◊◊◊ ◊◊◊◊◊ ◊◊◊◊◊	727,300	
◊◊◊◊◊ ◊◊◊◊◊ ◊◊◊◊◊ ◊◊◊◊◊ ◊◊◊◊◊ ◊◊◊◊◊ ◊◊◊◊◊ ◊◊◊◊◊ ◊◊◊◊◊ ◊◊◊◊◊	727,350	
◊◊◊◊◊ ◊◊◊◊◊ ◊◊◊◊◊ ◊◊◊◊◊ ◊◊◊◊◊ ◊◊◊◊◊ ◊◊◊◊◊ ◊◊◊◊◊ ◊◊◊◊◊ ◊◊◊◊◊	727,400	
◊◊◊◊◊ ◊◊◊◊◊ ◊◊◊◊◊ ◊◊◊◊◊ ◊◊◊◊◊ ◊◊◊◊◊ ◊◊◊◊◊ ◊◊◊◊◊ ◊◊◊◊◊ ◊◊◊◊◊	727,450	
◊◊◊◊◊ ◊◊◊◊◊ ◊◊◊◊◊ ◊◊◊◊◊ ◊◊◊◊◊ ◊◊◊◊◊ ◊◊◊◊◊ ◊◊◊◊◊ ◊◊◊◊◊ ◊◊◊◊◊	727,500	
◊◊◊◊◊ ◊◊◊◊◊ ◊◊◊◊◊ ◊◊◊◊◊ ◊◊◊◊◊ ◊◊◊◊◊ ◊◊◊◊◊ ◊◊◊◊◊ ◊◊◊◊◊ ◊◊◊◊◊	727,550	
◊◊◊◊◊ ◊◊◊◊◊ ◊◊◊◊◊ ◊◊◊◊◊ ◊◊◊◊◊ ◊◊◊◊◊ ◊◊◊◊◊ ◊◊◊◊◊ ◊◊◊◊◊ ◊◊◊◊◊	727,600	
◊◊◊◊◊ ◊◊◊◊◊ ◊◊◊◊◊ ◊◊◊◊◊ ◊◊◊◊◊ ◊◊◊◊◊ ◊◊◊◊◊ ◊◊◊◊◊ ◊◊◊◊◊ ◊◊◊◊◊	727,650	
◊◊◊◊◊ ◊◊◊◊◊ ◊◊◊◊◊ ◊◊◊◊◊ ◊◊◊◊◊ ◊◊◊◊◊ ◊◊◊◊◊ ◊◊◊◊◊ ◊◊◊◊◊ ◊◊◊◊◊	727,700	
◊◊◊◊◊ ◊◊◊◊◊ ◊◊◊◊◊ ◊◊◊◊◊ ◊◊◊◊◊ ◊◊◊◊◊ ◊◊◊◊◊ ◊◊◊◊◊ ◊◊◊◊◊ ◊◊◊◊◊	727,750	
◊◊◊◊◊ ◊◊◊◊◊ ◊◊◊◊◊ ◊◊◊◊◊ ◊◊◊◊◊ ◊◊◊◊◊ ◊◊◊◊◊ ◊◊◊◊◊ ◊◊◊◊◊ ◊◊◊◊◊	727,800	
◊◊◊◊◊ ◊◊◊◊◊ ◊◊◊◊◊ ◊◊◊◊◊ ◊◊◊◊◊ ◊◊◊◊◊ ◊◊◊◊◊ ◊◊◊◊◊ ◊◊◊◊◊ ◊◊◊◊◊	727,850	
◊◊◊◊◊ ◊◊◊◊◊ ◊◊◊◊◊ ◊◊◊◊◊ ◊◊◊◊◊ ◊◊◊◊◊ ◊◊◊◊◊ ◊◊◊◊◊ ◊◊◊◊◊ ◊◊◊◊◊	727,900	
◊◊◊◊◊ ◊◊◊◊◊ ◊◊◊◊◊ ◊◊◊◊◊ ◊◊◊◊◊ ◊◊◊◊◊ ◊◊◊◊◊ ◊◊◊◊◊ ◊◊◊◊◊ ◊◊◊◊◊	727,950	
◊◊◊◊◊ ◊◊◊◊◊ ◊◊◊◊◊ ◊◊◊◊◊ ◊◊◊◊◊ ◊◊◊◊◊ ◊◊◊◊◊ ◊◊◊◊◊ ◊◊◊◊◊ ◊◊◊◊◊	**728,000**	

The Number 728,001 to the Number 730,000

Start		Notes
728,050
728,100
728,150
728,200
728,250
728,300
728,350
728,400
728,450
728,500
728,550
728,600
728,650
728,700
728,750
728,800
728,850
728,900
728,950
729,000
729,050
729,100
729,150
729,200
729,250
729,300
729,350
729,400
729,450
729,500
729,550
729,600
729,650
729,700
729,750
729,800
729,850
729,900
729,950
730,000

The Number 730,001 to the Number 732,000

Start		Notes
730,050
730,100
730,150
730,200
730,250
730,300
730,350
730,400
730,450
730,500
730,550
730,600
730,650
730,700
730,750
730,800
730,850
730,900
730,950
731,000
731,050
731,100
731,150
731,200
731,250
731,300
731,350
731,400
731,450
731,500
731,550
731,600
731,650
731,700
731,750
731,800
731,850
731,900
731,950
732,000

The Number 732,001 to the Number 734,000

Start | | Notes

◊◊◊◊◊ ◊◊◊◊◊ ◊◊◊◊◊ ◊◊◊◊◊ ◊◊◊◊◊ ◊◊◊◊◊ ◊◊◊◊◊ ◊◊◊◊◊ ◊◊◊◊◊ ◊◊◊◊◊	732,050
◊◊◊◊◊ ◊◊◊◊◊ ◊◊◊◊◊ ◊◊◊◊◊ ◊◊◊◊◊ ◊◊◊◊◊ ◊◊◊◊◊ ◊◊◊◊◊ ◊◊◊◊◊ ◊◊◊◊◊	732,100
◊◊◊◊◊ ◊◊◊◊◊ ◊◊◊◊◊ ◊◊◊◊◊ ◊◊◊◊◊ ◊◊◊◊◊ ◊◊◊◊◊ ◊◊◊◊◊ ◊◊◊◊◊ ◊◊◊◊◊	732,150
◊◊◊◊◊ ◊◊◊◊◊ ◊◊◊◊◊ ◊◊◊◊◊ ◊◊◊◊◊ ◊◊◊◊◊ ◊◊◊◊◊ ◊◊◊◊◊ ◊◊◊◊◊ ◊◊◊◊◊	732,200
◊◊◊◊◊ ◊◊◊◊◊ ◊◊◊◊◊ ◊◊◊◊◊ ◊◊◊◊◊ ◊◊◊◊◊ ◊◊◊◊◊ ◊◊◊◊◊ ◊◊◊◊◊ ◊◊◊◊◊	732,250
◊◊◊◊◊ ◊◊◊◊◊ ◊◊◊◊◊ ◊◊◊◊◊ ◊◊◊◊◊ ◊◊◊◊◊ ◊◊◊◊◊ ◊◊◊◊◊ ◊◊◊◊◊ ◊◊◊◊◊	732,300
◊◊◊◊◊ ◊◊◊◊◊ ◊◊◊◊◊ ◊◊◊◊◊ ◊◊◊◊◊ ◊◊◊◊◊ ◊◊◊◊◊ ◊◊◊◊◊ ◊◊◊◊◊ ◊◊◊◊◊	732,350
◊◊◊◊◊ ◊◊◊◊◊ ◊◊◊◊◊ ◊◊◊◊◊ ◊◊◊◊◊ ◊◊◊◊◊ ◊◊◊◊◊ ◊◊◊◊◊ ◊◊◊◊◊ ◊◊◊◊◊	732,400
◊◊◊◊◊ ◊◊◊◊◊ ◊◊◊◊◊ ◊◊◊◊◊ ◊◊◊◊◊ ◊◊◊◊◊ ◊◊◊◊◊ ◊◊◊◊◊ ◊◊◊◊◊ ◊◊◊◊◊	732,450
◊◊◊◊◊ ◊◊◊◊◊ ◊◊◊◊◊ ◊◊◊◊◊ ◊◊◊◊◊ ◊◊◊◊◊ ◊◊◊◊◊ ◊◊◊◊◊ ◊◊◊◊◊ ◊◊◊◊◊	732,500
◊◊◊◊◊ ◊◊◊◊◊ ◊◊◊◊◊ ◊◊◊◊◊ ◊◊◊◊◊ ◊◊◊◊◊ ◊◊◊◊◊ ◊◊◊◊◊ ◊◊◊◊◊ ◊◊◊◊◊	732,550
◊◊◊◊◊ ◊◊◊◊◊ ◊◊◊◊◊ ◊◊◊◊◊ ◊◊◊◊◊ ◊◊◊◊◊ ◊◊◊◊◊ ◊◊◊◊◊ ◊◊◊◊◊ ◊◊◊◊◊	732,600
◊◊◊◊◊ ◊◊◊◊◊ ◊◊◊◊◊ ◊◊◊◊◊ ◊◊◊◊◊ ◊◊◊◊◊ ◊◊◊◊◊ ◊◊◊◊◊ ◊◊◊◊◊ ◊◊◊◊◊	732,650
◊◊◊◊◊ ◊◊◊◊◊ ◊◊◊◊◊ ◊◊◊◊◊ ◊◊◊◊◊ ◊◊◊◊◊ ◊◊◊◊◊ ◊◊◊◊◊ ◊◊◊◊◊ ◊◊◊◊◊	732,700
◊◊◊◊◊ ◊◊◊◊◊ ◊◊◊◊◊ ◊◊◊◊◊ ◊◊◊◊◊ ◊◊◊◊◊ ◊◊◊◊◊ ◊◊◊◊◊ ◊◊◊◊◊ ◊◊◊◊◊	732,750
◊◊◊◊◊ ◊◊◊◊◊ ◊◊◊◊◊ ◊◊◊◊◊ ◊◊◊◊◊ ◊◊◊◊◊ ◊◊◊◊◊ ◊◊◊◊◊ ◊◊◊◊◊ ◊◊◊◊◊	732,800
◊◊◊◊◊ ◊◊◊◊◊ ◊◊◊◊◊ ◊◊◊◊◊ ◊◊◊◊◊ ◊◊◊◊◊ ◊◊◊◊◊ ◊◊◊◊◊ ◊◊◊◊◊ ◊◊◊◊◊	732,850
◊◊◊◊◊ ◊◊◊◊◊ ◊◊◊◊◊ ◊◊◊◊◊ ◊◊◊◊◊ ◊◊◊◊◊ ◊◊◊◊◊ ◊◊◊◊◊ ◊◊◊◊◊ ◊◊◊◊◊	732,900
◊◊◊◊◊ ◊◊◊◊◊ ◊◊◊◊◊ ◊◊◊◊◊ ◊◊◊◊◊ ◊◊◊◊◊ ◊◊◊◊◊ ◊◊◊◊◊ ◊◊◊◊◊ ◊◊◊◊◊	732,950
◊◊◊◊◊ ◊◊◊◊◊ ◊◊◊◊◊ ◊◊◊◊◊ ◊◊◊◊◊ ◊◊◊◊◊ ◊◊◊◊◊ ◊◊◊◊◊ ◊◊◊◊◊ ◊◊◊◊◊	**733,000**
◊◊◊◊◊ ◊◊◊◊◊ ◊◊◊◊◊ ◊◊◊◊◊ ◊◊◊◊◊ ◊◊◊◊◊ ◊◊◊◊◊ ◊◊◊◊◊ ◊◊◊◊◊ ◊◊◊◊◊	733,050
◊◊◊◊◊ ◊◊◊◊◊ ◊◊◊◊◊ ◊◊◊◊◊ ◊◊◊◊◊ ◊◊◊◊◊ ◊◊◊◊◊ ◊◊◊◊◊ ◊◊◊◊◊ ◊◊◊◊◊	733,100
◊◊◊◊◊ ◊◊◊◊◊ ◊◊◊◊◊ ◊◊◊◊◊ ◊◊◊◊◊ ◊◊◊◊◊ ◊◊◊◊◊ ◊◊◊◊◊ ◊◊◊◊◊ ◊◊◊◊◊	733,150
◊◊◊◊◊ ◊◊◊◊◊ ◊◊◊◊◊ ◊◊◊◊◊ ◊◊◊◊◊ ◊◊◊◊◊ ◊◊◊◊◊ ◊◊◊◊◊ ◊◊◊◊◊ ◊◊◊◊◊	733,200
◊◊◊◊◊ ◊◊◊◊◊ ◊◊◊◊◊ ◊◊◊◊◊ ◊◊◊◊◊ ◊◊◊◊◊ ◊◊◊◊◊ ◊◊◊◊◊ ◊◊◊◊◊ ◊◊◊◊◊	733,250
◊◊◊◊◊ ◊◊◊◊◊ ◊◊◊◊◊ ◊◊◊◊◊ ◊◊◊◊◊ ◊◊◊◊◊ ◊◊◊◊◊ ◊◊◊◊◊ ◊◊◊◊◊ ◊◊◊◊◊	733,300
◊◊◊◊◊ ◊◊◊◊◊ ◊◊◊◊◊ ◊◊◊◊◊ ◊◊◊◊◊ ◊◊◊◊◊ ◊◊◊◊◊ ◊◊◊◊◊ ◊◊◊◊◊ ◊◊◊◊◊	733,350
◊◊◊◊◊ ◊◊◊◊◊ ◊◊◊◊◊ ◊◊◊◊◊ ◊◊◊◊◊ ◊◊◊◊◊ ◊◊◊◊◊ ◊◊◊◊◊ ◊◊◊◊◊ ◊◊◊◊◊	733,400
◊◊◊◊◊ ◊◊◊◊◊ ◊◊◊◊◊ ◊◊◊◊◊ ◊◊◊◊◊ ◊◊◊◊◊ ◊◊◊◊◊ ◊◊◊◊◊ ◊◊◊◊◊ ◊◊◊◊◊	733,450
◊◊◊◊◊ ◊◊◊◊◊ ◊◊◊◊◊ ◊◊◊◊◊ ◊◊◊◊◊ ◊◊◊◊◊ ◊◊◊◊◊ ◊◊◊◊◊ ◊◊◊◊◊ ◊◊◊◊◊	733,500
◊◊◊◊◊ ◊◊◊◊◊ ◊◊◊◊◊ ◊◊◊◊◊ ◊◊◊◊◊ ◊◊◊◊◊ ◊◊◊◊◊ ◊◊◊◊◊ ◊◊◊◊◊ ◊◊◊◊◊	733,550
◊◊◊◊◊ ◊◊◊◊◊ ◊◊◊◊◊ ◊◊◊◊◊ ◊◊◊◊◊ ◊◊◊◊◊ ◊◊◊◊◊ ◊◊◊◊◊ ◊◊◊◊◊ ◊◊◊◊◊	733,600
◊◊◊◊◊ ◊◊◊◊◊ ◊◊◊◊◊ ◊◊◊◊◊ ◊◊◊◊◊ ◊◊◊◊◊ ◊◊◊◊◊ ◊◊◊◊◊ ◊◊◊◊◊ ◊◊◊◊◊	733,650
◊◊◊◊◊ ◊◊◊◊◊ ◊◊◊◊◊ ◊◊◊◊◊ ◊◊◊◊◊ ◊◊◊◊◊ ◊◊◊◊◊ ◊◊◊◊◊ ◊◊◊◊◊ ◊◊◊◊◊	733,700
◊◊◊◊◊ ◊◊◊◊◊ ◊◊◊◊◊ ◊◊◊◊◊ ◊◊◊◊◊ ◊◊◊◊◊ ◊◊◊◊◊ ◊◊◊◊◊ ◊◊◊◊◊ ◊◊◊◊◊	733,750
◊◊◊◊◊ ◊◊◊◊◊ ◊◊◊◊◊ ◊◊◊◊◊ ◊◊◊◊◊ ◊◊◊◊◊ ◊◊◊◊◊ ◊◊◊◊◊ ◊◊◊◊◊ ◊◊◊◊◊	733,800
◊◊◊◊◊ ◊◊◊◊◊ ◊◊◊◊◊ ◊◊◊◊◊ ◊◊◊◊◊ ◊◊◊◊◊ ◊◊◊◊◊ ◊◊◊◊◊ ◊◊◊◊◊ ◊◊◊◊◊	733,850
◊◊◊◊◊ ◊◊◊◊◊ ◊◊◊◊◊ ◊◊◊◊◊ ◊◊◊◊◊ ◊◊◊◊◊ ◊◊◊◊◊ ◊◊◊◊◊ ◊◊◊◊◊ ◊◊◊◊◊	733,900
◊◊◊◊◊ ◊◊◊◊◊ ◊◊◊◊◊ ◊◊◊◊◊ ◊◊◊◊◊ ◊◊◊◊◊ ◊◊◊◊◊ ◊◊◊◊◊ ◊◊◊◊◊ ◊◊◊◊◊	733,950
◊◊◊◊◊ ◊◊◊◊◊ ◊◊◊◊◊ ◊◊◊◊◊ ◊◊◊◊◊ ◊◊◊◊◊ ◊◊◊◊◊ ◊◊◊◊◊ ◊◊◊◊◊ ◊◊◊◊◊	**734,000**

The Number 734,001 to the Number 736,000

Start

Notes

	734,050
	734,100
	734,150
	734,200
	734,250
	734,300
	734,350
	734,400
	734,450
	734,500
	734,550
	734,600
	734,650
	734,700
	734,750
	734,800
	734,850
	734,900
	734,950
	735,000
	735,050
	735,100
	735,150
	735,200
	735,250
	735,300
	735,350
	735,400
	735,450
	735,500
	735,550
	735,600
	735,650
	735,700
	735,750
	735,800
	735,850
	735,900
	735,950
	736,000

The Number 736,001 to the Number 738,000

Start

Notes

	736,050
	736,100
	736,150
	736,200
	736,250
	736,300
	736,350
	736,400
	736,450
	736,500
	736,550
	736,600
	736,650
	736,700
	736,750
	736,800
	736,850
	736,900
	736,950
	737,000
	737,050
	737,100
	737,150
	737,200
	737,250
	737,300
	737,350
	737,400
	737,450
	737,500
	737,550
	737,600
	737,650
	737,700
	737,750
	737,800
	737,850
	737,900
	737,950
	738,000

The Number 738,001 to the Number 740,000

Start		Notes
	738,050
	738,100
	738,150
	738,200
	738,250
	738,300
	738,350
	738,400
	738,450
	738,500
	738,550
	738,600
	738,650
	738,700
	738,750
	738,800
	738,850
	738,900
	738,950
	739,000
	739,050
	739,100
	739,150
	739,200
	739,250
	739,300
	739,350
	739,400
	739,450
	739,500
	739,550
	739,600
	739,650
	739,700
	739,750
	739,800
	739,850
	739,900
	739,950
	740,000

The Number 740,001 to the Number 742,000

Start | | Notes

	740,050
	740,100
	740,150
	740,200
	740,250
	740,300
	740,350
	740,400
	740,450
	740,500
	740,550
	740,600
	740,650
	740,700
	740,750
	740,800
	740,850
	740,900
	740,950
	741,000
	741,050
	741,100
	741,150
	741,200
	741,250
	741,300
	741,350
	741,400
	741,450
	741,500
	741,550
	741,600
	741,650
	741,700
	741,750
	741,800
	741,850
	741,900
	741,950
	742,000

The Number 742,001 to the Number 744,000

Start   Notes

◊◊◊◊◊ ◊◊◊◊◊ ◊◊◊◊◊ ◊◊◊◊◊ ◊◊◊◊◊ ◊◊◊◊◊ ◊◊◊◊◊ ◊◊◊◊◊ ◊◊◊◊◊ ◊◊◊◊◊	742,050
◊◊◊◊◊ ◊◊◊◊◊ ◊◊◊◊◊ ◊◊◊◊◊ ◊◊◊◊◊ ◊◊◊◊◊ ◊◊◊◊◊ ◊◊◊◊◊ ◊◊◊◊◊ ◊◊◊◊◊	742,100
◊◊◊◊◊ ◊◊◊◊◊ ◊◊◊◊◊ ◊◊◊◊◊ ◊◊◊◊◊ ◊◊◊◊◊ ◊◊◊◊◊ ◊◊◊◊◊ ◊◊◊◊◊ ◊◊◊◊◊	742,150
◊◊◊◊◊ ◊◊◊◊◊ ◊◊◊◊◊ ◊◊◊◊◊ ◊◊◊◊◊ ◊◊◊◊◊ ◊◊◊◊◊ ◊◊◊◊◊ ◊◊◊◊◊ ◊◊◊◊◊	742,200
◊◊◊◊◊ ◊◊◊◊◊ ◊◊◊◊◊ ◊◊◊◊◊ ◊◊◊◊◊ ◊◊◊◊◊ ◊◊◊◊◊ ◊◊◊◊◊ ◊◊◊◊◊ ◊◊◊◊◊	742,250
◊◊◊◊◊ ◊◊◊◊◊ ◊◊◊◊◊ ◊◊◊◊◊ ◊◊◊◊◊ ◊◊◊◊◊ ◊◊◊◊◊ ◊◊◊◊◊ ◊◊◊◊◊ ◊◊◊◊◊	742,300
◊◊◊◊◊ ◊◊◊◊◊ ◊◊◊◊◊ ◊◊◊◊◊ ◊◊◊◊◊ ◊◊◊◊◊ ◊◊◊◊◊ ◊◊◊◊◊ ◊◊◊◊◊ ◊◊◊◊◊	742,350
◊◊◊◊◊ ◊◊◊◊◊ ◊◊◊◊◊ ◊◊◊◊◊ ◊◊◊◊◊ ◊◊◊◊◊ ◊◊◊◊◊ ◊◊◊◊◊ ◊◊◊◊◊ ◊◊◊◊◊	742,400
◊◊◊◊◊ ◊◊◊◊◊ ◊◊◊◊◊ ◊◊◊◊◊ ◊◊◊◊◊ ◊◊◊◊◊ ◊◊◊◊◊ ◊◊◊◊◊ ◊◊◊◊◊ ◊◊◊◊◊	742,450
◊◊◊◊◊ ◊◊◊◊◊ ◊◊◊◊◊ ◊◊◊◊◊ ◊◊◊◊◊ ◊◊◊◊◊ ◊◊◊◊◊ ◊◊◊◊◊ ◊◊◊◊◊ ◊◊◊◊◊	742,500
◊◊◊◊◊ ◊◊◊◊◊ ◊◊◊◊◊ ◊◊◊◊◊ ◊◊◊◊◊ ◊◊◊◊◊ ◊◊◊◊◊ ◊◊◊◊◊ ◊◊◊◊◊ ◊◊◊◊◊	742,550
◊◊◊◊◊ ◊◊◊◊◊ ◊◊◊◊◊ ◊◊◊◊◊ ◊◊◊◊◊ ◊◊◊◊◊ ◊◊◊◊◊ ◊◊◊◊◊ ◊◊◊◊◊ ◊◊◊◊◊	742,600
◊◊◊◊◊ ◊◊◊◊◊ ◊◊◊◊◊ ◊◊◊◊◊ ◊◊◊◊◊ ◊◊◊◊◊ ◊◊◊◊◊ ◊◊◊◊◊ ◊◊◊◊◊ ◊◊◊◊◊	742,650
◊◊◊◊◊ ◊◊◊◊◊ ◊◊◊◊◊ ◊◊◊◊◊ ◊◊◊◊◊ ◊◊◊◊◊ ◊◊◊◊◊ ◊◊◊◊◊ ◊◊◊◊◊ ◊◊◊◊◊	742,700
◊◊◊◊◊ ◊◊◊◊◊ ◊◊◊◊◊ ◊◊◊◊◊ ◊◊◊◊◊ ◊◊◊◊◊ ◊◊◊◊◊ ◊◊◊◊◊ ◊◊◊◊◊ ◊◊◊◊◊	742,750
◊◊◊◊◊ ◊◊◊◊◊ ◊◊◊◊◊ ◊◊◊◊◊ ◊◊◊◊◊ ◊◊◊◊◊ ◊◊◊◊◊ ◊◊◊◊◊ ◊◊◊◊◊ ◊◊◊◊◊	742,800
◊◊◊◊◊ ◊◊◊◊◊ ◊◊◊◊◊ ◊◊◊◊◊ ◊◊◊◊◊ ◊◊◊◊◊ ◊◊◊◊◊ ◊◊◊◊◊ ◊◊◊◊◊ ◊◊◊◊◊	742,850
◊◊◊◊◊ ◊◊◊◊◊ ◊◊◊◊◊ ◊◊◊◊◊ ◊◊◊◊◊ ◊◊◊◊◊ ◊◊◊◊◊ ◊◊◊◊◊ ◊◊◊◊◊ ◊◊◊◊◊	742,900
◊◊◊◊◊ ◊◊◊◊◊ ◊◊◊◊◊ ◊◊◊◊◊ ◊◊◊◊◊ ◊◊◊◊◊ ◊◊◊◊◊ ◊◊◊◊◊ ◊◊◊◊◊ ◊◊◊◊◊	742,950
◊◊◊◊◊ ◊◊◊◊◊ ◊◊◊◊◊ ◊◊◊◊◊ ◊◊◊◊◊ ◊◊◊◊◊ ◊◊◊◊◊ ◊◊◊◊◊ ◊◊◊◊◊ ◊◊◊◊◊	**743,000**
◊◊◊◊◊ ◊◊◊◊◊ ◊◊◊◊◊ ◊◊◊◊◊ ◊◊◊◊◊ ◊◊◊◊◊ ◊◊◊◊◊ ◊◊◊◊◊ ◊◊◊◊◊ ◊◊◊◊◊	743,050
◊◊◊◊◊ ◊◊◊◊◊ ◊◊◊◊◊ ◊◊◊◊◊ ◊◊◊◊◊ ◊◊◊◊◊ ◊◊◊◊◊ ◊◊◊◊◊ ◊◊◊◊◊ ◊◊◊◊◊	743,100
◊◊◊◊◊ ◊◊◊◊◊ ◊◊◊◊◊ ◊◊◊◊◊ ◊◊◊◊◊ ◊◊◊◊◊ ◊◊◊◊◊ ◊◊◊◊◊ ◊◊◊◊◊ ◊◊◊◊◊	743,150
◊◊◊◊◊ ◊◊◊◊◊ ◊◊◊◊◊ ◊◊◊◊◊ ◊◊◊◊◊ ◊◊◊◊◊ ◊◊◊◊◊ ◊◊◊◊◊ ◊◊◊◊◊ ◊◊◊◊◊	743,200
◊◊◊◊◊ ◊◊◊◊◊ ◊◊◊◊◊ ◊◊◊◊◊ ◊◊◊◊◊ ◊◊◊◊◊ ◊◊◊◊◊ ◊◊◊◊◊ ◊◊◊◊◊ ◊◊◊◊◊	743,250
◊◊◊◊◊ ◊◊◊◊◊ ◊◊◊◊◊ ◊◊◊◊◊ ◊◊◊◊◊ ◊◊◊◊◊ ◊◊◊◊◊ ◊◊◊◊◊ ◊◊◊◊◊ ◊◊◊◊◊	743,300
◊◊◊◊◊ ◊◊◊◊◊ ◊◊◊◊◊ ◊◊◊◊◊ ◊◊◊◊◊ ◊◊◊◊◊ ◊◊◊◊◊ ◊◊◊◊◊ ◊◊◊◊◊ ◊◊◊◊◊	743,350
◊◊◊◊◊ ◊◊◊◊◊ ◊◊◊◊◊ ◊◊◊◊◊ ◊◊◊◊◊ ◊◊◊◊◊ ◊◊◊◊◊ ◊◊◊◊◊ ◊◊◊◊◊ ◊◊◊◊◊	743,400
◊◊◊◊◊ ◊◊◊◊◊ ◊◊◊◊◊ ◊◊◊◊◊ ◊◊◊◊◊ ◊◊◊◊◊ ◊◊◊◊◊ ◊◊◊◊◊ ◊◊◊◊◊ ◊◊◊◊◊	743,450
◊◊◊◊◊ ◊◊◊◊◊ ◊◊◊◊◊ ◊◊◊◊◊ ◊◊◊◊◊ ◊◊◊◊◊ ◊◊◊◊◊ ◊◊◊◊◊ ◊◊◊◊◊ ◊◊◊◊◊	743,500
◊◊◊◊◊ ◊◊◊◊◊ ◊◊◊◊◊ ◊◊◊◊◊ ◊◊◊◊◊ ◊◊◊◊◊ ◊◊◊◊◊ ◊◊◊◊◊ ◊◊◊◊◊ ◊◊◊◊◊	743,550
◊◊◊◊◊ ◊◊◊◊◊ ◊◊◊◊◊ ◊◊◊◊◊ ◊◊◊◊◊ ◊◊◊◊◊ ◊◊◊◊◊ ◊◊◊◊◊ ◊◊◊◊◊ ◊◊◊◊◊	743,600
◊◊◊◊◊ ◊◊◊◊◊ ◊◊◊◊◊ ◊◊◊◊◊ ◊◊◊◊◊ ◊◊◊◊◊ ◊◊◊◊◊ ◊◊◊◊◊ ◊◊◊◊◊ ◊◊◊◊◊	743,650
◊◊◊◊◊ ◊◊◊◊◊ ◊◊◊◊◊ ◊◊◊◊◊ ◊◊◊◊◊ ◊◊◊◊◊ ◊◊◊◊◊ ◊◊◊◊◊ ◊◊◊◊◊ ◊◊◊◊◊	743,700
◊◊◊◊◊ ◊◊◊◊◊ ◊◊◊◊◊ ◊◊◊◊◊ ◊◊◊◊◊ ◊◊◊◊◊ ◊◊◊◊◊ ◊◊◊◊◊ ◊◊◊◊◊ ◊◊◊◊◊	743,750
◊◊◊◊◊ ◊◊◊◊◊ ◊◊◊◊◊ ◊◊◊◊◊ ◊◊◊◊◊ ◊◊◊◊◊ ◊◊◊◊◊ ◊◊◊◊◊ ◊◊◊◊◊ ◊◊◊◊◊	743,800
◊◊◊◊◊ ◊◊◊◊◊ ◊◊◊◊◊ ◊◊◊◊◊ ◊◊◊◊◊ ◊◊◊◊◊ ◊◊◊◊◊ ◊◊◊◊◊ ◊◊◊◊◊ ◊◊◊◊◊	743,850
◊◊◊◊◊ ◊◊◊◊◊ ◊◊◊◊◊ ◊◊◊◊◊ ◊◊◊◊◊ ◊◊◊◊◊ ◊◊◊◊◊ ◊◊◊◊◊ ◊◊◊◊◊ ◊◊◊◊◊	743,900
◊◊◊◊◊ ◊◊◊◊◊ ◊◊◊◊◊ ◊◊◊◊◊ ◊◊◊◊◊ ◊◊◊◊◊ ◊◊◊◊◊ ◊◊◊◊◊ ◊◊◊◊◊ ◊◊◊◊◊	743,950
◊◊◊◊◊ ◊◊◊◊◊ ◊◊◊◊◊ ◊◊◊◊◊ ◊◊◊◊◊ ◊◊◊◊◊ ◊◊◊◊◊ ◊◊◊◊◊ ◊◊◊◊◊ ◊◊◊◊◊	**744,000**

◇ The Number 744,001 to the Number 746,000

Start
Notes

	744,050
	744,100
	744,150
	744,200
	744,250
	744,300
	744,350
	744,400
	744,450
	744,500
	744,550
	744,600
	744,650
	744,700
	744,750
	744,800
	744,850
	744,900
	744,950
	745,000
	745,050
	745,100
	745,150
	745,200
	745,250
	745,300
	745,350
	745,400
	745,450
	745,500
	745,550
	745,600
	745,650
	745,700
	745,750
	745,800
	745,850
	745,900
	745,950
	746,000

The Number 746,001 to the Number 748,000

Start

Notes

	746,050
	746,100
	746,150
	746,200
	746,250
	746,300
	746,350
	746,400
	746,450
	746,500
	746,550
	746,600
	746,650
	746,700
	746,750
	746,800
	746,850
	746,900
	746,950
	747,000
	747,050
	747,100
	747,150
	747,200
	747,250
	747,300
	747,350
	747,400
	747,450
	747,500
	747,550
	747,600
	747,650
	747,700
	747,750
	747,800
	747,850
	747,900
	747,950
	748,000

The Number 748,001 to the Number 750,000

Start

	748,050
	748,100
	748,150
	748,200
	748,250
	748,300
	748,350
	748,400
	748,450
	748,500
	748,550
	748,600
	748,650
	748,700
	748,750
	748,800
	748,850
	748,900
	748,950
	749,000
	749,050
	749,100
	749,150
	749,200
	749,250
	749,300
	749,350
	749,400
	749,450
	749,500
	749,550
	749,600
	749,650
	749,700
	749,750
	749,800
	749,850
	749,900
	749,950
	750,000

The Number 750,001 to the Number 752,000

Start

Notes

◊◊◊◊◊ ◊◊◊◊◊ ◊◊◊◊◊ ◊◊◊◊◊ ◊◊◊◊◊ ◊◊◊◊◊ ◊◊◊◊◊ ◊◊◊◊◊ ◊◊◊◊◊ ◊◊◊◊◊	750,050
◊◊◊◊◊ ◊◊◊◊◊ ◊◊◊◊◊ ◊◊◊◊◊ ◊◊◊◊◊ ◊◊◊◊◊ ◊◊◊◊◊ ◊◊◊◊◊ ◊◊◊◊◊ ◊◊◊◊◊	750,100
◊◊◊◊◊ ◊◊◊◊◊ ◊◊◊◊◊ ◊◊◊◊◊ ◊◊◊◊◊ ◊◊◊◊◊ ◊◊◊◊◊ ◊◊◊◊◊ ◊◊◊◊◊ ◊◊◊◊◊	750,150
◊◊◊◊◊ ◊◊◊◊◊ ◊◊◊◊◊ ◊◊◊◊◊ ◊◊◊◊◊ ◊◊◊◊◊ ◊◊◊◊◊ ◊◊◊◊◊ ◊◊◊◊◊ ◊◊◊◊◊	750,200
◊◊◊◊◊ ◊◊◊◊◊ ◊◊◊◊◊ ◊◊◊◊◊ ◊◊◊◊◊ ◊◊◊◊◊ ◊◊◊◊◊ ◊◊◊◊◊ ◊◊◊◊◊ ◊◊◊◊◊	750,250
◊◊◊◊◊ ◊◊◊◊◊ ◊◊◊◊◊ ◊◊◊◊◊ ◊◊◊◊◊ ◊◊◊◊◊ ◊◊◊◊◊ ◊◊◊◊◊ ◊◊◊◊◊ ◊◊◊◊◊	750,300
◊◊◊◊◊ ◊◊◊◊◊ ◊◊◊◊◊ ◊◊◊◊◊ ◊◊◊◊◊ ◊◊◊◊◊ ◊◊◊◊◊ ◊◊◊◊◊ ◊◊◊◊◊ ◊◊◊◊◊	750,350
◊◊◊◊◊ ◊◊◊◊◊ ◊◊◊◊◊ ◊◊◊◊◊ ◊◊◊◊◊ ◊◊◊◊◊ ◊◊◊◊◊ ◊◊◊◊◊ ◊◊◊◊◊ ◊◊◊◊◊	750,400
◊◊◊◊◊ ◊◊◊◊◊ ◊◊◊◊◊ ◊◊◊◊◊ ◊◊◊◊◊ ◊◊◊◊◊ ◊◊◊◊◊ ◊◊◊◊◊ ◊◊◊◊◊ ◊◊◊◊◊	750,450
◊◊◊◊◊ ◊◊◊◊◊ ◊◊◊◊◊ ◊◊◊◊◊ ◊◊◊◊◊ ◊◊◊◊◊ ◊◊◊◊◊ ◊◊◊◊◊ ◊◊◊◊◊ ◊◊◊◊◊	750,500
◊◊◊◊◊ ◊◊◊◊◊ ◊◊◊◊◊ ◊◊◊◊◊ ◊◊◊◊◊ ◊◊◊◊◊ ◊◊◊◊◊ ◊◊◊◊◊ ◊◊◊◊◊ ◊◊◊◊◊	750,550
◊◊◊◊◊ ◊◊◊◊◊ ◊◊◊◊◊ ◊◊◊◊◊ ◊◊◊◊◊ ◊◊◊◊◊ ◊◊◊◊◊ ◊◊◊◊◊ ◊◊◊◊◊ ◊◊◊◊◊	750,600
◊◊◊◊◊ ◊◊◊◊◊ ◊◊◊◊◊ ◊◊◊◊◊ ◊◊◊◊◊ ◊◊◊◊◊ ◊◊◊◊◊ ◊◊◊◊◊ ◊◊◊◊◊ ◊◊◊◊◊	750,650
◊◊◊◊◊ ◊◊◊◊◊ ◊◊◊◊◊ ◊◊◊◊◊ ◊◊◊◊◊ ◊◊◊◊◊ ◊◊◊◊◊ ◊◊◊◊◊ ◊◊◊◊◊ ◊◊◊◊◊	750,700
◊◊◊◊◊ ◊◊◊◊◊ ◊◊◊◊◊ ◊◊◊◊◊ ◊◊◊◊◊ ◊◊◊◊◊ ◊◊◊◊◊ ◊◊◊◊◊ ◊◊◊◊◊ ◊◊◊◊◊	750,750
◊◊◊◊◊ ◊◊◊◊◊ ◊◊◊◊◊ ◊◊◊◊◊ ◊◊◊◊◊ ◊◊◊◊◊ ◊◊◊◊◊ ◊◊◊◊◊ ◊◊◊◊◊ ◊◊◊◊◊	750,800
◊◊◊◊◊ ◊◊◊◊◊ ◊◊◊◊◊ ◊◊◊◊◊ ◊◊◊◊◊ ◊◊◊◊◊ ◊◊◊◊◊ ◊◊◊◊◊ ◊◊◊◊◊ ◊◊◊◊◊	750,850
◊◊◊◊◊ ◊◊◊◊◊ ◊◊◊◊◊ ◊◊◊◊◊ ◊◊◊◊◊ ◊◊◊◊◊ ◊◊◊◊◊ ◊◊◊◊◊ ◊◊◊◊◊ ◊◊◊◊◊	750,900
◊◊◊◊◊ ◊◊◊◊◊ ◊◊◊◊◊ ◊◊◊◊◊ ◊◊◊◊◊ ◊◊◊◊◊ ◊◊◊◊◊ ◊◊◊◊◊ ◊◊◊◊◊ ◊◊◊◊◊	750,950
◊◊◊◊◊ ◊◊◊◊◊ ◊◊◊◊◊ ◊◊◊◊◊ ◊◊◊◊◊ ◊◊◊◊◊ ◊◊◊◊◊ ◊◊◊◊◊ ◊◊◊◊◊ ◊◊◊◊◊	**751,000**
◊◊◊◊◊ ◊◊◊◊◊ ◊◊◊◊◊ ◊◊◊◊◊ ◊◊◊◊◊ ◊◊◊◊◊ ◊◊◊◊◊ ◊◊◊◊◊ ◊◊◊◊◊ ◊◊◊◊◊	751,050
◊◊◊◊◊ ◊◊◊◊◊ ◊◊◊◊◊ ◊◊◊◊◊ ◊◊◊◊◊ ◊◊◊◊◊ ◊◊◊◊◊ ◊◊◊◊◊ ◊◊◊◊◊ ◊◊◊◊◊	751,100
◊◊◊◊◊ ◊◊◊◊◊ ◊◊◊◊◊ ◊◊◊◊◊ ◊◊◊◊◊ ◊◊◊◊◊ ◊◊◊◊◊ ◊◊◊◊◊ ◊◊◊◊◊ ◊◊◊◊◊	751,150
◊◊◊◊◊ ◊◊◊◊◊ ◊◊◊◊◊ ◊◊◊◊◊ ◊◊◊◊◊ ◊◊◊◊◊ ◊◊◊◊◊ ◊◊◊◊◊ ◊◊◊◊◊ ◊◊◊◊◊	751,200
◊◊◊◊◊ ◊◊◊◊◊ ◊◊◊◊◊ ◊◊◊◊◊ ◊◊◊◊◊ ◊◊◊◊◊ ◊◊◊◊◊ ◊◊◊◊◊ ◊◊◊◊◊ ◊◊◊◊◊	751,250
◊◊◊◊◊ ◊◊◊◊◊ ◊◊◊◊◊ ◊◊◊◊◊ ◊◊◊◊◊ ◊◊◊◊◊ ◊◊◊◊◊ ◊◊◊◊◊ ◊◊◊◊◊ ◊◊◊◊◊	751,300
◊◊◊◊◊ ◊◊◊◊◊ ◊◊◊◊◊ ◊◊◊◊◊ ◊◊◊◊◊ ◊◊◊◊◊ ◊◊◊◊◊ ◊◊◊◊◊ ◊◊◊◊◊ ◊◊◊◊◊	751,350
◊◊◊◊◊ ◊◊◊◊◊ ◊◊◊◊◊ ◊◊◊◊◊ ◊◊◊◊◊ ◊◊◊◊◊ ◊◊◊◊◊ ◊◊◊◊◊ ◊◊◊◊◊ ◊◊◊◊◊	751,400
◊◊◊◊◊ ◊◊◊◊◊ ◊◊◊◊◊ ◊◊◊◊◊ ◊◊◊◊◊ ◊◊◊◊◊ ◊◊◊◊◊ ◊◊◊◊◊ ◊◊◊◊◊ ◊◊◊◊◊	751,450
◊◊◊◊◊ ◊◊◊◊◊ ◊◊◊◊◊ ◊◊◊◊◊ ◊◊◊◊◊ ◊◊◊◊◊ ◊◊◊◊◊ ◊◊◊◊◊ ◊◊◊◊◊ ◊◊◊◊◊	751,500
◊◊◊◊◊ ◊◊◊◊◊ ◊◊◊◊◊ ◊◊◊◊◊ ◊◊◊◊◊ ◊◊◊◊◊ ◊◊◊◊◊ ◊◊◊◊◊ ◊◊◊◊◊ ◊◊◊◊◊	751,550
◊◊◊◊◊ ◊◊◊◊◊ ◊◊◊◊◊ ◊◊◊◊◊ ◊◊◊◊◊ ◊◊◊◊◊ ◊◊◊◊◊ ◊◊◊◊◊ ◊◊◊◊◊ ◊◊◊◊◊	751,600
◊◊◊◊◊ ◊◊◊◊◊ ◊◊◊◊◊ ◊◊◊◊◊ ◊◊◊◊◊ ◊◊◊◊◊ ◊◊◊◊◊ ◊◊◊◊◊ ◊◊◊◊◊ ◊◊◊◊◊	751,650
◊◊◊◊◊ ◊◊◊◊◊ ◊◊◊◊◊ ◊◊◊◊◊ ◊◊◊◊◊ ◊◊◊◊◊ ◊◊◊◊◊ ◊◊◊◊◊ ◊◊◊◊◊ ◊◊◊◊◊	751,700
◊◊◊◊◊ ◊◊◊◊◊ ◊◊◊◊◊ ◊◊◊◊◊ ◊◊◊◊◊ ◊◊◊◊◊ ◊◊◊◊◊ ◊◊◊◊◊ ◊◊◊◊◊ ◊◊◊◊◊	751,750
◊◊◊◊◊ ◊◊◊◊◊ ◊◊◊◊◊ ◊◊◊◊◊ ◊◊◊◊◊ ◊◊◊◊◊ ◊◊◊◊◊ ◊◊◊◊◊ ◊◊◊◊◊ ◊◊◊◊◊	751,800
◊◊◊◊◊ ◊◊◊◊◊ ◊◊◊◊◊ ◊◊◊◊◊ ◊◊◊◊◊ ◊◊◊◊◊ ◊◊◊◊◊ ◊◊◊◊◊ ◊◊◊◊◊ ◊◊◊◊◊	751,850
◊◊◊◊◊ ◊◊◊◊◊ ◊◊◊◊◊ ◊◊◊◊◊ ◊◊◊◊◊ ◊◊◊◊◊ ◊◊◊◊◊ ◊◊◊◊◊ ◊◊◊◊◊ ◊◊◊◊◊	751,900
◊◊◊◊◊ ◊◊◊◊◊ ◊◊◊◊◊ ◊◊◊◊◊ ◊◊◊◊◊ ◊◊◊◊◊ ◊◊◊◊◊ ◊◊◊◊◊ ◊◊◊◊◊ ◊◊◊◊◊	751,950
◊◊◊◊◊ ◊◊◊◊◊ ◊◊◊◊◊ ◊◊◊◊◊ ◊◊◊◊◊ ◊◊◊◊◊ ◊◊◊◊◊ ◊◊◊◊◊ ◊◊◊◊◊ ◊◊◊◊◊	**752,000**

The Number 752,001 to the Number 754,000

Start

	752,050
	752,100
	752,150
	752,200
	752,250
	752,300
	752,350
	752,400
	752,450
	752,500
	752,550
	752,600
	752,650
	752,700
	752,750
	752,800
	752,850
	752,900
	752,950
	753,000
	753,050
	753,100
	753,150
	753,200
	753,250
	753,300
	753,350
	753,400
	753,450
	753,500
	753,550
	753,600
	753,650
	753,700
	753,750
	753,800
	753,850
	753,900
	753,950
	754,000

The Number 754,001 to the Number 756,000

Start

Notes

	754,050
	754,100
	754,150
	754,200
	754,250
	754,300
	754,350
	754,400
	754,450
	754,500
	754,550
	754,600
	754,650
	754,700
	754,750
	754,800
	754,850
	754,900
	754,950
	755,000
	755,050
	755,100
	755,150
	755,200
	755,250
	755,300
	755,350
	755,400
	755,450
	755,500
	755,550
	755,600
	755,650
	755,700
	755,750
	755,800
	755,850
	755,900
	755,950
	756,000

◇ The Number 756,001 to the Number 758,000

Start | | Notes

	Number	Notes
	756,050	
	756,100	
	756,150	
	756,200	
	756,250	
	756,300	
	756,350	
	756,400	
	756,450	
	756,500	
	756,550	
	756,600	
	756,650	
	756,700	
	756,750	
	756,800	
	756,850	
	756,900	
	756,950	
	757,000	
	757,050	
	757,100	
	757,150	
	757,200	
	757,250	
	757,300	
	757,350	
	757,400	
	757,450	
	757,500	
	757,550	
	757,600	
	757,650	
	757,700	
	757,750	
	757,800	
	757,850	
	757,900	
	757,950	
	758,000	

The Number 758,001 to the Number 760,000

Start

	Notes
758,050	
758,100	
758,150	
758,200	
758,250	
758,300	
758,350	
758,400	
758,450	
758,500	
758,550	
758,600	
758,650	
758,700	
758,750	
758,800	
758,850	
758,900	
758,950	
759,000	
759,050	
759,100	
759,150	
759,200	
759,250	
759,300	
759,350	
759,400	
759,450	
759,500	
759,550	
759,600	
759,650	
759,700	
759,750	
759,800	
759,850	
759,900	
759,950	
760,000	

The Number 760,001 to the Number 762,000

Start

Notes

	760,050
	760,100
	760,150
	760,200
	760,250
	760,300
	760,350
	760,400
	760,450
	760,500
	760,550
	760,600
	760,650
	760,700
	760,750
	760,800
	760,850
	760,900
	760,950
	761,000
	761,050
	761,100
	761,150
	761,200
	761,250
	761,300
	761,350
	761,400
	761,450
	761,500
	761,550
	761,600
	761,650
	761,700
	761,750
	761,800
	761,850
	761,900
	761,950
	762,000

The Number 762,001 to the Number 764,000

Start

Notes

◊◊◊◊◊ ◊◊◊◊◊ ◊◊◊◊◊ ◊◊◊◊◊ ◊◊◊◊◊ ◊◊◊◊◊ ◊◊◊◊◊ ◊◊◊◊◊ ◊◊◊◊◊ ◊◊◊◊◊	762,050	
◊◊◊◊◊ ◊◊◊◊◊ ◊◊◊◊◊ ◊◊◊◊◊ ◊◊◊◊◊ ◊◊◊◊◊ ◊◊◊◊◊ ◊◊◊◊◊ ◊◊◊◊◊ ◊◊◊◊◊	762,100	
◊◊◊◊◊ ◊◊◊◊◊ ◊◊◊◊◊ ◊◊◊◊◊ ◊◊◊◊◊ ◊◊◊◊◊ ◊◊◊◊◊ ◊◊◊◊◊ ◊◊◊◊◊ ◊◊◊◊◊	762,150	
◊◊◊◊◊ ◊◊◊◊◊ ◊◊◊◊◊ ◊◊◊◊◊ ◊◊◊◊◊ ◊◊◊◊◊ ◊◊◊◊◊ ◊◊◊◊◊ ◊◊◊◊◊ ◊◊◊◊◊	762,200	
◊◊◊◊◊ ◊◊◊◊◊ ◊◊◊◊◊ ◊◊◊◊◊ ◊◊◊◊◊ ◊◊◊◊◊ ◊◊◊◊◊ ◊◊◊◊◊ ◊◊◊◊◊ ◊◊◊◊◊	762,250	
◊◊◊◊◊ ◊◊◊◊◊ ◊◊◊◊◊ ◊◊◊◊◊ ◊◊◊◊◊ ◊◊◊◊◊ ◊◊◊◊◊ ◊◊◊◊◊ ◊◊◊◊◊ ◊◊◊◊◊	762,300	
◊◊◊◊◊ ◊◊◊◊◊ ◊◊◊◊◊ ◊◊◊◊◊ ◊◊◊◊◊ ◊◊◊◊◊ ◊◊◊◊◊ ◊◊◊◊◊ ◊◊◊◊◊ ◊◊◊◊◊	762,350	
◊◊◊◊◊ ◊◊◊◊◊ ◊◊◊◊◊ ◊◊◊◊◊ ◊◊◊◊◊ ◊◊◊◊◊ ◊◊◊◊◊ ◊◊◊◊◊ ◊◊◊◊◊ ◊◊◊◊◊	762,400	
◊◊◊◊◊ ◊◊◊◊◊ ◊◊◊◊◊ ◊◊◊◊◊ ◊◊◊◊◊ ◊◊◊◊◊ ◊◊◊◊◊ ◊◊◊◊◊ ◊◊◊◊◊ ◊◊◊◊◊	762,450	
◊◊◊◊◊ ◊◊◊◊◊ ◊◊◊◊◊ ◊◊◊◊◊ ◊◊◊◊◊ ◊◊◊◊◊ ◊◊◊◊◊ ◊◊◊◊◊ ◊◊◊◊◊ ◊◊◊◊◊	762,500	
◊◊◊◊◊ ◊◊◊◊◊ ◊◊◊◊◊ ◊◊◊◊◊ ◊◊◊◊◊ ◊◊◊◊◊ ◊◊◊◊◊ ◊◊◊◊◊ ◊◊◊◊◊ ◊◊◊◊◊	762,550	
◊◊◊◊◊ ◊◊◊◊◊ ◊◊◊◊◊ ◊◊◊◊◊ ◊◊◊◊◊ ◊◊◊◊◊ ◊◊◊◊◊ ◊◊◊◊◊ ◊◊◊◊◊ ◊◊◊◊◊	762,600	
◊◊◊◊◊ ◊◊◊◊◊ ◊◊◊◊◊ ◊◊◊◊◊ ◊◊◊◊◊ ◊◊◊◊◊ ◊◊◊◊◊ ◊◊◊◊◊ ◊◊◊◊◊ ◊◊◊◊◊	762,650	
◊◊◊◊◊ ◊◊◊◊◊ ◊◊◊◊◊ ◊◊◊◊◊ ◊◊◊◊◊ ◊◊◊◊◊ ◊◊◊◊◊ ◊◊◊◊◊ ◊◊◊◊◊ ◊◊◊◊◊	762,700	
◊◊◊◊◊ ◊◊◊◊◊ ◊◊◊◊◊ ◊◊◊◊◊ ◊◊◊◊◊ ◊◊◊◊◊ ◊◊◊◊◊ ◊◊◊◊◊ ◊◊◊◊◊ ◊◊◊◊◊	762,750	
◊◊◊◊◊ ◊◊◊◊◊ ◊◊◊◊◊ ◊◊◊◊◊ ◊◊◊◊◊ ◊◊◊◊◊ ◊◊◊◊◊ ◊◊◊◊◊ ◊◊◊◊◊ ◊◊◊◊◊	762,800	
◊◊◊◊◊ ◊◊◊◊◊ ◊◊◊◊◊ ◊◊◊◊◊ ◊◊◊◊◊ ◊◊◊◊◊ ◊◊◊◊◊ ◊◊◊◊◊ ◊◊◊◊◊ ◊◊◊◊◊	762,850	
◊◊◊◊◊ ◊◊◊◊◊ ◊◊◊◊◊ ◊◊◊◊◊ ◊◊◊◊◊ ◊◊◊◊◊ ◊◊◊◊◊ ◊◊◊◊◊ ◊◊◊◊◊ ◊◊◊◊◊	762,900	
◊◊◊◊◊ ◊◊◊◊◊ ◊◊◊◊◊ ◊◊◊◊◊ ◊◊◊◊◊ ◊◊◊◊◊ ◊◊◊◊◊ ◊◊◊◊◊ ◊◊◊◊◊ ◊◊◊◊◊	762,950	
◊◊◊◊◊ ◊◊◊◊◊ ◊◊◊◊◊ ◊◊◊◊◊ ◊◊◊◊◊ ◊◊◊◊◊ ◊◊◊◊◊ ◊◊◊◊◊ ◊◊◊◊◊ ◊◊◊◊◊	**763,000**	
◊◊◊◊◊ ◊◊◊◊◊ ◊◊◊◊◊ ◊◊◊◊◊ ◊◊◊◊◊ ◊◊◊◊◊ ◊◊◊◊◊ ◊◊◊◊◊ ◊◊◊◊◊ ◊◊◊◊◊	763,050	
◊◊◊◊◊ ◊◊◊◊◊ ◊◊◊◊◊ ◊◊◊◊◊ ◊◊◊◊◊ ◊◊◊◊◊ ◊◊◊◊◊ ◊◊◊◊◊ ◊◊◊◊◊ ◊◊◊◊◊	763,100	
◊◊◊◊◊ ◊◊◊◊◊ ◊◊◊◊◊ ◊◊◊◊◊ ◊◊◊◊◊ ◊◊◊◊◊ ◊◊◊◊◊ ◊◊◊◊◊ ◊◊◊◊◊ ◊◊◊◊◊	763,150	
◊◊◊◊◊ ◊◊◊◊◊ ◊◊◊◊◊ ◊◊◊◊◊ ◊◊◊◊◊ ◊◊◊◊◊ ◊◊◊◊◊ ◊◊◊◊◊ ◊◊◊◊◊ ◊◊◊◊◊	763,200	
◊◊◊◊◊ ◊◊◊◊◊ ◊◊◊◊◊ ◊◊◊◊◊ ◊◊◊◊◊ ◊◊◊◊◊ ◊◊◊◊◊ ◊◊◊◊◊ ◊◊◊◊◊ ◊◊◊◊◊	763,250	
◊◊◊◊◊ ◊◊◊◊◊ ◊◊◊◊◊ ◊◊◊◊◊ ◊◊◊◊◊ ◊◊◊◊◊ ◊◊◊◊◊ ◊◊◊◊◊ ◊◊◊◊◊ ◊◊◊◊◊	763,300	
◊◊◊◊◊ ◊◊◊◊◊ ◊◊◊◊◊ ◊◊◊◊◊ ◊◊◊◊◊ ◊◊◊◊◊ ◊◊◊◊◊ ◊◊◊◊◊ ◊◊◊◊◊ ◊◊◊◊◊	763,350	
◊◊◊◊◊ ◊◊◊◊◊ ◊◊◊◊◊ ◊◊◊◊◊ ◊◊◊◊◊ ◊◊◊◊◊ ◊◊◊◊◊ ◊◊◊◊◊ ◊◊◊◊◊ ◊◊◊◊◊	763,400	
◊◊◊◊◊ ◊◊◊◊◊ ◊◊◊◊◊ ◊◊◊◊◊ ◊◊◊◊◊ ◊◊◊◊◊ ◊◊◊◊◊ ◊◊◊◊◊ ◊◊◊◊◊ ◊◊◊◊◊	763,450	
◊◊◊◊◊ ◊◊◊◊◊ ◊◊◊◊◊ ◊◊◊◊◊ ◊◊◊◊◊ ◊◊◊◊◊ ◊◊◊◊◊ ◊◊◊◊◊ ◊◊◊◊◊ ◊◊◊◊◊	763,500	
◊◊◊◊◊ ◊◊◊◊◊ ◊◊◊◊◊ ◊◊◊◊◊ ◊◊◊◊◊ ◊◊◊◊◊ ◊◊◊◊◊ ◊◊◊◊◊ ◊◊◊◊◊ ◊◊◊◊◊	763,550	
◊◊◊◊◊ ◊◊◊◊◊ ◊◊◊◊◊ ◊◊◊◊◊ ◊◊◊◊◊ ◊◊◊◊◊ ◊◊◊◊◊ ◊◊◊◊◊ ◊◊◊◊◊ ◊◊◊◊◊	763,600	
◊◊◊◊◊ ◊◊◊◊◊ ◊◊◊◊◊ ◊◊◊◊◊ ◊◊◊◊◊ ◊◊◊◊◊ ◊◊◊◊◊ ◊◊◊◊◊ ◊◊◊◊◊ ◊◊◊◊◊	763,650	
◊◊◊◊◊ ◊◊◊◊◊ ◊◊◊◊◊ ◊◊◊◊◊ ◊◊◊◊◊ ◊◊◊◊◊ ◊◊◊◊◊ ◊◊◊◊◊ ◊◊◊◊◊ ◊◊◊◊◊	763,700	
◊◊◊◊◊ ◊◊◊◊◊ ◊◊◊◊◊ ◊◊◊◊◊ ◊◊◊◊◊ ◊◊◊◊◊ ◊◊◊◊◊ ◊◊◊◊◊ ◊◊◊◊◊ ◊◊◊◊◊	763,750	
◊◊◊◊◊ ◊◊◊◊◊ ◊◊◊◊◊ ◊◊◊◊◊ ◊◊◊◊◊ ◊◊◊◊◊ ◊◊◊◊◊ ◊◊◊◊◊ ◊◊◊◊◊ ◊◊◊◊◊	763,800	
◊◊◊◊◊ ◊◊◊◊◊ ◊◊◊◊◊ ◊◊◊◊◊ ◊◊◊◊◊ ◊◊◊◊◊ ◊◊◊◊◊ ◊◊◊◊◊ ◊◊◊◊◊ ◊◊◊◊◊	763,850	
◊◊◊◊◊ ◊◊◊◊◊ ◊◊◊◊◊ ◊◊◊◊◊ ◊◊◊◊◊ ◊◊◊◊◊ ◊◊◊◊◊ ◊◊◊◊◊ ◊◊◊◊◊ ◊◊◊◊◊	763,900	
◊◊◊◊◊ ◊◊◊◊◊ ◊◊◊◊◊ ◊◊◊◊◊ ◊◊◊◊◊ ◊◊◊◊◊ ◊◊◊◊◊ ◊◊◊◊◊ ◊◊◊◊◊ ◊◊◊◊◊	763,950	
◊◊◊◊◊ ◊◊◊◊◊ ◊◊◊◊◊ ◊◊◊◊◊ ◊◊◊◊◊ ◊◊◊◊◊ ◊◊◊◊◊ ◊◊◊◊◊ ◊◊◊◊◊ ◊◊◊◊◊	**764,000**	

The Number 764,001 to the Number 766,000

	Notes
764,050	
764,100	
764,150	
764,200	
764,250	
764,300	
764,350	
764,400	
764,450	
764,500	
764,550	
764,600	
764,650	
764,700	
764,750	
764,800	
764,850	
764,900	
764,950	
765,000	
765,050	
765,100	
765,150	
765,200	
765,250	
765,300	
765,350	
765,400	
765,450	
765,500	
765,550	
765,600	
765,650	
765,700	
765,750	
765,800	
765,850	
765,900	
765,950	
766,000	

The Number 766,001 to the Number 768,000

Start

Notes

	766,050
	766,100
	766,150
	766,200
	766,250
	766,300
	766,350
	766,400
	766,450
	766,500
	766,550
	766,600
	766,650
	766,700
	766,750
	766,800
	766,850
	766,900
	766,950
	767,000
	767,050
	767,100
	767,150
	767,200
	767,250
	767,300
	767,350
	767,400
	767,450
	767,500
	767,550
	767,600
	767,650
	767,700
	767,750
	767,800
	767,850
	767,900
	767,950
	768,000

The Number 768,001 to the Number 770,000

Start

Notes

	768,050
	768,100
	768,150
	768,200
	768,250
	768,300
	768,350
	768,400
	768,450
	768,500
	768,550
	768,600
	768,650
	768,700
	768,750
	768,800
	768,850
	768,900
	768,950
	769,000
	769,050
	769,100
	769,150
	769,200
	769,250
	769,300
	769,350
	769,400
	769,450
	769,500
	769,550
	769,600
	769,650
	769,700
	769,750
	769,800
	769,850
	769,900
	769,950
	770,000

The Number 770,001 to the Number 772,000

Start

Notes

770,050
770,100
770,150
770,200
770,250
770,300
770,350
770,400
770,450
770,500
770,550
770,600
770,650
770,700
770,750
770,800
770,850
770,900
770,950
771,000
771,050
771,100
771,150
771,200
771,250
771,300
771,350
771,400
771,450
771,500
771,550
771,600
771,650
771,700
771,750
771,800
771,850
771,900
771,950
772,000

The Number 772,001 to the Number 774,000

Start

Notes

	772,050
	772,100
	772,150
	772,200
	772,250
	772,300
	772,350
	772,400
	772,450
	772,500
	772,550
	772,600
	772,650
	772,700
	772,750
	772,800
	772,850
	772,900
	772,950
	773,000
	773,050
	773,100
	773,150
	773,200
	773,250
	773,300
	773,350
	773,400
	773,450
	773,500
	773,550
	773,600
	773,650
	773,700
	773,750
	773,800
	773,850
	773,900
	773,950
	774,000

The Number 774,001 to the Number 776,000

Start

Notes

	774,050
	774,100
	774,150
	774,200
	774,250
	774,300
	774,350
	774,400
	774,450
	774,500
	774,550
	774,600
	774,650
	774,700
	774,750
	774,800
	774,850
	774,900
	774,950
	775,000
	775,050
	775,100
	775,150
	775,200
	775,250
	775,300
	775,350
	775,400
	775,450
	775,500
	775,550
	775,600
	775,650
	775,700
	775,750
	775,800
	775,850
	775,900
	775,950
	776,000

The Number 776,001 to the Number 778,000

Start

Notes

	776,050
	776,100
	776,150
	776,200
	776,250
	776,300
	776,350
	776,400
	776,450
	776,500
	776,550
	776,600
	776,650
	776,700
	776,750
	776,800
	776,850
	776,900
	776,950
	777,000
	777,050
	777,100
	777,150
	777,200
	777,250
	777,300
	777,350
	777,400
	777,450
	777,500
	777,550
	777,600
	777,650
	777,700
	777,750
	777,800
	777,850
	777,900
	777,950
	778,000

The Number 778,001 to the Number 780,000

Start | Notes

	778,050
	778,100
	778,150
	778,200
	778,250
	778,300
	778,350
	778,400
	778,450
	778,500
	778,550
	778,600
	778,650
	778,700
	778,750
	778,800
	778,850
	778,900
	778,950
	779,000
	779,050
	779,100
	779,150
	779,200
	779,250
	779,300
	779,350
	779,400
	779,450
	779,500
	779,550
	779,600
	779,650
	779,700
	779,750
	779,800
	779,850
	779,900
	779,950
	780,000

The Number 780,001 to the Number 782,000

Start

Notes

	780,050
	780,100
	780,150
	780,200
	780,250
	780,300
	780,350
	780,400
	780,450
	780,500
	780,550
	780,600
	780,650
	780,700
	780,750
	780,800
	780,850
	780,900
	780,950
	781,000
	781,050
	781,100
	781,150
	781,200
	781,250
	781,300
	781,350
	781,400
	781,450
	781,500
	781,550
	781,600
	781,650
	781,700
	781,750
	781,800
	781,850
	781,900
	781,950
	782,000

The Number 782,001 to the Number 784,000

Start   Notes

	782,050
	782,100
	782,150
	782,200
	782,250
	782,300
	782,350
	782,400
	782,450
	782,500
	782,550
	782,600
	782,650
	782,700
	782,750
	782,800
	782,850
	782,900
	782,950
	783,000
	783,050
	783,100
	783,150
	783,200
	783,250
	783,300
	783,350
	783,400
	783,450
	783,500
	783,550
	783,600
	783,650
	783,700
	783,750
	783,800
	783,850
	783,900
	783,950
	784,000

◇ The Number 784,001 to the Number 786,000

Start   Notes

	784,050
	784,100
	784,150
	784,200
	784,250
	784,300
	784,350
	784,400
	784,450
	784,500
	784,550
	784,600
	784,650
	784,700
	784,750
	784,800
	784,850
	784,900
	784,950
	785,000
	785,050
	785,100
	785,150
	785,200
	785,250
	785,300
	785,350
	785,400
	785,450
	785,500
	785,550
	785,600
	785,650
	785,700
	785,750
	785,800
	785,850
	785,900
	785,950
	786,000

The Number 786,001 to the Number 788,000

Start		Notes
	786,050
	786,100
	786,150
	786,200
	786,250
	786,300
	786,350
	786,400
	786,450
	786,500
	786,550
	786,600
	786,650
	786,700
	786,750
	786,800
	786,850
	786,900
	786,950
	787,000
	787,050
	787,100
	787,150
	787,200
	787,250
	787,300
	787,350
	787,400
	787,450
	787,500
	787,550
	787,600
	787,650
	787,700
	787,750
	787,800
	787,850
	787,900
	787,950
	788,000

The Number 788,001 to the Number 790,000

Start   **Notes**

	788,050
	788,100
	788,150
	788,200
	788,250
	788,300
	788,350
	788,400
	788,450
	788,500
	788,550
	788,600
	788,650
	788,700
	788,750
	788,800
	788,850
	788,900
	788,950
	789,000
	789,050
	789,100
	789,150
	789,200
	789,250
	789,300
	789,350
	789,400
	789,450
	789,500
	789,550
	789,600
	789,650
	789,700
	789,750
	789,800
	789,850
	789,900
	789,950
	790,000

The Number 790,001 to the Number 792,000

Start		Notes
790,050
790,100
790,150
790,200
790,250
790,300
790,350
790,400
790,450
790,500
790,550
790,600
790,650
790,700
790,750
790,800
790,850
790,900
790,950
791,000
791,050
791,100
791,150
791,200
791,250
791,300
791,350
791,400
791,450
791,500
791,550
791,600
791,650
791,700
791,750
791,800
791,850
791,900
791,950
792,000

The Number 792,001 to the Number 794,000

Start   **Notes**

Number	Notes
792,050	
792,100	
792,150	
792,200	
792,250	
792,300	
792,350	
792,400	
792,450	
792,500	
792,550	
792,600	
792,650	
792,700	
792,750	
792,800	
792,850	
792,900	
792,950	
793,000	
793,050	
793,100	
793,150	
793,200	
793,250	
793,300	
793,350	
793,400	
793,450	
793,500	
793,550	
793,600	
793,650	
793,700	
793,750	
793,800	
793,850	
793,900	
793,950	
794,000	

The Number 794,001 to the Number 796,000

Start

Notes

	794,050
	794,100
	794,150
	794,200
	794,250
	794,300
	794,350
	794,400
	794,450
	794,500
	794,550
	794,600
	794,650
	794,700
	794,750
	794,800
	794,850
	794,900
	794,950
	795,000
	795,050
	795,100
	795,150
	795,200
	795,250
	795,300
	795,350
	795,400
	795,450
	795,500
	795,550
	795,600
	795,650
	795,700
	795,750
	795,800
	795,850
	795,900
	795,950
	796,000

The Number 796,001 to the Number 798,000

Start

	Number
	796,050
	796,100
	796,150
	796,200
	796,250
	796,300
	796,350
	796,400
	796,450
	796,500
	796,550
	796,600
	796,650
	796,700
	796,750
	796,800
	796,850
	796,900
	796,950
	797,000
	797,050
	797,100
	797,150
	797,200
	797,250
	797,300
	797,350
	797,400
	797,450
	797,500
	797,550
	797,600
	797,650
	797,700
	797,750
	797,800
	797,850
	797,900
	797,950
	798,000

The Number 798,001 to the Number 800,000

Start

	Number
	798,050
	798,100
	798,150
	798,200
	798,250
	798,300
	798,350
	798,400
	798,450
	798,500
	798,550
	798,600
	798,650
	798,700
	798,750
	798,800
	798,850
	798,900
	798,950
	799,000
	799,050
	799,100
	799,150
	799,200
	799,250
	799,300
	799,350
	799,400
	799,450
	799,500
	799,550
	799,600
	799,650
	799,700
	799,750
	799,800
	799,850
	799,900
	799,950
	800,000

The Number 800,001 to the Number 802,000

Start

Notes

	800,050
	800,100
	800,150
	800,200
	800,250
	800,300
	800,350
	800,400
	800,450
	800,500
	800,550
	800,600
	800,650
	800,700
	800,750
	800,800
	800,850
	800,900
	800,950
	801,000
	801,050
	801,100
	801,150
	801,200
	801,250
	801,300
	801,350
	801,400
	801,450
	801,500
	801,550
	801,600
	801,650
	801,700
	801,750
	801,800
	801,850
	801,900
	801,950
	802,000

The Number 802,001 to the Number 804,000

Start   Notes

	802,050
	802,100
	802,150
	802,200
	802,250
	802,300
	802,350
	802,400
	802,450
	802,500
	802,550
	802,600
	802,650
	802,700
	802,750
	802,800
	802,850
	802,900
	802,950
	803,000
	803,050
	803,100
	803,150
	803,200
	803,250
	803,300
	803,350
	803,400
	803,450
	803,500
	803,550
	803,600
	803,650
	803,700
	803,750
	803,800
	803,850
	803,900
	803,950
	804,000

The Number 804,001 to the Number 806,000

Start		Notes
	804,050	
	804,100	
	804,150	
	804,200	
	804,250	
	804,300	
	804,350	
	804,400	
	804,450	
	804,500	
	804,550	
	804,600	
	804,650	
	804,700	
	804,750	
	804,800	
	804,850	
	804,900	
	804,950	
	805,000	
	805,050	
	805,100	
	805,150	
	805,200	
	805,250	
	805,300	
	805,350	
	805,400	
	805,450	
	805,500	
	805,550	
	805,600	
	805,650	
	805,700	
	805,750	
	805,800	
	805,850	
	805,900	
	805,950	
	806,000	

The Number 806,001 to the Number 808,000

Start

Notes

◊◊◊◊◊	806,050
◊◊◊◊◊	806,100
◊◊◊◊◊	806,150
◊◊◊◊◊	806,200
◊◊◊◊◊	806,250
◊◊◊◊◊	806,300
◊◊◊◊◊	806,350
◊◊◊◊◊	806,400
◊◊◊◊◊	806,450
◊◊◊◊◊	806,500
◊◊◊◊◊	806,550
◊◊◊◊◊	806,600
◊◊◊◊◊	806,650
◊◊◊◊◊	806,700
◊◊◊◊◊	806,750
◊◊◊◊◊	806,800
◊◊◊◊◊	806,850
◊◊◊◊◊	806,900
◊◊◊◊◊	806,950
◊◊◊◊◊	**807,000**
◊◊◊◊◊	807,050
◊◊◊◊◊	807,100
◊◊◊◊◊	807,150
◊◊◊◊◊	807,200
◊◊◊◊◊	807,250
◊◊◊◊◊	807,300
◊◊◊◊◊	807,350
◊◊◊◊◊	807,400
◊◊◊◊◊	807,450
◊◊◊◊◊	807,500
◊◊◊◊◊	807,550
◊◊◊◊◊	807,600
◊◊◊◊◊	807,650
◊◊◊◊◊	807,700
◊◊◊◊◊	807,750
◊◊◊◊◊	807,800
◊◊◊◊◊	807,850
◊◊◊◊◊	807,900
◊◊◊◊◊	807,950
◊◊◊◊◊	**808,000**

◇ The Number 808,001 to the Number 810,000

Start | | **Notes**

◇◇◇◇◇ ◇◇◇◇◇ ◇◇◇◇◇ ◇◇◇◇◇ ◇◇◇◇◇ ◇◇◇◇◇ ◇◇◇◇◇ ◇◇◇◇◇ ◇◇◇◇◇ ◇◇◇◇◇	808,050
◇◇◇◇◇ ◇◇◇◇◇ ◇◇◇◇◇ ◇◇◇◇◇ ◇◇◇◇◇ ◇◇◇◇◇ ◇◇◇◇◇ ◇◇◇◇◇ ◇◇◇◇◇ ◇◇◇◇◇	808,100
◇◇◇◇◇ ◇◇◇◇◇ ◇◇◇◇◇ ◇◇◇◇◇ ◇◇◇◇◇ ◇◇◇◇◇ ◇◇◇◇◇ ◇◇◇◇◇ ◇◇◇◇◇ ◇◇◇◇◇	808,150
◇◇◇◇◇ ◇◇◇◇◇ ◇◇◇◇◇ ◇◇◇◇◇ ◇◇◇◇◇ ◇◇◇◇◇ ◇◇◇◇◇ ◇◇◇◇◇ ◇◇◇◇◇ ◇◇◇◇◇	808,200
◇◇◇◇◇ ◇◇◇◇◇ ◇◇◇◇◇ ◇◇◇◇◇ ◇◇◇◇◇ ◇◇◇◇◇ ◇◇◇◇◇ ◇◇◇◇◇ ◇◇◇◇◇ ◇◇◇◇◇	808,250
◇◇◇◇◇ ◇◇◇◇◇ ◇◇◇◇◇ ◇◇◇◇◇ ◇◇◇◇◇ ◇◇◇◇◇ ◇◇◇◇◇ ◇◇◇◇◇ ◇◇◇◇◇ ◇◇◇◇◇	808,300
◇◇◇◇◇ ◇◇◇◇◇ ◇◇◇◇◇ ◇◇◇◇◇ ◇◇◇◇◇ ◇◇◇◇◇ ◇◇◇◇◇ ◇◇◇◇◇ ◇◇◇◇◇ ◇◇◇◇◇	808,350
◇◇◇◇◇ ◇◇◇◇◇ ◇◇◇◇◇ ◇◇◇◇◇ ◇◇◇◇◇ ◇◇◇◇◇ ◇◇◇◇◇ ◇◇◇◇◇ ◇◇◇◇◇ ◇◇◇◇◇	808,400
◇◇◇◇◇ ◇◇◇◇◇ ◇◇◇◇◇ ◇◇◇◇◇ ◇◇◇◇◇ ◇◇◇◇◇ ◇◇◇◇◇ ◇◇◇◇◇ ◇◇◇◇◇ ◇◇◇◇◇	808,450
◇◇◇◇◇ ◇◇◇◇◇ ◇◇◇◇◇ ◇◇◇◇◇ ◇◇◇◇◇ ◇◇◇◇◇ ◇◇◇◇◇ ◇◇◇◇◇ ◇◇◇◇◇ ◇◇◇◇◇	808,500
◇◇◇◇◇ ◇◇◇◇◇ ◇◇◇◇◇ ◇◇◇◇◇ ◇◇◇◇◇ ◇◇◇◇◇ ◇◇◇◇◇ ◇◇◇◇◇ ◇◇◇◇◇ ◇◇◇◇◇	808,550
◇◇◇◇◇ ◇◇◇◇◇ ◇◇◇◇◇ ◇◇◇◇◇ ◇◇◇◇◇ ◇◇◇◇◇ ◇◇◇◇◇ ◇◇◇◇◇ ◇◇◇◇◇ ◇◇◇◇◇	808,600
◇◇◇◇◇ ◇◇◇◇◇ ◇◇◇◇◇ ◇◇◇◇◇ ◇◇◇◇◇ ◇◇◇◇◇ ◇◇◇◇◇ ◇◇◇◇◇ ◇◇◇◇◇ ◇◇◇◇◇	808,650
◇◇◇◇◇ ◇◇◇◇◇ ◇◇◇◇◇ ◇◇◇◇◇ ◇◇◇◇◇ ◇◇◇◇◇ ◇◇◇◇◇ ◇◇◇◇◇ ◇◇◇◇◇ ◇◇◇◇◇	808,700
◇◇◇◇◇ ◇◇◇◇◇ ◇◇◇◇◇ ◇◇◇◇◇ ◇◇◇◇◇ ◇◇◇◇◇ ◇◇◇◇◇ ◇◇◇◇◇ ◇◇◇◇◇ ◇◇◇◇◇	808,750
◇◇◇◇◇ ◇◇◇◇◇ ◇◇◇◇◇ ◇◇◇◇◇ ◇◇◇◇◇ ◇◇◇◇◇ ◇◇◇◇◇ ◇◇◇◇◇ ◇◇◇◇◇ ◇◇◇◇◇	808,800
◇◇◇◇◇ ◇◇◇◇◇ ◇◇◇◇◇ ◇◇◇◇◇ ◇◇◇◇◇ ◇◇◇◇◇ ◇◇◇◇◇ ◇◇◇◇◇ ◇◇◇◇◇ ◇◇◇◇◇	808,850
◇◇◇◇◇ ◇◇◇◇◇ ◇◇◇◇◇ ◇◇◇◇◇ ◇◇◇◇◇ ◇◇◇◇◇ ◇◇◇◇◇ ◇◇◇◇◇ ◇◇◇◇◇ ◇◇◇◇◇	808,900
◇◇◇◇◇ ◇◇◇◇◇ ◇◇◇◇◇ ◇◇◇◇◇ ◇◇◇◇◇ ◇◇◇◇◇ ◇◇◇◇◇ ◇◇◇◇◇ ◇◇◇◇◇ ◇◇◇◇◇	808,950
◇◇◇◇◇ ◇◇◇◇◇ ◇◇◇◇◇ ◇◇◇◇◇ ◇◇◇◇◇ ◇◇◇◇◇ ◇◇◇◇◇ ◇◇◇◇◇ ◇◇◇◇◇ ◇◇◇◇◇	**809,000**
◇◇◇◇◇ ◇◇◇◇◇ ◇◇◇◇◇ ◇◇◇◇◇ ◇◇◇◇◇ ◇◇◇◇◇ ◇◇◇◇◇ ◇◇◇◇◇ ◇◇◇◇◇ ◇◇◇◇◇	809,050
◇◇◇◇◇ ◇◇◇◇◇ ◇◇◇◇◇ ◇◇◇◇◇ ◇◇◇◇◇ ◇◇◇◇◇ ◇◇◇◇◇ ◇◇◇◇◇ ◇◇◇◇◇ ◇◇◇◇◇	809,100
◇◇◇◇◇ ◇◇◇◇◇ ◇◇◇◇◇ ◇◇◇◇◇ ◇◇◇◇◇ ◇◇◇◇◇ ◇◇◇◇◇ ◇◇◇◇◇ ◇◇◇◇◇ ◇◇◇◇◇	809,150
◇◇◇◇◇ ◇◇◇◇◇ ◇◇◇◇◇ ◇◇◇◇◇ ◇◇◇◇◇ ◇◇◇◇◇ ◇◇◇◇◇ ◇◇◇◇◇ ◇◇◇◇◇ ◇◇◇◇◇	809,200
◇◇◇◇◇ ◇◇◇◇◇ ◇◇◇◇◇ ◇◇◇◇◇ ◇◇◇◇◇ ◇◇◇◇◇ ◇◇◇◇◇ ◇◇◇◇◇ ◇◇◇◇◇ ◇◇◇◇◇	809,250
◇◇◇◇◇ ◇◇◇◇◇ ◇◇◇◇◇ ◇◇◇◇◇ ◇◇◇◇◇ ◇◇◇◇◇ ◇◇◇◇◇ ◇◇◇◇◇ ◇◇◇◇◇ ◇◇◇◇◇	809,300
◇◇◇◇◇ ◇◇◇◇◇ ◇◇◇◇◇ ◇◇◇◇◇ ◇◇◇◇◇ ◇◇◇◇◇ ◇◇◇◇◇ ◇◇◇◇◇ ◇◇◇◇◇ ◇◇◇◇◇	809,350
◇◇◇◇◇ ◇◇◇◇◇ ◇◇◇◇◇ ◇◇◇◇◇ ◇◇◇◇◇ ◇◇◇◇◇ ◇◇◇◇◇ ◇◇◇◇◇ ◇◇◇◇◇ ◇◇◇◇◇	809,400
◇◇◇◇◇ ◇◇◇◇◇ ◇◇◇◇◇ ◇◇◇◇◇ ◇◇◇◇◇ ◇◇◇◇◇ ◇◇◇◇◇ ◇◇◇◇◇ ◇◇◇◇◇ ◇◇◇◇◇	809,450
◇◇◇◇◇ ◇◇◇◇◇ ◇◇◇◇◇ ◇◇◇◇◇ ◇◇◇◇◇ ◇◇◇◇◇ ◇◇◇◇◇ ◇◇◇◇◇ ◇◇◇◇◇ ◇◇◇◇◇	809,500
◇◇◇◇◇ ◇◇◇◇◇ ◇◇◇◇◇ ◇◇◇◇◇ ◇◇◇◇◇ ◇◇◇◇◇ ◇◇◇◇◇ ◇◇◇◇◇ ◇◇◇◇◇ ◇◇◇◇◇	809,550
◇◇◇◇◇ ◇◇◇◇◇ ◇◇◇◇◇ ◇◇◇◇◇ ◇◇◇◇◇ ◇◇◇◇◇ ◇◇◇◇◇ ◇◇◇◇◇ ◇◇◇◇◇ ◇◇◇◇◇	809,600
◇◇◇◇◇ ◇◇◇◇◇ ◇◇◇◇◇ ◇◇◇◇◇ ◇◇◇◇◇ ◇◇◇◇◇ ◇◇◇◇◇ ◇◇◇◇◇ ◇◇◇◇◇ ◇◇◇◇◇	809,650
◇◇◇◇◇ ◇◇◇◇◇ ◇◇◇◇◇ ◇◇◇◇◇ ◇◇◇◇◇ ◇◇◇◇◇ ◇◇◇◇◇ ◇◇◇◇◇ ◇◇◇◇◇ ◇◇◇◇◇	809,700
◇◇◇◇◇ ◇◇◇◇◇ ◇◇◇◇◇ ◇◇◇◇◇ ◇◇◇◇◇ ◇◇◇◇◇ ◇◇◇◇◇ ◇◇◇◇◇ ◇◇◇◇◇ ◇◇◇◇◇	809,750
◇◇◇◇◇ ◇◇◇◇◇ ◇◇◇◇◇ ◇◇◇◇◇ ◇◇◇◇◇ ◇◇◇◇◇ ◇◇◇◇◇ ◇◇◇◇◇ ◇◇◇◇◇ ◇◇◇◇◇	809,800
◇◇◇◇◇ ◇◇◇◇◇ ◇◇◇◇◇ ◇◇◇◇◇ ◇◇◇◇◇ ◇◇◇◇◇ ◇◇◇◇◇ ◇◇◇◇◇ ◇◇◇◇◇ ◇◇◇◇◇	809,850
◇◇◇◇◇ ◇◇◇◇◇ ◇◇◇◇◇ ◇◇◇◇◇ ◇◇◇◇◇ ◇◇◇◇◇ ◇◇◇◇◇ ◇◇◇◇◇ ◇◇◇◇◇ ◇◇◇◇◇	809,900
◇◇◇◇◇ ◇◇◇◇◇ ◇◇◇◇◇ ◇◇◇◇◇ ◇◇◇◇◇ ◇◇◇◇◇ ◇◇◇◇◇ ◇◇◇◇◇ ◇◇◇◇◇ ◇◇◇◇◇	809,950
◇◇◇◇◇ ◇◇◇◇◇ ◇◇◇◇◇ ◇◇◇◇◇ ◇◇◇◇◇ ◇◇◇◇◇ ◇◇◇◇◇ ◇◇◇◇◇ ◇◇◇◇◇ ◇◇◇◇◇	**810,000**

The Number 810,001 to the Number 812,000

Start		Notes

Number	Notes
810,050	
810,100	
810,150	
810,200	
810,250	
810,300	
810,350	
810,400	
810,450	
810,500	
810,550	
810,600	
810,650	
810,700	
810,750	
810,800	
810,850	
810,900	
810,950	
811,000	
811,050	
811,100	
811,150	
811,200	
811,250	
811,300	
811,350	
811,400	
811,450	
811,500	
811,550	
811,600	
811,650	
811,700	
811,750	
811,800	
811,850	
811,900	
811,950	
812,000	

The Number 812,001 to the Number 814,000

Start   Notes

	812,050
	812,100
	812,150
	812,200
	812,250
	812,300
	812,350
	812,400
	812,450
	812,500
	812,550
	812,600
	812,650
	812,700
	812,750
	812,800
	812,850
	812,900
	812,950
	813,000
	813,050
	813,100
	813,150
	813,200
	813,250
	813,300
	813,350
	813,400
	813,450
	813,500
	813,550
	813,600
	813,650
	813,700
	813,750
	813,800
	813,850
	813,900
	813,950
	814,000

The Number 814,001 to the Number 816,000

Start | Notes

	814,050
	814,100
	814,150
	814,200
	814,250
	814,300
	814,350
	814,400
	814,450
	814,500
	814,550
	814,600
	814,650
	814,700
	814,750
	814,800
	814,850
	814,900
	814,950
	815,000
	815,050
	815,100
	815,150
	815,200
	815,250
	815,300
	815,350
	815,400
	815,450
	815,500
	815,550
	815,600
	815,650
	815,700
	815,750
	815,800
	815,850
	815,900
	815,950
	816,000

The Number 816,001 to the Number 818,000

Start   Notes

◇◇◇◇◇ ◇◇◇◇◇ ◇◇◇◇◇ ◇◇◇◇◇ ◇◇◇◇◇ ◇◇◇◇◇ ◇◇◇◇◇ ◇◇◇◇◇ ◇◇◇◇◇ ◇◇◇◇◇	816,050
◇◇◇◇◇ ◇◇◇◇◇ ◇◇◇◇◇ ◇◇◇◇◇ ◇◇◇◇◇ ◇◇◇◇◇ ◇◇◇◇◇ ◇◇◇◇◇ ◇◇◇◇◇ ◇◇◇◇◇	816,100
◇◇◇◇◇ ◇◇◇◇◇ ◇◇◇◇◇ ◇◇◇◇◇ ◇◇◇◇◇ ◇◇◇◇◇ ◇◇◇◇◇ ◇◇◇◇◇ ◇◇◇◇◇ ◇◇◇◇◇	816,150
◇◇◇◇◇ ◇◇◇◇◇ ◇◇◇◇◇ ◇◇◇◇◇ ◇◇◇◇◇ ◇◇◇◇◇ ◇◇◇◇◇ ◇◇◇◇◇ ◇◇◇◇◇ ◇◇◇◇◇	816,200
◇◇◇◇◇ ◇◇◇◇◇ ◇◇◇◇◇ ◇◇◇◇◇ ◇◇◇◇◇ ◇◇◇◇◇ ◇◇◇◇◇ ◇◇◇◇◇ ◇◇◇◇◇ ◇◇◇◇◇	816,250
◇◇◇◇◇ ◇◇◇◇◇ ◇◇◇◇◇ ◇◇◇◇◇ ◇◇◇◇◇ ◇◇◇◇◇ ◇◇◇◇◇ ◇◇◇◇◇ ◇◇◇◇◇ ◇◇◇◇◇	816,300
◇◇◇◇◇ ◇◇◇◇◇ ◇◇◇◇◇ ◇◇◇◇◇ ◇◇◇◇◇ ◇◇◇◇◇ ◇◇◇◇◇ ◇◇◇◇◇ ◇◇◇◇◇ ◇◇◇◇◇	816,350
◇◇◇◇◇ ◇◇◇◇◇ ◇◇◇◇◇ ◇◇◇◇◇ ◇◇◇◇◇ ◇◇◇◇◇ ◇◇◇◇◇ ◇◇◇◇◇ ◇◇◇◇◇ ◇◇◇◇◇	816,400
◇◇◇◇◇ ◇◇◇◇◇ ◇◇◇◇◇ ◇◇◇◇◇ ◇◇◇◇◇ ◇◇◇◇◇ ◇◇◇◇◇ ◇◇◇◇◇ ◇◇◇◇◇ ◇◇◇◇◇	816,450
◇◇◇◇◇ ◇◇◇◇◇ ◇◇◇◇◇ ◇◇◇◇◇ ◇◇◇◇◇ ◇◇◇◇◇ ◇◇◇◇◇ ◇◇◇◇◇ ◇◇◇◇◇ ◇◇◇◇◇	816,500
◇◇◇◇◇ ◇◇◇◇◇ ◇◇◇◇◇ ◇◇◇◇◇ ◇◇◇◇◇ ◇◇◇◇◇ ◇◇◇◇◇ ◇◇◇◇◇ ◇◇◇◇◇ ◇◇◇◇◇	816,550
◇◇◇◇◇ ◇◇◇◇◇ ◇◇◇◇◇ ◇◇◇◇◇ ◇◇◇◇◇ ◇◇◇◇◇ ◇◇◇◇◇ ◇◇◇◇◇ ◇◇◇◇◇ ◇◇◇◇◇	816,600
◇◇◇◇◇ ◇◇◇◇◇ ◇◇◇◇◇ ◇◇◇◇◇ ◇◇◇◇◇ ◇◇◇◇◇ ◇◇◇◇◇ ◇◇◇◇◇ ◇◇◇◇◇ ◇◇◇◇◇	816,650
◇◇◇◇◇ ◇◇◇◇◇ ◇◇◇◇◇ ◇◇◇◇◇ ◇◇◇◇◇ ◇◇◇◇◇ ◇◇◇◇◇ ◇◇◇◇◇ ◇◇◇◇◇ ◇◇◇◇◇	816,700
◇◇◇◇◇ ◇◇◇◇◇ ◇◇◇◇◇ ◇◇◇◇◇ ◇◇◇◇◇ ◇◇◇◇◇ ◇◇◇◇◇ ◇◇◇◇◇ ◇◇◇◇◇ ◇◇◇◇◇	816,750
◇◇◇◇◇ ◇◇◇◇◇ ◇◇◇◇◇ ◇◇◇◇◇ ◇◇◇◇◇ ◇◇◇◇◇ ◇◇◇◇◇ ◇◇◇◇◇ ◇◇◇◇◇ ◇◇◇◇◇	816,800
◇◇◇◇◇ ◇◇◇◇◇ ◇◇◇◇◇ ◇◇◇◇◇ ◇◇◇◇◇ ◇◇◇◇◇ ◇◇◇◇◇ ◇◇◇◇◇ ◇◇◇◇◇ ◇◇◇◇◇	816,850
◇◇◇◇◇ ◇◇◇◇◇ ◇◇◇◇◇ ◇◇◇◇◇ ◇◇◇◇◇ ◇◇◇◇◇ ◇◇◇◇◇ ◇◇◇◇◇ ◇◇◇◇◇ ◇◇◇◇◇	816,900
◇◇◇◇◇ ◇◇◇◇◇ ◇◇◇◇◇ ◇◇◇◇◇ ◇◇◇◇◇ ◇◇◇◇◇ ◇◇◇◇◇ ◇◇◇◇◇ ◇◇◇◇◇ ◇◇◇◇◇	816,950
◇◇◇◇◇ ◇◇◇◇◇ ◇◇◇◇◇ ◇◇◇◇◇ ◇◇◇◇◇ ◇◇◇◇◇ ◇◇◇◇◇ ◇◇◇◇◇ ◇◇◇◇◇ ◇◇◇◇◇	**817,000**
◇◇◇◇◇ ◇◇◇◇◇ ◇◇◇◇◇ ◇◇◇◇◇ ◇◇◇◇◇ ◇◇◇◇◇ ◇◇◇◇◇ ◇◇◇◇◇ ◇◇◇◇◇ ◇◇◇◇◇	817,050
◇◇◇◇◇ ◇◇◇◇◇ ◇◇◇◇◇ ◇◇◇◇◇ ◇◇◇◇◇ ◇◇◇◇◇ ◇◇◇◇◇ ◇◇◇◇◇ ◇◇◇◇◇ ◇◇◇◇◇	817,100
◇◇◇◇◇ ◇◇◇◇◇ ◇◇◇◇◇ ◇◇◇◇◇ ◇◇◇◇◇ ◇◇◇◇◇ ◇◇◇◇◇ ◇◇◇◇◇ ◇◇◇◇◇ ◇◇◇◇◇	817,150
◇◇◇◇◇ ◇◇◇◇◇ ◇◇◇◇◇ ◇◇◇◇◇ ◇◇◇◇◇ ◇◇◇◇◇ ◇◇◇◇◇ ◇◇◇◇◇ ◇◇◇◇◇ ◇◇◇◇◇	817,200
◇◇◇◇◇ ◇◇◇◇◇ ◇◇◇◇◇ ◇◇◇◇◇ ◇◇◇◇◇ ◇◇◇◇◇ ◇◇◇◇◇ ◇◇◇◇◇ ◇◇◇◇◇ ◇◇◇◇◇	817,250
◇◇◇◇◇ ◇◇◇◇◇ ◇◇◇◇◇ ◇◇◇◇◇ ◇◇◇◇◇ ◇◇◇◇◇ ◇◇◇◇◇ ◇◇◇◇◇ ◇◇◇◇◇ ◇◇◇◇◇	817,300
◇◇◇◇◇ ◇◇◇◇◇ ◇◇◇◇◇ ◇◇◇◇◇ ◇◇◇◇◇ ◇◇◇◇◇ ◇◇◇◇◇ ◇◇◇◇◇ ◇◇◇◇◇ ◇◇◇◇◇	817,350
◇◇◇◇◇ ◇◇◇◇◇ ◇◇◇◇◇ ◇◇◇◇◇ ◇◇◇◇◇ ◇◇◇◇◇ ◇◇◇◇◇ ◇◇◇◇◇ ◇◇◇◇◇ ◇◇◇◇◇	817,400
◇◇◇◇◇ ◇◇◇◇◇ ◇◇◇◇◇ ◇◇◇◇◇ ◇◇◇◇◇ ◇◇◇◇◇ ◇◇◇◇◇ ◇◇◇◇◇ ◇◇◇◇◇ ◇◇◇◇◇	817,450
◇◇◇◇◇ ◇◇◇◇◇ ◇◇◇◇◇ ◇◇◇◇◇ ◇◇◇◇◇ ◇◇◇◇◇ ◇◇◇◇◇ ◇◇◇◇◇ ◇◇◇◇◇ ◇◇◇◇◇	817,500
◇◇◇◇◇ ◇◇◇◇◇ ◇◇◇◇◇ ◇◇◇◇◇ ◇◇◇◇◇ ◇◇◇◇◇ ◇◇◇◇◇ ◇◇◇◇◇ ◇◇◇◇◇ ◇◇◇◇◇	817,550
◇◇◇◇◇ ◇◇◇◇◇ ◇◇◇◇◇ ◇◇◇◇◇ ◇◇◇◇◇ ◇◇◇◇◇ ◇◇◇◇◇ ◇◇◇◇◇ ◇◇◇◇◇ ◇◇◇◇◇	817,600
◇◇◇◇◇ ◇◇◇◇◇ ◇◇◇◇◇ ◇◇◇◇◇ ◇◇◇◇◇ ◇◇◇◇◇ ◇◇◇◇◇ ◇◇◇◇◇ ◇◇◇◇◇ ◇◇◇◇◇	817,650
◇◇◇◇◇ ◇◇◇◇◇ ◇◇◇◇◇ ◇◇◇◇◇ ◇◇◇◇◇ ◇◇◇◇◇ ◇◇◇◇◇ ◇◇◇◇◇ ◇◇◇◇◇ ◇◇◇◇◇	817,700
◇◇◇◇◇ ◇◇◇◇◇ ◇◇◇◇◇ ◇◇◇◇◇ ◇◇◇◇◇ ◇◇◇◇◇ ◇◇◇◇◇ ◇◇◇◇◇ ◇◇◇◇◇ ◇◇◇◇◇	817,750
◇◇◇◇◇ ◇◇◇◇◇ ◇◇◇◇◇ ◇◇◇◇◇ ◇◇◇◇◇ ◇◇◇◇◇ ◇◇◇◇◇ ◇◇◇◇◇ ◇◇◇◇◇ ◇◇◇◇◇	817,800
◇◇◇◇◇ ◇◇◇◇◇ ◇◇◇◇◇ ◇◇◇◇◇ ◇◇◇◇◇ ◇◇◇◇◇ ◇◇◇◇◇ ◇◇◇◇◇ ◇◇◇◇◇ ◇◇◇◇◇	817,850
◇◇◇◇◇ ◇◇◇◇◇ ◇◇◇◇◇ ◇◇◇◇◇ ◇◇◇◇◇ ◇◇◇◇◇ ◇◇◇◇◇ ◇◇◇◇◇ ◇◇◇◇◇ ◇◇◇◇◇	817,900
◇◇◇◇◇ ◇◇◇◇◇ ◇◇◇◇◇ ◇◇◇◇◇ ◇◇◇◇◇ ◇◇◇◇◇ ◇◇◇◇◇ ◇◇◇◇◇ ◇◇◇◇◇ ◇◇◇◇◇	817,950
◇◇◇◇◇ ◇◇◇◇◇ ◇◇◇◇◇ ◇◇◇◇◇ ◇◇◇◇◇ ◇◇◇◇◇ ◇◇◇◇◇ ◇◇◇◇◇ ◇◇◇◇◇ ◇◇◇◇◇	**818,000**

The Number 818,001 to the Number 820,000

Start

	Notes
	818,050
	818,100
	818,150
	818,200
	818,250
	818,300
	818,350
	818,400
	818,450
	818,500
	818,550
	818,600
	818,650
	818,700
	818,750
	818,800
	818,850
	818,900
	818,950
	819,000
	819,050
	819,100
	819,150
	819,200
	819,250
	819,300
	819,350
	819,400
	819,450
	819,500
	819,550
	819,600
	819,650
	819,700
	819,750
	819,800
	819,850
	819,900
	819,950
	820,000

The Number 820,001 to the Number 822,000

Start

	Notes
	820,050
	820,100
	820,150
	820,200
	820,250
	820,300
	820,350
	820,400
	820,450
	820,500
	820,550
	820,600
	820,650
	820,700
	820,750
	820,800
	820,850
	820,900
	820,950
	821,000
	821,050
	821,100
	821,150
	821,200
	821,250
	821,300
	821,350
	821,400
	821,450
	821,500
	821,550
	821,600
	821,650
	821,700
	821,750
	821,800
	821,850
	821,900
	821,950
	822,000

The Number 822,001 to the Number 824,000

Start	Notes
	822,050
	822,100
	822,150
	822,200
	822,250
	822,300
	822,350
	822,400
	822,450
	822,500
	822,550
	822,600
	822,650
	822,700
	822,750
	822,800
	822,850
	822,900
	822,950
	823,000
	823,050
	823,100
	823,150
	823,200
	823,250
	823,300
	823,350
	823,400
	823,450
	823,500
	823,550
	823,600
	823,650
	823,700
	823,750
	823,800
	823,850
	823,900
	823,950
	824,000

The Number 824,001 to the Number 826,000

Start **Notes**

	824,050	
	824,100	
	824,150	
	824,200	
	824,250	
	824,300	
	824,350	
	824,400	
	824,450	
	824,500	
	824,550	
	824,600	
	824,650	
	824,700	
	824,750	
	824,800	
	824,850	
	824,900	
	824,950	
	825,000	
	825,050	
	825,100	
	825,150	
	825,200	
	825,250	
	825,300	
	825,350	
	825,400	
	825,450	
	825,500	
	825,550	
	825,600	
	825,650	
	825,700	
	825,750	
	825,800	
	825,850	
	825,900	
	825,950	
	826,000	

The Number 826,001 to the Number 828,000

Start

Notes

	826,050
	826,100
	826,150
	826,200
	826,250
	826,300
	826,350
	826,400
	826,450
	826,500
	826,550
	826,600
	826,650
	826,700
	826,750
	826,800
	826,850
	826,900
	826,950
	827,000
	827,050
	827,100
	827,150
	827,200
	827,250
	827,300
	827,350
	827,400
	827,450
	827,500
	827,550
	827,600
	827,650
	827,700
	827,750
	827,800
	827,850
	827,900
	827,950
	828,000

The Number 828,001 to the Number 830,000

Start   Notes

◊◊◊◊◊ ◊◊◊◊◊ ◊◊◊◊◊ ◊◊◊◊◊ ◊◊◊◊◊ ◊◊◊◊◊ ◊◊◊◊◊ ◊◊◊◊◊ ◊◊◊◊◊ ◊◊◊◊◊	828,050
◊◊◊◊◊ ◊◊◊◊◊ ◊◊◊◊◊ ◊◊◊◊◊ ◊◊◊◊◊ ◊◊◊◊◊ ◊◊◊◊◊ ◊◊◊◊◊ ◊◊◊◊◊ ◊◊◊◊◊	828,100
◊◊◊◊◊ ◊◊◊◊◊ ◊◊◊◊◊ ◊◊◊◊◊ ◊◊◊◊◊ ◊◊◊◊◊ ◊◊◊◊◊ ◊◊◊◊◊ ◊◊◊◊◊ ◊◊◊◊◊	828,150
◊◊◊◊◊ ◊◊◊◊◊ ◊◊◊◊◊ ◊◊◊◊◊ ◊◊◊◊◊ ◊◊◊◊◊ ◊◊◊◊◊ ◊◊◊◊◊ ◊◊◊◊◊ ◊◊◊◊◊	828,200
◊◊◊◊◊ ◊◊◊◊◊ ◊◊◊◊◊ ◊◊◊◊◊ ◊◊◊◊◊ ◊◊◊◊◊ ◊◊◊◊◊ ◊◊◊◊◊ ◊◊◊◊◊ ◊◊◊◊◊	828,250
◊◊◊◊◊ ◊◊◊◊◊ ◊◊◊◊◊ ◊◊◊◊◊ ◊◊◊◊◊ ◊◊◊◊◊ ◊◊◊◊◊ ◊◊◊◊◊ ◊◊◊◊◊ ◊◊◊◊◊	828,300
◊◊◊◊◊ ◊◊◊◊◊ ◊◊◊◊◊ ◊◊◊◊◊ ◊◊◊◊◊ ◊◊◊◊◊ ◊◊◊◊◊ ◊◊◊◊◊ ◊◊◊◊◊ ◊◊◊◊◊	828,350
◊◊◊◊◊ ◊◊◊◊◊ ◊◊◊◊◊ ◊◊◊◊◊ ◊◊◊◊◊ ◊◊◊◊◊ ◊◊◊◊◊ ◊◊◊◊◊ ◊◊◊◊◊ ◊◊◊◊◊	828,400
◊◊◊◊◊ ◊◊◊◊◊ ◊◊◊◊◊ ◊◊◊◊◊ ◊◊◊◊◊ ◊◊◊◊◊ ◊◊◊◊◊ ◊◊◊◊◊ ◊◊◊◊◊ ◊◊◊◊◊	828,450
◊◊◊◊◊ ◊◊◊◊◊ ◊◊◊◊◊ ◊◊◊◊◊ ◊◊◊◊◊ ◊◊◊◊◊ ◊◊◊◊◊ ◊◊◊◊◊ ◊◊◊◊◊ ◊◊◊◊◊	828,500
◊◊◊◊◊ ◊◊◊◊◊ ◊◊◊◊◊ ◊◊◊◊◊ ◊◊◊◊◊ ◊◊◊◊◊ ◊◊◊◊◊ ◊◊◊◊◊ ◊◊◊◊◊ ◊◊◊◊◊	828,550
◊◊◊◊◊ ◊◊◊◊◊ ◊◊◊◊◊ ◊◊◊◊◊ ◊◊◊◊◊ ◊◊◊◊◊ ◊◊◊◊◊ ◊◊◊◊◊ ◊◊◊◊◊ ◊◊◊◊◊	828,600
◊◊◊◊◊ ◊◊◊◊◊ ◊◊◊◊◊ ◊◊◊◊◊ ◊◊◊◊◊ ◊◊◊◊◊ ◊◊◊◊◊ ◊◊◊◊◊ ◊◊◊◊◊ ◊◊◊◊◊	828,650
◊◊◊◊◊ ◊◊◊◊◊ ◊◊◊◊◊ ◊◊◊◊◊ ◊◊◊◊◊ ◊◊◊◊◊ ◊◊◊◊◊ ◊◊◊◊◊ ◊◊◊◊◊ ◊◊◊◊◊	828,700
◊◊◊◊◊ ◊◊◊◊◊ ◊◊◊◊◊ ◊◊◊◊◊ ◊◊◊◊◊ ◊◊◊◊◊ ◊◊◊◊◊ ◊◊◊◊◊ ◊◊◊◊◊ ◊◊◊◊◊	828,750
◊◊◊◊◊ ◊◊◊◊◊ ◊◊◊◊◊ ◊◊◊◊◊ ◊◊◊◊◊ ◊◊◊◊◊ ◊◊◊◊◊ ◊◊◊◊◊ ◊◊◊◊◊ ◊◊◊◊◊	828,800
◊◊◊◊◊ ◊◊◊◊◊ ◊◊◊◊◊ ◊◊◊◊◊ ◊◊◊◊◊ ◊◊◊◊◊ ◊◊◊◊◊ ◊◊◊◊◊ ◊◊◊◊◊ ◊◊◊◊◊	828,850
◊◊◊◊◊ ◊◊◊◊◊ ◊◊◊◊◊ ◊◊◊◊◊ ◊◊◊◊◊ ◊◊◊◊◊ ◊◊◊◊◊ ◊◊◊◊◊ ◊◊◊◊◊ ◊◊◊◊◊	828,900
◊◊◊◊◊ ◊◊◊◊◊ ◊◊◊◊◊ ◊◊◊◊◊ ◊◊◊◊◊ ◊◊◊◊◊ ◊◊◊◊◊ ◊◊◊◊◊ ◊◊◊◊◊ ◊◊◊◊◊	828,950
◊◊◊◊◊ ◊◊◊◊◊ ◊◊◊◊◊ ◊◊◊◊◊ ◊◊◊◊◊ ◊◊◊◊◊ ◊◊◊◊◊ ◊◊◊◊◊ ◊◊◊◊◊ ◊◊◊◊◊	**829,000**
◊◊◊◊◊ ◊◊◊◊◊ ◊◊◊◊◊ ◊◊◊◊◊ ◊◊◊◊◊ ◊◊◊◊◊ ◊◊◊◊◊ ◊◊◊◊◊ ◊◊◊◊◊ ◊◊◊◊◊	829,050
◊◊◊◊◊ ◊◊◊◊◊ ◊◊◊◊◊ ◊◊◊◊◊ ◊◊◊◊◊ ◊◊◊◊◊ ◊◊◊◊◊ ◊◊◊◊◊ ◊◊◊◊◊ ◊◊◊◊◊	829,100
◊◊◊◊◊ ◊◊◊◊◊ ◊◊◊◊◊ ◊◊◊◊◊ ◊◊◊◊◊ ◊◊◊◊◊ ◊◊◊◊◊ ◊◊◊◊◊ ◊◊◊◊◊ ◊◊◊◊◊	829,150
◊◊◊◊◊ ◊◊◊◊◊ ◊◊◊◊◊ ◊◊◊◊◊ ◊◊◊◊◊ ◊◊◊◊◊ ◊◊◊◊◊ ◊◊◊◊◊ ◊◊◊◊◊ ◊◊◊◊◊	829,200
◊◊◊◊◊ ◊◊◊◊◊ ◊◊◊◊◊ ◊◊◊◊◊ ◊◊◊◊◊ ◊◊◊◊◊ ◊◊◊◊◊ ◊◊◊◊◊ ◊◊◊◊◊ ◊◊◊◊◊	829,250
◊◊◊◊◊ ◊◊◊◊◊ ◊◊◊◊◊ ◊◊◊◊◊ ◊◊◊◊◊ ◊◊◊◊◊ ◊◊◊◊◊ ◊◊◊◊◊ ◊◊◊◊◊ ◊◊◊◊◊	829,300
◊◊◊◊◊ ◊◊◊◊◊ ◊◊◊◊◊ ◊◊◊◊◊ ◊◊◊◊◊ ◊◊◊◊◊ ◊◊◊◊◊ ◊◊◊◊◊ ◊◊◊◊◊ ◊◊◊◊◊	829,350
◊◊◊◊◊ ◊◊◊◊◊ ◊◊◊◊◊ ◊◊◊◊◊ ◊◊◊◊◊ ◊◊◊◊◊ ◊◊◊◊◊ ◊◊◊◊◊ ◊◊◊◊◊ ◊◊◊◊◊	829,400
◊◊◊◊◊ ◊◊◊◊◊ ◊◊◊◊◊ ◊◊◊◊◊ ◊◊◊◊◊ ◊◊◊◊◊ ◊◊◊◊◊ ◊◊◊◊◊ ◊◊◊◊◊ ◊◊◊◊◊	829,450
◊◊◊◊◊ ◊◊◊◊◊ ◊◊◊◊◊ ◊◊◊◊◊ ◊◊◊◊◊ ◊◊◊◊◊ ◊◊◊◊◊ ◊◊◊◊◊ ◊◊◊◊◊ ◊◊◊◊◊	829,500
◊◊◊◊◊ ◊◊◊◊◊ ◊◊◊◊◊ ◊◊◊◊◊ ◊◊◊◊◊ ◊◊◊◊◊ ◊◊◊◊◊ ◊◊◊◊◊ ◊◊◊◊◊ ◊◊◊◊◊	829,550
◊◊◊◊◊ ◊◊◊◊◊ ◊◊◊◊◊ ◊◊◊◊◊ ◊◊◊◊◊ ◊◊◊◊◊ ◊◊◊◊◊ ◊◊◊◊◊ ◊◊◊◊◊ ◊◊◊◊◊	829,600
◊◊◊◊◊ ◊◊◊◊◊ ◊◊◊◊◊ ◊◊◊◊◊ ◊◊◊◊◊ ◊◊◊◊◊ ◊◊◊◊◊ ◊◊◊◊◊ ◊◊◊◊◊ ◊◊◊◊◊	829,650
◊◊◊◊◊ ◊◊◊◊◊ ◊◊◊◊◊ ◊◊◊◊◊ ◊◊◊◊◊ ◊◊◊◊◊ ◊◊◊◊◊ ◊◊◊◊◊ ◊◊◊◊◊ ◊◊◊◊◊	829,700
◊◊◊◊◊ ◊◊◊◊◊ ◊◊◊◊◊ ◊◊◊◊◊ ◊◊◊◊◊ ◊◊◊◊◊ ◊◊◊◊◊ ◊◊◊◊◊ ◊◊◊◊◊ ◊◊◊◊◊	829,750
◊◊◊◊◊ ◊◊◊◊◊ ◊◊◊◊◊ ◊◊◊◊◊ ◊◊◊◊◊ ◊◊◊◊◊ ◊◊◊◊◊ ◊◊◊◊◊ ◊◊◊◊◊ ◊◊◊◊◊	829,800
◊◊◊◊◊ ◊◊◊◊◊ ◊◊◊◊◊ ◊◊◊◊◊ ◊◊◊◊◊ ◊◊◊◊◊ ◊◊◊◊◊ ◊◊◊◊◊ ◊◊◊◊◊ ◊◊◊◊◊	829,850
◊◊◊◊◊ ◊◊◊◊◊ ◊◊◊◊◊ ◊◊◊◊◊ ◊◊◊◊◊ ◊◊◊◊◊ ◊◊◊◊◊ ◊◊◊◊◊ ◊◊◊◊◊ ◊◊◊◊◊	829,900
◊◊◊◊◊ ◊◊◊◊◊ ◊◊◊◊◊ ◊◊◊◊◊ ◊◊◊◊◊ ◊◊◊◊◊ ◊◊◊◊◊ ◊◊◊◊◊ ◊◊◊◊◊ ◊◊◊◊◊	829,950
◊◊◊◊◊ ◊◊◊◊◊ ◊◊◊◊◊ ◊◊◊◊◊ ◊◊◊◊◊ ◊◊◊◊◊ ◊◊◊◊◊ ◊◊◊◊◊ ◊◊◊◊◊ ◊◊◊◊◊	**830,000**

The Number 830,001 to the Number 832,000

Start		Notes
	830,050
	830,100
	830,150
	830,200
	830,250
	830,300
	830,350
	830,400
	830,450
	830,500
	830,550
	830,600
	830,650
	830,700
	830,750
	830,800
	830,850
	830,900
	830,950
	831,000
	831,050
	831,100
	831,150
	831,200
	831,250
	831,300
	831,350
	831,400
	831,450
	831,500
	831,550
	831,600
	831,650
	831,700
	831,750
	831,800
	831,850
	831,900
	831,950
	832,000

The Number 832,001 to the Number 834,000

Start   Notes

◇◇◇◇◇ ◇◇◇◇◇ ◇◇◇◇◇ ◇◇◇◇◇ ◇◇◇◇◇ ◇◇◇◇◇ ◇◇◇◇◇ ◇◇◇◇◇ ◇◇◇◇◇ ◇◇◇◇◇	832,050
◇◇◇◇◇ ◇◇◇◇◇ ◇◇◇◇◇ ◇◇◇◇◇ ◇◇◇◇◇ ◇◇◇◇◇ ◇◇◇◇◇ ◇◇◇◇◇ ◇◇◇◇◇ ◇◇◇◇◇	832,100
◇◇◇◇◇ ◇◇◇◇◇ ◇◇◇◇◇ ◇◇◇◇◇ ◇◇◇◇◇ ◇◇◇◇◇ ◇◇◇◇◇ ◇◇◇◇◇ ◇◇◇◇◇ ◇◇◇◇◇	832,150
◇◇◇◇◇ ◇◇◇◇◇ ◇◇◇◇◇ ◇◇◇◇◇ ◇◇◇◇◇ ◇◇◇◇◇ ◇◇◇◇◇ ◇◇◇◇◇ ◇◇◇◇◇ ◇◇◇◇◇	832,200
◇◇◇◇◇ ◇◇◇◇◇ ◇◇◇◇◇ ◇◇◇◇◇ ◇◇◇◇◇ ◇◇◇◇◇ ◇◇◇◇◇ ◇◇◇◇◇ ◇◇◇◇◇ ◇◇◇◇◇	832,250
◇◇◇◇◇ ◇◇◇◇◇ ◇◇◇◇◇ ◇◇◇◇◇ ◇◇◇◇◇ ◇◇◇◇◇ ◇◇◇◇◇ ◇◇◇◇◇ ◇◇◇◇◇ ◇◇◇◇◇	832,300
◇◇◇◇◇ ◇◇◇◇◇ ◇◇◇◇◇ ◇◇◇◇◇ ◇◇◇◇◇ ◇◇◇◇◇ ◇◇◇◇◇ ◇◇◇◇◇ ◇◇◇◇◇ ◇◇◇◇◇	832,350
◇◇◇◇◇ ◇◇◇◇◇ ◇◇◇◇◇ ◇◇◇◇◇ ◇◇◇◇◇ ◇◇◇◇◇ ◇◇◇◇◇ ◇◇◇◇◇ ◇◇◇◇◇ ◇◇◇◇◇	832,400
◇◇◇◇◇ ◇◇◇◇◇ ◇◇◇◇◇ ◇◇◇◇◇ ◇◇◇◇◇ ◇◇◇◇◇ ◇◇◇◇◇ ◇◇◇◇◇ ◇◇◇◇◇ ◇◇◇◇◇	832,450
◇◇◇◇◇ ◇◇◇◇◇ ◇◇◇◇◇ ◇◇◇◇◇ ◇◇◇◇◇ ◇◇◇◇◇ ◇◇◇◇◇ ◇◇◇◇◇ ◇◇◇◇◇ ◇◇◇◇◇	832,500
◇◇◇◇◇ ◇◇◇◇◇ ◇◇◇◇◇ ◇◇◇◇◇ ◇◇◇◇◇ ◇◇◇◇◇ ◇◇◇◇◇ ◇◇◇◇◇ ◇◇◇◇◇ ◇◇◇◇◇	832,550
◇◇◇◇◇ ◇◇◇◇◇ ◇◇◇◇◇ ◇◇◇◇◇ ◇◇◇◇◇ ◇◇◇◇◇ ◇◇◇◇◇ ◇◇◇◇◇ ◇◇◇◇◇ ◇◇◇◇◇	832,600
◇◇◇◇◇ ◇◇◇◇◇ ◇◇◇◇◇ ◇◇◇◇◇ ◇◇◇◇◇ ◇◇◇◇◇ ◇◇◇◇◇ ◇◇◇◇◇ ◇◇◇◇◇ ◇◇◇◇◇	832,650
◇◇◇◇◇ ◇◇◇◇◇ ◇◇◇◇◇ ◇◇◇◇◇ ◇◇◇◇◇ ◇◇◇◇◇ ◇◇◇◇◇ ◇◇◇◇◇ ◇◇◇◇◇ ◇◇◇◇◇	832,700
◇◇◇◇◇ ◇◇◇◇◇ ◇◇◇◇◇ ◇◇◇◇◇ ◇◇◇◇◇ ◇◇◇◇◇ ◇◇◇◇◇ ◇◇◇◇◇ ◇◇◇◇◇ ◇◇◇◇◇	832,750
◇◇◇◇◇ ◇◇◇◇◇ ◇◇◇◇◇ ◇◇◇◇◇ ◇◇◇◇◇ ◇◇◇◇◇ ◇◇◇◇◇ ◇◇◇◇◇ ◇◇◇◇◇ ◇◇◇◇◇	832,800
◇◇◇◇◇ ◇◇◇◇◇ ◇◇◇◇◇ ◇◇◇◇◇ ◇◇◇◇◇ ◇◇◇◇◇ ◇◇◇◇◇ ◇◇◇◇◇ ◇◇◇◇◇ ◇◇◇◇◇	832,850
◇◇◇◇◇ ◇◇◇◇◇ ◇◇◇◇◇ ◇◇◇◇◇ ◇◇◇◇◇ ◇◇◇◇◇ ◇◇◇◇◇ ◇◇◇◇◇ ◇◇◇◇◇ ◇◇◇◇◇	832,900
◇◇◇◇◇ ◇◇◇◇◇ ◇◇◇◇◇ ◇◇◇◇◇ ◇◇◇◇◇ ◇◇◇◇◇ ◇◇◇◇◇ ◇◇◇◇◇ ◇◇◇◇◇ ◇◇◇◇◇	832,950
◇◇◇◇◇ ◇◇◇◇◇ ◇◇◇◇◇ ◇◇◇◇◇ ◇◇◇◇◇ ◇◇◇◇◇ ◇◇◇◇◇ ◇◇◇◇◇ ◇◇◇◇◇ ◇◇◇◇◇	**833,000**
◇◇◇◇◇ ◇◇◇◇◇ ◇◇◇◇◇ ◇◇◇◇◇ ◇◇◇◇◇ ◇◇◇◇◇ ◇◇◇◇◇ ◇◇◇◇◇ ◇◇◇◇◇ ◇◇◇◇◇	833,050
◇◇◇◇◇ ◇◇◇◇◇ ◇◇◇◇◇ ◇◇◇◇◇ ◇◇◇◇◇ ◇◇◇◇◇ ◇◇◇◇◇ ◇◇◇◇◇ ◇◇◇◇◇ ◇◇◇◇◇	833,100
◇◇◇◇◇ ◇◇◇◇◇ ◇◇◇◇◇ ◇◇◇◇◇ ◇◇◇◇◇ ◇◇◇◇◇ ◇◇◇◇◇ ◇◇◇◇◇ ◇◇◇◇◇ ◇◇◇◇◇	833,150
◇◇◇◇◇ ◇◇◇◇◇ ◇◇◇◇◇ ◇◇◇◇◇ ◇◇◇◇◇ ◇◇◇◇◇ ◇◇◇◇◇ ◇◇◇◇◇ ◇◇◇◇◇ ◇◇◇◇◇	833,200
◇◇◇◇◇ ◇◇◇◇◇ ◇◇◇◇◇ ◇◇◇◇◇ ◇◇◇◇◇ ◇◇◇◇◇ ◇◇◇◇◇ ◇◇◇◇◇ ◇◇◇◇◇ ◇◇◇◇◇	833,250
◇◇◇◇◇ ◇◇◇◇◇ ◇◇◇◇◇ ◇◇◇◇◇ ◇◇◇◇◇ ◇◇◇◇◇ ◇◇◇◇◇ ◇◇◇◇◇ ◇◇◇◇◇ ◇◇◇◇◇	833,300
◇◇◇◇◇ ◇◇◇◇◇ ◇◇◇◇◇ ◇◇◇◇◇ ◇◇◇◇◇ ◇◇◇◇◇ ◇◇◇◇◇ ◇◇◇◇◇ ◇◇◇◇◇ ◇◇◇◇◇	833,350
◇◇◇◇◇ ◇◇◇◇◇ ◇◇◇◇◇ ◇◇◇◇◇ ◇◇◇◇◇ ◇◇◇◇◇ ◇◇◇◇◇ ◇◇◇◇◇ ◇◇◇◇◇ ◇◇◇◇◇	833,400
◇◇◇◇◇ ◇◇◇◇◇ ◇◇◇◇◇ ◇◇◇◇◇ ◇◇◇◇◇ ◇◇◇◇◇ ◇◇◇◇◇ ◇◇◇◇◇ ◇◇◇◇◇ ◇◇◇◇◇	833,450
◇◇◇◇◇ ◇◇◇◇◇ ◇◇◇◇◇ ◇◇◇◇◇ ◇◇◇◇◇ ◇◇◇◇◇ ◇◇◇◇◇ ◇◇◇◇◇ ◇◇◇◇◇ ◇◇◇◇◇	833,500
◇◇◇◇◇ ◇◇◇◇◇ ◇◇◇◇◇ ◇◇◇◇◇ ◇◇◇◇◇ ◇◇◇◇◇ ◇◇◇◇◇ ◇◇◇◇◇ ◇◇◇◇◇ ◇◇◇◇◇	833,550
◇◇◇◇◇ ◇◇◇◇◇ ◇◇◇◇◇ ◇◇◇◇◇ ◇◇◇◇◇ ◇◇◇◇◇ ◇◇◇◇◇ ◇◇◇◇◇ ◇◇◇◇◇ ◇◇◇◇◇	833,600
◇◇◇◇◇ ◇◇◇◇◇ ◇◇◇◇◇ ◇◇◇◇◇ ◇◇◇◇◇ ◇◇◇◇◇ ◇◇◇◇◇ ◇◇◇◇◇ ◇◇◇◇◇ ◇◇◇◇◇	833,650
◇◇◇◇◇ ◇◇◇◇◇ ◇◇◇◇◇ ◇◇◇◇◇ ◇◇◇◇◇ ◇◇◇◇◇ ◇◇◇◇◇ ◇◇◇◇◇ ◇◇◇◇◇ ◇◇◇◇◇	833,700
◇◇◇◇◇ ◇◇◇◇◇ ◇◇◇◇◇ ◇◇◇◇◇ ◇◇◇◇◇ ◇◇◇◇◇ ◇◇◇◇◇ ◇◇◇◇◇ ◇◇◇◇◇ ◇◇◇◇◇	833,750
◇◇◇◇◇ ◇◇◇◇◇ ◇◇◇◇◇ ◇◇◇◇◇ ◇◇◇◇◇ ◇◇◇◇◇ ◇◇◇◇◇ ◇◇◇◇◇ ◇◇◇◇◇ ◇◇◇◇◇	833,800
◇◇◇◇◇ ◇◇◇◇◇ ◇◇◇◇◇ ◇◇◇◇◇ ◇◇◇◇◇ ◇◇◇◇◇ ◇◇◇◇◇ ◇◇◇◇◇ ◇◇◇◇◇ ◇◇◇◇◇	833,850
◇◇◇◇◇ ◇◇◇◇◇ ◇◇◇◇◇ ◇◇◇◇◇ ◇◇◇◇◇ ◇◇◇◇◇ ◇◇◇◇◇ ◇◇◇◇◇ ◇◇◇◇◇ ◇◇◇◇◇	833,900
◇◇◇◇◇ ◇◇◇◇◇ ◇◇◇◇◇ ◇◇◇◇◇ ◇◇◇◇◇ ◇◇◇◇◇ ◇◇◇◇◇ ◇◇◇◇◇ ◇◇◇◇◇ ◇◇◇◇◇	833,950
◇◇◇◇◇ ◇◇◇◇◇ ◇◇◇◇◇ ◇◇◇◇◇ ◇◇◇◇◇ ◇◇◇◇◇ ◇◇◇◇◇ ◇◇◇◇◇ ◇◇◇◇◇ ◇◇◇◇◇	**834,000**

The Number 834,001 to the Number 836,000

Start   Notes

	834,050
	834,100
	834,150
	834,200
	834,250
	834,300
	834,350
	834,400
	834,450
	834,500
	834,550
	834,600
	834,650
	834,700
	834,750
	834,800
	834,850
	834,900
	834,950
	835,000
	835,050
	835,100
	835,150
	835,200
	835,250
	835,300
	835,350
	835,400
	835,450
	835,500
	835,550
	835,600
	835,650
	835,700
	835,750
	835,800
	835,850
	835,900
	835,950
	836,000

The Number 836,001 to the Number 838,000

Start   Notes

◊◊◊◊◊ ◊◊◊◊◊ ◊◊◊◊◊ ◊◊◊◊◊ ◊◊◊◊◊ ◊◊◊◊◊ ◊◊◊◊◊ ◊◊◊◊◊ ◊◊◊◊◊ ◊◊◊◊◊	836,050
◊◊◊◊◊ ◊◊◊◊◊ ◊◊◊◊◊ ◊◊◊◊◊ ◊◊◊◊◊ ◊◊◊◊◊ ◊◊◊◊◊ ◊◊◊◊◊ ◊◊◊◊◊ ◊◊◊◊◊	836,100
◊◊◊◊◊ ◊◊◊◊◊ ◊◊◊◊◊ ◊◊◊◊◊ ◊◊◊◊◊ ◊◊◊◊◊ ◊◊◊◊◊ ◊◊◊◊◊ ◊◊◊◊◊ ◊◊◊◊◊	836,150
◊◊◊◊◊ ◊◊◊◊◊ ◊◊◊◊◊ ◊◊◊◊◊ ◊◊◊◊◊ ◊◊◊◊◊ ◊◊◊◊◊ ◊◊◊◊◊ ◊◊◊◊◊ ◊◊◊◊◊	836,200
◊◊◊◊◊ ◊◊◊◊◊ ◊◊◊◊◊ ◊◊◊◊◊ ◊◊◊◊◊ ◊◊◊◊◊ ◊◊◊◊◊ ◊◊◊◊◊ ◊◊◊◊◊ ◊◊◊◊◊	836,250
◊◊◊◊◊ ◊◊◊◊◊ ◊◊◊◊◊ ◊◊◊◊◊ ◊◊◊◊◊ ◊◊◊◊◊ ◊◊◊◊◊ ◊◊◊◊◊ ◊◊◊◊◊ ◊◊◊◊◊	836,300
◊◊◊◊◊ ◊◊◊◊◊ ◊◊◊◊◊ ◊◊◊◊◊ ◊◊◊◊◊ ◊◊◊◊◊ ◊◊◊◊◊ ◊◊◊◊◊ ◊◊◊◊◊ ◊◊◊◊◊	836,350
◊◊◊◊◊ ◊◊◊◊◊ ◊◊◊◊◊ ◊◊◊◊◊ ◊◊◊◊◊ ◊◊◊◊◊ ◊◊◊◊◊ ◊◊◊◊◊ ◊◊◊◊◊ ◊◊◊◊◊	836,400
◊◊◊◊◊ ◊◊◊◊◊ ◊◊◊◊◊ ◊◊◊◊◊ ◊◊◊◊◊ ◊◊◊◊◊ ◊◊◊◊◊ ◊◊◊◊◊ ◊◊◊◊◊ ◊◊◊◊◊	836,450
◊◊◊◊◊ ◊◊◊◊◊ ◊◊◊◊◊ ◊◊◊◊◊ ◊◊◊◊◊ ◊◊◊◊◊ ◊◊◊◊◊ ◊◊◊◊◊ ◊◊◊◊◊ ◊◊◊◊◊	836,500
◊◊◊◊◊ ◊◊◊◊◊ ◊◊◊◊◊ ◊◊◊◊◊ ◊◊◊◊◊ ◊◊◊◊◊ ◊◊◊◊◊ ◊◊◊◊◊ ◊◊◊◊◊ ◊◊◊◊◊	836,550
◊◊◊◊◊ ◊◊◊◊◊ ◊◊◊◊◊ ◊◊◊◊◊ ◊◊◊◊◊ ◊◊◊◊◊ ◊◊◊◊◊ ◊◊◊◊◊ ◊◊◊◊◊ ◊◊◊◊◊	836,600
◊◊◊◊◊ ◊◊◊◊◊ ◊◊◊◊◊ ◊◊◊◊◊ ◊◊◊◊◊ ◊◊◊◊◊ ◊◊◊◊◊ ◊◊◊◊◊ ◊◊◊◊◊ ◊◊◊◊◊	836,650
◊◊◊◊◊ ◊◊◊◊◊ ◊◊◊◊◊ ◊◊◊◊◊ ◊◊◊◊◊ ◊◊◊◊◊ ◊◊◊◊◊ ◊◊◊◊◊ ◊◊◊◊◊ ◊◊◊◊◊	836,700
◊◊◊◊◊ ◊◊◊◊◊ ◊◊◊◊◊ ◊◊◊◊◊ ◊◊◊◊◊ ◊◊◊◊◊ ◊◊◊◊◊ ◊◊◊◊◊ ◊◊◊◊◊ ◊◊◊◊◊	836,750
◊◊◊◊◊ ◊◊◊◊◊ ◊◊◊◊◊ ◊◊◊◊◊ ◊◊◊◊◊ ◊◊◊◊◊ ◊◊◊◊◊ ◊◊◊◊◊ ◊◊◊◊◊ ◊◊◊◊◊	836,800
◊◊◊◊◊ ◊◊◊◊◊ ◊◊◊◊◊ ◊◊◊◊◊ ◊◊◊◊◊ ◊◊◊◊◊ ◊◊◊◊◊ ◊◊◊◊◊ ◊◊◊◊◊ ◊◊◊◊◊	836,850
◊◊◊◊◊ ◊◊◊◊◊ ◊◊◊◊◊ ◊◊◊◊◊ ◊◊◊◊◊ ◊◊◊◊◊ ◊◊◊◊◊ ◊◊◊◊◊ ◊◊◊◊◊ ◊◊◊◊◊	836,900
◊◊◊◊◊ ◊◊◊◊◊ ◊◊◊◊◊ ◊◊◊◊◊ ◊◊◊◊◊ ◊◊◊◊◊ ◊◊◊◊◊ ◊◊◊◊◊ ◊◊◊◊◊ ◊◊◊◊◊	836,950
◊◊◊◊◊ ◊◊◊◊◊ ◊◊◊◊◊ ◊◊◊◊◊ ◊◊◊◊◊ ◊◊◊◊◊ ◊◊◊◊◊ ◊◊◊◊◊ ◊◊◊◊◊ ◊◊◊◊◊	**837,000**
◊◊◊◊◊ ◊◊◊◊◊ ◊◊◊◊◊ ◊◊◊◊◊ ◊◊◊◊◊ ◊◊◊◊◊ ◊◊◊◊◊ ◊◊◊◊◊ ◊◊◊◊◊ ◊◊◊◊◊	837,050
◊◊◊◊◊ ◊◊◊◊◊ ◊◊◊◊◊ ◊◊◊◊◊ ◊◊◊◊◊ ◊◊◊◊◊ ◊◊◊◊◊ ◊◊◊◊◊ ◊◊◊◊◊ ◊◊◊◊◊	837,100
◊◊◊◊◊ ◊◊◊◊◊ ◊◊◊◊◊ ◊◊◊◊◊ ◊◊◊◊◊ ◊◊◊◊◊ ◊◊◊◊◊ ◊◊◊◊◊ ◊◊◊◊◊ ◊◊◊◊◊	837,150
◊◊◊◊◊ ◊◊◊◊◊ ◊◊◊◊◊ ◊◊◊◊◊ ◊◊◊◊◊ ◊◊◊◊◊ ◊◊◊◊◊ ◊◊◊◊◊ ◊◊◊◊◊ ◊◊◊◊◊	837,200
◊◊◊◊◊ ◊◊◊◊◊ ◊◊◊◊◊ ◊◊◊◊◊ ◊◊◊◊◊ ◊◊◊◊◊ ◊◊◊◊◊ ◊◊◊◊◊ ◊◊◊◊◊ ◊◊◊◊◊	837,250
◊◊◊◊◊ ◊◊◊◊◊ ◊◊◊◊◊ ◊◊◊◊◊ ◊◊◊◊◊ ◊◊◊◊◊ ◊◊◊◊◊ ◊◊◊◊◊ ◊◊◊◊◊ ◊◊◊◊◊	837,300
◊◊◊◊◊ ◊◊◊◊◊ ◊◊◊◊◊ ◊◊◊◊◊ ◊◊◊◊◊ ◊◊◊◊◊ ◊◊◊◊◊ ◊◊◊◊◊ ◊◊◊◊◊ ◊◊◊◊◊	837,350
◊◊◊◊◊ ◊◊◊◊◊ ◊◊◊◊◊ ◊◊◊◊◊ ◊◊◊◊◊ ◊◊◊◊◊ ◊◊◊◊◊ ◊◊◊◊◊ ◊◊◊◊◊ ◊◊◊◊◊	837,400
◊◊◊◊◊ ◊◊◊◊◊ ◊◊◊◊◊ ◊◊◊◊◊ ◊◊◊◊◊ ◊◊◊◊◊ ◊◊◊◊◊ ◊◊◊◊◊ ◊◊◊◊◊ ◊◊◊◊◊	837,450
◊◊◊◊◊ ◊◊◊◊◊ ◊◊◊◊◊ ◊◊◊◊◊ ◊◊◊◊◊ ◊◊◊◊◊ ◊◊◊◊◊ ◊◊◊◊◊ ◊◊◊◊◊ ◊◊◊◊◊	837,500
◊◊◊◊◊ ◊◊◊◊◊ ◊◊◊◊◊ ◊◊◊◊◊ ◊◊◊◊◊ ◊◊◊◊◊ ◊◊◊◊◊ ◊◊◊◊◊ ◊◊◊◊◊ ◊◊◊◊◊	837,550
◊◊◊◊◊ ◊◊◊◊◊ ◊◊◊◊◊ ◊◊◊◊◊ ◊◊◊◊◊ ◊◊◊◊◊ ◊◊◊◊◊ ◊◊◊◊◊ ◊◊◊◊◊ ◊◊◊◊◊	837,600
◊◊◊◊◊ ◊◊◊◊◊ ◊◊◊◊◊ ◊◊◊◊◊ ◊◊◊◊◊ ◊◊◊◊◊ ◊◊◊◊◊ ◊◊◊◊◊ ◊◊◊◊◊ ◊◊◊◊◊	837,650
◊◊◊◊◊ ◊◊◊◊◊ ◊◊◊◊◊ ◊◊◊◊◊ ◊◊◊◊◊ ◊◊◊◊◊ ◊◊◊◊◊ ◊◊◊◊◊ ◊◊◊◊◊ ◊◊◊◊◊	837,700
◊◊◊◊◊ ◊◊◊◊◊ ◊◊◊◊◊ ◊◊◊◊◊ ◊◊◊◊◊ ◊◊◊◊◊ ◊◊◊◊◊ ◊◊◊◊◊ ◊◊◊◊◊ ◊◊◊◊◊	837,750
◊◊◊◊◊ ◊◊◊◊◊ ◊◊◊◊◊ ◊◊◊◊◊ ◊◊◊◊◊ ◊◊◊◊◊ ◊◊◊◊◊ ◊◊◊◊◊ ◊◊◊◊◊ ◊◊◊◊◊	837,800
◊◊◊◊◊ ◊◊◊◊◊ ◊◊◊◊◊ ◊◊◊◊◊ ◊◊◊◊◊ ◊◊◊◊◊ ◊◊◊◊◊ ◊◊◊◊◊ ◊◊◊◊◊ ◊◊◊◊◊	837,850
◊◊◊◊◊ ◊◊◊◊◊ ◊◊◊◊◊ ◊◊◊◊◊ ◊◊◊◊◊ ◊◊◊◊◊ ◊◊◊◊◊ ◊◊◊◊◊ ◊◊◊◊◊ ◊◊◊◊◊	837,900
◊◊◊◊◊ ◊◊◊◊◊ ◊◊◊◊◊ ◊◊◊◊◊ ◊◊◊◊◊ ◊◊◊◊◊ ◊◊◊◊◊ ◊◊◊◊◊ ◊◊◊◊◊ ◊◊◊◊◊	837,950
◊◊◊◊◊ ◊◊◊◊◊ ◊◊◊◊◊ ◊◊◊◊◊ ◊◊◊◊◊ ◊◊◊◊◊ ◊◊◊◊◊ ◊◊◊◊◊ ◊◊◊◊◊ ◊◊◊◊◊	**838,000**

◇ The Number 838,001 to the Number 840,000

Start   **Notes**

	838,050
	838,100
	838,150
	838,200
	838,250
	838,300
	838,350
	838,400
	838,450
	838,500
	838,550
	838,600
	838,650
	838,700
	838,750
	838,800
	838,850
	838,900
	838,950
	839,000
	839,050
	839,100
	839,150
	839,200
	839,250
	839,300
	839,350
	839,400
	839,450
	839,500
	839,550
	839,600
	839,650
	839,700
	839,750
	839,800
	839,850
	839,900
	839,950
	840,000

◊ The Number 840,001 to the Number 842,000

Start | | Notes

840,050	
840,100	
840,150	
840,200	
840,250	
840,300	
840,350	
840,400	
840,450	
840,500	
840,550	
840,600	
840,650	
840,700	
840,750	
840,800	
840,850	
840,900	
840,950	
841,000	
841,050	
841,100	
841,150	
841,200	
841,250	
841,300	
841,350	
841,400	
841,450	
841,500	
841,550	
841,600	
841,650	
841,700	
841,750	
841,800	
841,850	
841,900	
841,950	
842,000	

The Number 842,001 to the Number 844,000

Start		Notes
◊◊◊◊◊ ◊◊◊◊◊ ◊◊◊◊◊ ◊◊◊◊◊ ◊◊◊◊◊ ◊◊◊◊◊ ◊◊◊◊◊ ◊◊◊◊◊ ◊◊◊◊◊ ◊◊◊◊◊	842,050	
◊◊◊◊◊ ◊◊◊◊◊ ◊◊◊◊◊ ◊◊◊◊◊ ◊◊◊◊◊ ◊◊◊◊◊ ◊◊◊◊◊ ◊◊◊◊◊ ◊◊◊◊◊ ◊◊◊◊◊	842,100	
◊◊◊◊◊ ◊◊◊◊◊ ◊◊◊◊◊ ◊◊◊◊◊ ◊◊◊◊◊ ◊◊◊◊◊ ◊◊◊◊◊ ◊◊◊◊◊ ◊◊◊◊◊ ◊◊◊◊◊	842,150	
◊◊◊◊◊ ◊◊◊◊◊ ◊◊◊◊◊ ◊◊◊◊◊ ◊◊◊◊◊ ◊◊◊◊◊ ◊◊◊◊◊ ◊◊◊◊◊ ◊◊◊◊◊ ◊◊◊◊◊	842,200	
◊◊◊◊◊ ◊◊◊◊◊ ◊◊◊◊◊ ◊◊◊◊◊ ◊◊◊◊◊ ◊◊◊◊◊ ◊◊◊◊◊ ◊◊◊◊◊ ◊◊◊◊◊ ◊◊◊◊◊	842,250	
◊◊◊◊◊ ◊◊◊◊◊ ◊◊◊◊◊ ◊◊◊◊◊ ◊◊◊◊◊ ◊◊◊◊◊ ◊◊◊◊◊ ◊◊◊◊◊ ◊◊◊◊◊ ◊◊◊◊◊	842,300	
◊◊◊◊◊ ◊◊◊◊◊ ◊◊◊◊◊ ◊◊◊◊◊ ◊◊◊◊◊ ◊◊◊◊◊ ◊◊◊◊◊ ◊◊◊◊◊ ◊◊◊◊◊ ◊◊◊◊◊	842,350	
◊◊◊◊◊ ◊◊◊◊◊ ◊◊◊◊◊ ◊◊◊◊◊ ◊◊◊◊◊ ◊◊◊◊◊ ◊◊◊◊◊ ◊◊◊◊◊ ◊◊◊◊◊ ◊◊◊◊◊	842,400	
◊◊◊◊◊ ◊◊◊◊◊ ◊◊◊◊◊ ◊◊◊◊◊ ◊◊◊◊◊ ◊◊◊◊◊ ◊◊◊◊◊ ◊◊◊◊◊ ◊◊◊◊◊ ◊◊◊◊◊	842,450	
◊◊◊◊◊ ◊◊◊◊◊ ◊◊◊◊◊ ◊◊◊◊◊ ◊◊◊◊◊ ◊◊◊◊◊ ◊◊◊◊◊ ◊◊◊◊◊ ◊◊◊◊◊ ◊◊◊◊◊	842,500	
◊◊◊◊◊ ◊◊◊◊◊ ◊◊◊◊◊ ◊◊◊◊◊ ◊◊◊◊◊ ◊◊◊◊◊ ◊◊◊◊◊ ◊◊◊◊◊ ◊◊◊◊◊ ◊◊◊◊◊	842,550	
◊◊◊◊◊ ◊◊◊◊◊ ◊◊◊◊◊ ◊◊◊◊◊ ◊◊◊◊◊ ◊◊◊◊◊ ◊◊◊◊◊ ◊◊◊◊◊ ◊◊◊◊◊ ◊◊◊◊◊	842,600	
◊◊◊◊◊ ◊◊◊◊◊ ◊◊◊◊◊ ◊◊◊◊◊ ◊◊◊◊◊ ◊◊◊◊◊ ◊◊◊◊◊ ◊◊◊◊◊ ◊◊◊◊◊ ◊◊◊◊◊	842,650	
◊◊◊◊◊ ◊◊◊◊◊ ◊◊◊◊◊ ◊◊◊◊◊ ◊◊◊◊◊ ◊◊◊◊◊ ◊◊◊◊◊ ◊◊◊◊◊ ◊◊◊◊◊ ◊◊◊◊◊	842,700	
◊◊◊◊◊ ◊◊◊◊◊ ◊◊◊◊◊ ◊◊◊◊◊ ◊◊◊◊◊ ◊◊◊◊◊ ◊◊◊◊◊ ◊◊◊◊◊ ◊◊◊◊◊ ◊◊◊◊◊	842,750	
◊◊◊◊◊ ◊◊◊◊◊ ◊◊◊◊◊ ◊◊◊◊◊ ◊◊◊◊◊ ◊◊◊◊◊ ◊◊◊◊◊ ◊◊◊◊◊ ◊◊◊◊◊ ◊◊◊◊◊	842,800	
◊◊◊◊◊ ◊◊◊◊◊ ◊◊◊◊◊ ◊◊◊◊◊ ◊◊◊◊◊ ◊◊◊◊◊ ◊◊◊◊◊ ◊◊◊◊◊ ◊◊◊◊◊ ◊◊◊◊◊	842,850	
◊◊◊◊◊ ◊◊◊◊◊ ◊◊◊◊◊ ◊◊◊◊◊ ◊◊◊◊◊ ◊◊◊◊◊ ◊◊◊◊◊ ◊◊◊◊◊ ◊◊◊◊◊ ◊◊◊◊◊	842,900	
◊◊◊◊◊ ◊◊◊◊◊ ◊◊◊◊◊ ◊◊◊◊◊ ◊◊◊◊◊ ◊◊◊◊◊ ◊◊◊◊◊ ◊◊◊◊◊ ◊◊◊◊◊ ◊◊◊◊◊	842,950	
◊◊◊◊◊ ◊◊◊◊◊ ◊◊◊◊◊ ◊◊◊◊◊ ◊◊◊◊◊ ◊◊◊◊◊ ◊◊◊◊◊ ◊◊◊◊◊ ◊◊◊◊◊ ◊◊◊◊◊	**843,000**	
◊◊◊◊◊ ◊◊◊◊◊ ◊◊◊◊◊ ◊◊◊◊◊ ◊◊◊◊◊ ◊◊◊◊◊ ◊◊◊◊◊ ◊◊◊◊◊ ◊◊◊◊◊ ◊◊◊◊◊	843,050	
◊◊◊◊◊ ◊◊◊◊◊ ◊◊◊◊◊ ◊◊◊◊◊ ◊◊◊◊◊ ◊◊◊◊◊ ◊◊◊◊◊ ◊◊◊◊◊ ◊◊◊◊◊ ◊◊◊◊◊	843,100	
◊◊◊◊◊ ◊◊◊◊◊ ◊◊◊◊◊ ◊◊◊◊◊ ◊◊◊◊◊ ◊◊◊◊◊ ◊◊◊◊◊ ◊◊◊◊◊ ◊◊◊◊◊ ◊◊◊◊◊	843,150	
◊◊◊◊◊ ◊◊◊◊◊ ◊◊◊◊◊ ◊◊◊◊◊ ◊◊◊◊◊ ◊◊◊◊◊ ◊◊◊◊◊ ◊◊◊◊◊ ◊◊◊◊◊ ◊◊◊◊◊	843,200	
◊◊◊◊◊ ◊◊◊◊◊ ◊◊◊◊◊ ◊◊◊◊◊ ◊◊◊◊◊ ◊◊◊◊◊ ◊◊◊◊◊ ◊◊◊◊◊ ◊◊◊◊◊ ◊◊◊◊◊	843,250	
◊◊◊◊◊ ◊◊◊◊◊ ◊◊◊◊◊ ◊◊◊◊◊ ◊◊◊◊◊ ◊◊◊◊◊ ◊◊◊◊◊ ◊◊◊◊◊ ◊◊◊◊◊ ◊◊◊◊◊	843,300	
◊◊◊◊◊ ◊◊◊◊◊ ◊◊◊◊◊ ◊◊◊◊◊ ◊◊◊◊◊ ◊◊◊◊◊ ◊◊◊◊◊ ◊◊◊◊◊ ◊◊◊◊◊ ◊◊◊◊◊	843,350	
◊◊◊◊◊ ◊◊◊◊◊ ◊◊◊◊◊ ◊◊◊◊◊ ◊◊◊◊◊ ◊◊◊◊◊ ◊◊◊◊◊ ◊◊◊◊◊ ◊◊◊◊◊ ◊◊◊◊◊	843,400	
◊◊◊◊◊ ◊◊◊◊◊ ◊◊◊◊◊ ◊◊◊◊◊ ◊◊◊◊◊ ◊◊◊◊◊ ◊◊◊◊◊ ◊◊◊◊◊ ◊◊◊◊◊ ◊◊◊◊◊	843,450	
◊◊◊◊◊ ◊◊◊◊◊ ◊◊◊◊◊ ◊◊◊◊◊ ◊◊◊◊◊ ◊◊◊◊◊ ◊◊◊◊◊ ◊◊◊◊◊ ◊◊◊◊◊ ◊◊◊◊◊	843,500	
◊◊◊◊◊ ◊◊◊◊◊ ◊◊◊◊◊ ◊◊◊◊◊ ◊◊◊◊◊ ◊◊◊◊◊ ◊◊◊◊◊ ◊◊◊◊◊ ◊◊◊◊◊ ◊◊◊◊◊	843,550	
◊◊◊◊◊ ◊◊◊◊◊ ◊◊◊◊◊ ◊◊◊◊◊ ◊◊◊◊◊ ◊◊◊◊◊ ◊◊◊◊◊ ◊◊◊◊◊ ◊◊◊◊◊ ◊◊◊◊◊	843,600	
◊◊◊◊◊ ◊◊◊◊◊ ◊◊◊◊◊ ◊◊◊◊◊ ◊◊◊◊◊ ◊◊◊◊◊ ◊◊◊◊◊ ◊◊◊◊◊ ◊◊◊◊◊ ◊◊◊◊◊	843,650	
◊◊◊◊◊ ◊◊◊◊◊ ◊◊◊◊◊ ◊◊◊◊◊ ◊◊◊◊◊ ◊◊◊◊◊ ◊◊◊◊◊ ◊◊◊◊◊ ◊◊◊◊◊ ◊◊◊◊◊	843,700	
◊◊◊◊◊ ◊◊◊◊◊ ◊◊◊◊◊ ◊◊◊◊◊ ◊◊◊◊◊ ◊◊◊◊◊ ◊◊◊◊◊ ◊◊◊◊◊ ◊◊◊◊◊ ◊◊◊◊◊	843,750	
◊◊◊◊◊ ◊◊◊◊◊ ◊◊◊◊◊ ◊◊◊◊◊ ◊◊◊◊◊ ◊◊◊◊◊ ◊◊◊◊◊ ◊◊◊◊◊ ◊◊◊◊◊ ◊◊◊◊◊	843,800	
◊◊◊◊◊ ◊◊◊◊◊ ◊◊◊◊◊ ◊◊◊◊◊ ◊◊◊◊◊ ◊◊◊◊◊ ◊◊◊◊◊ ◊◊◊◊◊ ◊◊◊◊◊ ◊◊◊◊◊	843,850	
◊◊◊◊◊ ◊◊◊◊◊ ◊◊◊◊◊ ◊◊◊◊◊ ◊◊◊◊◊ ◊◊◊◊◊ ◊◊◊◊◊ ◊◊◊◊◊ ◊◊◊◊◊ ◊◊◊◊◊	843,900	
◊◊◊◊◊ ◊◊◊◊◊ ◊◊◊◊◊ ◊◊◊◊◊ ◊◊◊◊◊ ◊◊◊◊◊ ◊◊◊◊◊ ◊◊◊◊◊ ◊◊◊◊◊ ◊◊◊◊◊	843,950	
◊◊◊◊◊ ◊◊◊◊◊ ◊◊◊◊◊ ◊◊◊◊◊ ◊◊◊◊◊ ◊◊◊◊◊ ◊◊◊◊◊ ◊◊◊◊◊ ◊◊◊◊◊ ◊◊◊◊◊	**844,000**	

The Number 844,001 to the Number 846,000

Start

Notes

	844,050
	844,100
	844,150
	844,200
	844,250
	844,300
	844,350
	844,400
	844,450
	844,500
	844,550
	844,600
	844,650
	844,700
	844,750
	844,800
	844,850
	844,900
	844,950
	845,000
	845,050
	845,100
	845,150
	845,200
	845,250
	845,300
	845,350
	845,400
	845,450
	845,500
	845,550
	845,600
	845,650
	845,700
	845,750
	845,800
	845,850
	845,900
	845,950
	846,000

The Number 846,001 to the Number 848,000

Start		Notes
	846,050	
	846,100	
	846,150	
	846,200	
	846,250	
	846,300	
	846,350	
	846,400	
	846,450	
	846,500	
	846,550	
	846,600	
	846,650	
	846,700	
	846,750	
	846,800	
	846,850	
	846,900	
	846,950	
	847,000	
	847,050	
	847,100	
	847,150	
	847,200	
	847,250	
	847,300	
	847,350	
	847,400	
	847,450	
	847,500	
	847,550	
	847,600	
	847,650	
	847,700	
	847,750	
	847,800	
	847,850	
	847,900	
	847,950	
	848,000	

The Number 848,001 to the Number 850,000

Start

Notes

	848,050
	848,100
	848,150
	848,200
	848,250
	848,300
	848,350
	848,400
	848,450
	848,500
	848,550
	848,600
	848,650
	848,700
	848,750
	848,800
	848,850
	848,900
	848,950
	849,000
	849,050
	849,100
	849,150
	849,200
	849,250
	849,300
	849,350
	849,400
	849,450
	849,500
	849,550
	849,600
	849,650
	849,700
	849,750
	849,800
	849,850
	849,900
	849,950
	850,000

◇ The Number 850,001 to the Number 852,000

Start	Notes
◊◊◊◊◊ ◊◊◊◊◊ ◊◊◊◊◊ ◊◊◊◊◊ ◊◊◊◊◊ ◊◊◊◊◊ ◊◊◊◊◊ ◊◊◊◊◊ ◊◊◊◊◊ ◊◊◊◊◊	850,050
◊◊◊◊◊ ◊◊◊◊◊ ◊◊◊◊◊ ◊◊◊◊◊ ◊◊◊◊◊ ◊◊◊◊◊ ◊◊◊◊◊ ◊◊◊◊◊ ◊◊◊◊◊ ◊◊◊◊◊	850,100
◊◊◊◊◊ ◊◊◊◊◊ ◊◊◊◊◊ ◊◊◊◊◊ ◊◊◊◊◊ ◊◊◊◊◊ ◊◊◊◊◊ ◊◊◊◊◊ ◊◊◊◊◊ ◊◊◊◊◊	850,150
◊◊◊◊◊ ◊◊◊◊◊ ◊◊◊◊◊ ◊◊◊◊◊ ◊◊◊◊◊ ◊◊◊◊◊ ◊◊◊◊◊ ◊◊◊◊◊ ◊◊◊◊◊ ◊◊◊◊◊	850,200
◊◊◊◊◊ ◊◊◊◊◊ ◊◊◊◊◊ ◊◊◊◊◊ ◊◊◊◊◊ ◊◊◊◊◊ ◊◊◊◊◊ ◊◊◊◊◊ ◊◊◊◊◊ ◊◊◊◊◊	850,250
◊◊◊◊◊ ◊◊◊◊◊ ◊◊◊◊◊ ◊◊◊◊◊ ◊◊◊◊◊ ◊◊◊◊◊ ◊◊◊◊◊ ◊◊◊◊◊ ◊◊◊◊◊ ◊◊◊◊◊	850,300
◊◊◊◊◊ ◊◊◊◊◊ ◊◊◊◊◊ ◊◊◊◊◊ ◊◊◊◊◊ ◊◊◊◊◊ ◊◊◊◊◊ ◊◊◊◊◊ ◊◊◊◊◊ ◊◊◊◊◊	850,350
◊◊◊◊◊ ◊◊◊◊◊ ◊◊◊◊◊ ◊◊◊◊◊ ◊◊◊◊◊ ◊◊◊◊◊ ◊◊◊◊◊ ◊◊◊◊◊ ◊◊◊◊◊ ◊◊◊◊◊	850,400
◊◊◊◊◊ ◊◊◊◊◊ ◊◊◊◊◊ ◊◊◊◊◊ ◊◊◊◊◊ ◊◊◊◊◊ ◊◊◊◊◊ ◊◊◊◊◊ ◊◊◊◊◊ ◊◊◊◊◊	850,450
◊◊◊◊◊ ◊◊◊◊◊ ◊◊◊◊◊ ◊◊◊◊◊ ◊◊◊◊◊ ◊◊◊◊◊ ◊◊◊◊◊ ◊◊◊◊◊ ◊◊◊◊◊ ◊◊◊◊◊	850,500
◊◊◊◊◊ ◊◊◊◊◊ ◊◊◊◊◊ ◊◊◊◊◊ ◊◊◊◊◊ ◊◊◊◊◊ ◊◊◊◊◊ ◊◊◊◊◊ ◊◊◊◊◊ ◊◊◊◊◊	850,550
◊◊◊◊◊ ◊◊◊◊◊ ◊◊◊◊◊ ◊◊◊◊◊ ◊◊◊◊◊ ◊◊◊◊◊ ◊◊◊◊◊ ◊◊◊◊◊ ◊◊◊◊◊ ◊◊◊◊◊	850,600
◊◊◊◊◊ ◊◊◊◊◊ ◊◊◊◊◊ ◊◊◊◊◊ ◊◊◊◊◊ ◊◊◊◊◊ ◊◊◊◊◊ ◊◊◊◊◊ ◊◊◊◊◊ ◊◊◊◊◊	850,650
◊◊◊◊◊ ◊◊◊◊◊ ◊◊◊◊◊ ◊◊◊◊◊ ◊◊◊◊◊ ◊◊◊◊◊ ◊◊◊◊◊ ◊◊◊◊◊ ◊◊◊◊◊ ◊◊◊◊◊	850,700
◊◊◊◊◊ ◊◊◊◊◊ ◊◊◊◊◊ ◊◊◊◊◊ ◊◊◊◊◊ ◊◊◊◊◊ ◊◊◊◊◊ ◊◊◊◊◊ ◊◊◊◊◊ ◊◊◊◊◊	850,750
◊◊◊◊◊ ◊◊◊◊◊ ◊◊◊◊◊ ◊◊◊◊◊ ◊◊◊◊◊ ◊◊◊◊◊ ◊◊◊◊◊ ◊◊◊◊◊ ◊◊◊◊◊ ◊◊◊◊◊	850,800
◊◊◊◊◊ ◊◊◊◊◊ ◊◊◊◊◊ ◊◊◊◊◊ ◊◊◊◊◊ ◊◊◊◊◊ ◊◊◊◊◊ ◊◊◊◊◊ ◊◊◊◊◊ ◊◊◊◊◊	850,850
◊◊◊◊◊ ◊◊◊◊◊ ◊◊◊◊◊ ◊◊◊◊◊ ◊◊◊◊◊ ◊◊◊◊◊ ◊◊◊◊◊ ◊◊◊◊◊ ◊◊◊◊◊ ◊◊◊◊◊	850,900
◊◊◊◊◊ ◊◊◊◊◊ ◊◊◊◊◊ ◊◊◊◊◊ ◊◊◊◊◊ ◊◊◊◊◊ ◊◊◊◊◊ ◊◊◊◊◊ ◊◊◊◊◊ ◊◊◊◊◊	850,950
◊◊◊◊◊ ◊◊◊◊◊ ◊◊◊◊◊ ◊◊◊◊◊ ◊◊◊◊◊ ◊◊◊◊◊ ◊◊◊◊◊ ◊◊◊◊◊ ◊◊◊◊◊ ◊◊◊◊◊	**851,000**
◊◊◊◊◊ ◊◊◊◊◊ ◊◊◊◊◊ ◊◊◊◊◊ ◊◊◊◊◊ ◊◊◊◊◊ ◊◊◊◊◊ ◊◊◊◊◊ ◊◊◊◊◊ ◊◊◊◊◊	851,050
◊◊◊◊◊ ◊◊◊◊◊ ◊◊◊◊◊ ◊◊◊◊◊ ◊◊◊◊◊ ◊◊◊◊◊ ◊◊◊◊◊ ◊◊◊◊◊ ◊◊◊◊◊ ◊◊◊◊◊	851,100
◊◊◊◊◊ ◊◊◊◊◊ ◊◊◊◊◊ ◊◊◊◊◊ ◊◊◊◊◊ ◊◊◊◊◊ ◊◊◊◊◊ ◊◊◊◊◊ ◊◊◊◊◊ ◊◊◊◊◊	851,150
◊◊◊◊◊ ◊◊◊◊◊ ◊◊◊◊◊ ◊◊◊◊◊ ◊◊◊◊◊ ◊◊◊◊◊ ◊◊◊◊◊ ◊◊◊◊◊ ◊◊◊◊◊ ◊◊◊◊◊	851,200
◊◊◊◊◊ ◊◊◊◊◊ ◊◊◊◊◊ ◊◊◊◊◊ ◊◊◊◊◊ ◊◊◊◊◊ ◊◊◊◊◊ ◊◊◊◊◊ ◊◊◊◊◊ ◊◊◊◊◊	851,250
◊◊◊◊◊ ◊◊◊◊◊ ◊◊◊◊◊ ◊◊◊◊◊ ◊◊◊◊◊ ◊◊◊◊◊ ◊◊◊◊◊ ◊◊◊◊◊ ◊◊◊◊◊ ◊◊◊◊◊	851,300
◊◊◊◊◊ ◊◊◊◊◊ ◊◊◊◊◊ ◊◊◊◊◊ ◊◊◊◊◊ ◊◊◊◊◊ ◊◊◊◊◊ ◊◊◊◊◊ ◊◊◊◊◊ ◊◊◊◊◊	851,350
◊◊◊◊◊ ◊◊◊◊◊ ◊◊◊◊◊ ◊◊◊◊◊ ◊◊◊◊◊ ◊◊◊◊◊ ◊◊◊◊◊ ◊◊◊◊◊ ◊◊◊◊◊ ◊◊◊◊◊	851,400
◊◊◊◊◊ ◊◊◊◊◊ ◊◊◊◊◊ ◊◊◊◊◊ ◊◊◊◊◊ ◊◊◊◊◊ ◊◊◊◊◊ ◊◊◊◊◊ ◊◊◊◊◊ ◊◊◊◊◊	851,450
◊◊◊◊◊ ◊◊◊◊◊ ◊◊◊◊◊ ◊◊◊◊◊ ◊◊◊◊◊ ◊◊◊◊◊ ◊◊◊◊◊ ◊◊◊◊◊ ◊◊◊◊◊ ◊◊◊◊◊	851,500
◊◊◊◊◊ ◊◊◊◊◊ ◊◊◊◊◊ ◊◊◊◊◊ ◊◊◊◊◊ ◊◊◊◊◊ ◊◊◊◊◊ ◊◊◊◊◊ ◊◊◊◊◊ ◊◊◊◊◊	851,550
◊◊◊◊◊ ◊◊◊◊◊ ◊◊◊◊◊ ◊◊◊◊◊ ◊◊◊◊◊ ◊◊◊◊◊ ◊◊◊◊◊ ◊◊◊◊◊ ◊◊◊◊◊ ◊◊◊◊◊	851,600
◊◊◊◊◊ ◊◊◊◊◊ ◊◊◊◊◊ ◊◊◊◊◊ ◊◊◊◊◊ ◊◊◊◊◊ ◊◊◊◊◊ ◊◊◊◊◊ ◊◊◊◊◊ ◊◊◊◊◊	851,650
◊◊◊◊◊ ◊◊◊◊◊ ◊◊◊◊◊ ◊◊◊◊◊ ◊◊◊◊◊ ◊◊◊◊◊ ◊◊◊◊◊ ◊◊◊◊◊ ◊◊◊◊◊ ◊◊◊◊◊	851,700
◊◊◊◊◊ ◊◊◊◊◊ ◊◊◊◊◊ ◊◊◊◊◊ ◊◊◊◊◊ ◊◊◊◊◊ ◊◊◊◊◊ ◊◊◊◊◊ ◊◊◊◊◊ ◊◊◊◊◊	851,750
◊◊◊◊◊ ◊◊◊◊◊ ◊◊◊◊◊ ◊◊◊◊◊ ◊◊◊◊◊ ◊◊◊◊◊ ◊◊◊◊◊ ◊◊◊◊◊ ◊◊◊◊◊ ◊◊◊◊◊	851,800
◊◊◊◊◊ ◊◊◊◊◊ ◊◊◊◊◊ ◊◊◊◊◊ ◊◊◊◊◊ ◊◊◊◊◊ ◊◊◊◊◊ ◊◊◊◊◊ ◊◊◊◊◊ ◊◊◊◊◊	851,850
◊◊◊◊◊ ◊◊◊◊◊ ◊◊◊◊◊ ◊◊◊◊◊ ◊◊◊◊◊ ◊◊◊◊◊ ◊◊◊◊◊ ◊◊◊◊◊ ◊◊◊◊◊ ◊◊◊◊◊	851,900
◊◊◊◊◊ ◊◊◊◊◊ ◊◊◊◊◊ ◊◊◊◊◊ ◊◊◊◊◊ ◊◊◊◊◊ ◊◊◊◊◊ ◊◊◊◊◊ ◊◊◊◊◊ ◊◊◊◊◊	851,950
◊◊◊◊◊ ◊◊◊◊◊ ◊◊◊◊◊ ◊◊◊◊◊ ◊◊◊◊◊ ◊◊◊◊◊ ◊◊◊◊◊ ◊◊◊◊◊ ◊◊◊◊◊ ◊◊◊◊◊	**852,000**

◇ The Number 852,001 to the Number 854,000

Start

Notes

	852,050
	852,100
	852,150
	852,200
	852,250
	852,300
	852,350
	852,400
	852,450
	852,500
	852,550
	852,600
	852,650
	852,700
	852,750
	852,800
	852,850
	852,900
	852,950
	853,000
	853,050
	853,100
	853,150
	853,200
	853,250
	853,300
	853,350
	853,400
	853,450
	853,500
	853,550
	853,600
	853,650
	853,700
	853,750
	853,800
	853,850
	853,900
	853,950
	854,000

The Number 854,001 to the Number 856,000

Start   Notes

	854,050
	854,100
	854,150
	854,200
	854,250
	854,300
	854,350
	854,400
	854,450
	854,500
	854,550
	854,600
	854,650
	854,700
	854,750
	854,800
	854,850
	854,900
	854,950
	855,000
	855,050
	855,100
	855,150
	855,200
	855,250
	855,300
	855,350
	855,400
	855,450
	855,500
	855,550
	855,600
	855,650
	855,700
	855,750
	855,800
	855,850
	855,900
	855,950
	856,000

The Number 856,001 to the Number 858,000

Start

Notes

	856,050
	856,100
	856,150
	856,200
	856,250
	856,300
	856,350
	856,400
	856,450
	856,500
	856,550
	856,600
	856,650
	856,700
	856,750
	856,800
	856,850
	856,900
	856,950
	857,000
	857,050
	857,100
	857,150
	857,200
	857,250
	857,300
	857,350
	857,400
	857,450
	857,500
	857,550
	857,600
	857,650
	857,700
	857,750
	857,800
	857,850
	857,900
	857,950
	858,000

The Number 858,001 to the Number 860,000

Start   Notes

	858,050
	858,100
	858,150
	858,200
	858,250
	858,300
	858,350
	858,400
	858,450
	858,500
	858,550
	858,600
	858,650
	858,700
	858,750
	858,800
	858,850
	858,900
	858,950
	859,000
	859,050
	859,100
	859,150
	859,200
	859,250
	859,300
	859,350
	859,400
	859,450
	859,500
	859,550
	859,600
	859,650
	859,700
	859,750
	859,800
	859,850
	859,900
	859,950
	860,000

The Number 860,001 to the Number 862,000

Start

Notes

	860,050
	860,100
	860,150
	860,200
	860,250
	860,300
	860,350
	860,400
	860,450
	860,500
	860,550
	860,600
	860,650
	860,700
	860,750
	860,800
	860,850
	860,900
	860,950
	861,000
	861,050
	861,100
	861,150
	861,200
	861,250
	861,300
	861,350
	861,400
	861,450
	861,500
	861,550
	861,600
	861,650
	861,700
	861,750
	861,800
	861,850
	861,900
	861,950
	862,000

The Number 862,001 to the Number 864,000

Start   Notes

	862,050
	862,100
	862,150
	862,200
	862,250
	862,300
	862,350
	862,400
	862,450
	862,500
	862,550
	862,600
	862,650
	862,700
	862,750
	862,800
	862,850
	862,900
	862,950
	863,000
	863,050
	863,100
	863,150
	863,200
	863,250
	863,300
	863,350
	863,400
	863,450
	863,500
	863,550
	863,600
	863,650
	863,700
	863,750
	863,800
	863,850
	863,900
	863,950
	864,000

The Number 864,001 to the Number 866,000

Start

Notes

◇◇◇◇◇ ◇◇◇◇◇ ◇◇◇◇◇ ◇◇◇◇◇ ◇◇◇◇◇ ◇◇◇◇◇ ◇◇◇◇◇ ◇◇◇◇◇ ◇◇◇◇◇ ◇◇◇◇◇	864,050
◇◇◇◇◇ ◇◇◇◇◇ ◇◇◇◇◇ ◇◇◇◇◇ ◇◇◇◇◇ ◇◇◇◇◇ ◇◇◇◇◇ ◇◇◇◇◇ ◇◇◇◇◇ ◇◇◇◇◇	864,100
◇◇◇◇◇ ◇◇◇◇◇ ◇◇◇◇◇ ◇◇◇◇◇ ◇◇◇◇◇ ◇◇◇◇◇ ◇◇◇◇◇ ◇◇◇◇◇ ◇◇◇◇◇ ◇◇◇◇◇	864,150
◇◇◇◇◇ ◇◇◇◇◇ ◇◇◇◇◇ ◇◇◇◇◇ ◇◇◇◇◇ ◇◇◇◇◇ ◇◇◇◇◇ ◇◇◇◇◇ ◇◇◇◇◇ ◇◇◇◇◇	864,200
◇◇◇◇◇ ◇◇◇◇◇ ◇◇◇◇◇ ◇◇◇◇◇ ◇◇◇◇◇ ◇◇◇◇◇ ◇◇◇◇◇ ◇◇◇◇◇ ◇◇◇◇◇ ◇◇◇◇◇	864,250
◇◇◇◇◇ ◇◇◇◇◇ ◇◇◇◇◇ ◇◇◇◇◇ ◇◇◇◇◇ ◇◇◇◇◇ ◇◇◇◇◇ ◇◇◇◇◇ ◇◇◇◇◇ ◇◇◇◇◇	864,300
◇◇◇◇◇ ◇◇◇◇◇ ◇◇◇◇◇ ◇◇◇◇◇ ◇◇◇◇◇ ◇◇◇◇◇ ◇◇◇◇◇ ◇◇◇◇◇ ◇◇◇◇◇ ◇◇◇◇◇	864,350
◇◇◇◇◇ ◇◇◇◇◇ ◇◇◇◇◇ ◇◇◇◇◇ ◇◇◇◇◇ ◇◇◇◇◇ ◇◇◇◇◇ ◇◇◇◇◇ ◇◇◇◇◇ ◇◇◇◇◇	864,400
◇◇◇◇◇ ◇◇◇◇◇ ◇◇◇◇◇ ◇◇◇◇◇ ◇◇◇◇◇ ◇◇◇◇◇ ◇◇◇◇◇ ◇◇◇◇◇ ◇◇◇◇◇ ◇◇◇◇◇	864,450
◇◇◇◇◇ ◇◇◇◇◇ ◇◇◇◇◇ ◇◇◇◇◇ ◇◇◇◇◇ ◇◇◇◇◇ ◇◇◇◇◇ ◇◇◇◇◇ ◇◇◇◇◇ ◇◇◇◇◇	864,500
◇◇◇◇◇ ◇◇◇◇◇ ◇◇◇◇◇ ◇◇◇◇◇ ◇◇◇◇◇ ◇◇◇◇◇ ◇◇◇◇◇ ◇◇◇◇◇ ◇◇◇◇◇ ◇◇◇◇◇	864,550
◇◇◇◇◇ ◇◇◇◇◇ ◇◇◇◇◇ ◇◇◇◇◇ ◇◇◇◇◇ ◇◇◇◇◇ ◇◇◇◇◇ ◇◇◇◇◇ ◇◇◇◇◇ ◇◇◇◇◇	864,600
◇◇◇◇◇ ◇◇◇◇◇ ◇◇◇◇◇ ◇◇◇◇◇ ◇◇◇◇◇ ◇◇◇◇◇ ◇◇◇◇◇ ◇◇◇◇◇ ◇◇◇◇◇ ◇◇◇◇◇	864,650
◇◇◇◇◇ ◇◇◇◇◇ ◇◇◇◇◇ ◇◇◇◇◇ ◇◇◇◇◇ ◇◇◇◇◇ ◇◇◇◇◇ ◇◇◇◇◇ ◇◇◇◇◇ ◇◇◇◇◇	864,700
◇◇◇◇◇ ◇◇◇◇◇ ◇◇◇◇◇ ◇◇◇◇◇ ◇◇◇◇◇ ◇◇◇◇◇ ◇◇◇◇◇ ◇◇◇◇◇ ◇◇◇◇◇ ◇◇◇◇◇	864,750
◇◇◇◇◇ ◇◇◇◇◇ ◇◇◇◇◇ ◇◇◇◇◇ ◇◇◇◇◇ ◇◇◇◇◇ ◇◇◇◇◇ ◇◇◇◇◇ ◇◇◇◇◇ ◇◇◇◇◇	864,800
◇◇◇◇◇ ◇◇◇◇◇ ◇◇◇◇◇ ◇◇◇◇◇ ◇◇◇◇◇ ◇◇◇◇◇ ◇◇◇◇◇ ◇◇◇◇◇ ◇◇◇◇◇ ◇◇◇◇◇	864,850
◇◇◇◇◇ ◇◇◇◇◇ ◇◇◇◇◇ ◇◇◇◇◇ ◇◇◇◇◇ ◇◇◇◇◇ ◇◇◇◇◇ ◇◇◇◇◇ ◇◇◇◇◇ ◇◇◇◇◇	864,900
◇◇◇◇◇ ◇◇◇◇◇ ◇◇◇◇◇ ◇◇◇◇◇ ◇◇◇◇◇ ◇◇◇◇◇ ◇◇◇◇◇ ◇◇◇◇◇ ◇◇◇◇◇ ◇◇◇◇◇	864,950
◇◇◇◇◇ ◇◇◇◇◇ ◇◇◇◇◇ ◇◇◇◇◇ ◇◇◇◇◇ ◇◇◇◇◇ ◇◇◇◇◇ ◇◇◇◇◇ ◇◇◇◇◇ ◇◇◇◇◇	**865,000**
◇◇◇◇◇ ◇◇◇◇◇ ◇◇◇◇◇ ◇◇◇◇◇ ◇◇◇◇◇ ◇◇◇◇◇ ◇◇◇◇◇ ◇◇◇◇◇ ◇◇◇◇◇ ◇◇◇◇◇	865,050
◇◇◇◇◇ ◇◇◇◇◇ ◇◇◇◇◇ ◇◇◇◇◇ ◇◇◇◇◇ ◇◇◇◇◇ ◇◇◇◇◇ ◇◇◇◇◇ ◇◇◇◇◇ ◇◇◇◇◇	865,100
◇◇◇◇◇ ◇◇◇◇◇ ◇◇◇◇◇ ◇◇◇◇◇ ◇◇◇◇◇ ◇◇◇◇◇ ◇◇◇◇◇ ◇◇◇◇◇ ◇◇◇◇◇ ◇◇◇◇◇	865,150
◇◇◇◇◇ ◇◇◇◇◇ ◇◇◇◇◇ ◇◇◇◇◇ ◇◇◇◇◇ ◇◇◇◇◇ ◇◇◇◇◇ ◇◇◇◇◇ ◇◇◇◇◇ ◇◇◇◇◇	865,200
◇◇◇◇◇ ◇◇◇◇◇ ◇◇◇◇◇ ◇◇◇◇◇ ◇◇◇◇◇ ◇◇◇◇◇ ◇◇◇◇◇ ◇◇◇◇◇ ◇◇◇◇◇ ◇◇◇◇◇	865,250
◇◇◇◇◇ ◇◇◇◇◇ ◇◇◇◇◇ ◇◇◇◇◇ ◇◇◇◇◇ ◇◇◇◇◇ ◇◇◇◇◇ ◇◇◇◇◇ ◇◇◇◇◇ ◇◇◇◇◇	865,300
◇◇◇◇◇ ◇◇◇◇◇ ◇◇◇◇◇ ◇◇◇◇◇ ◇◇◇◇◇ ◇◇◇◇◇ ◇◇◇◇◇ ◇◇◇◇◇ ◇◇◇◇◇ ◇◇◇◇◇	865,350
◇◇◇◇◇ ◇◇◇◇◇ ◇◇◇◇◇ ◇◇◇◇◇ ◇◇◇◇◇ ◇◇◇◇◇ ◇◇◇◇◇ ◇◇◇◇◇ ◇◇◇◇◇ ◇◇◇◇◇	865,400
◇◇◇◇◇ ◇◇◇◇◇ ◇◇◇◇◇ ◇◇◇◇◇ ◇◇◇◇◇ ◇◇◇◇◇ ◇◇◇◇◇ ◇◇◇◇◇ ◇◇◇◇◇ ◇◇◇◇◇	865,450
◇◇◇◇◇ ◇◇◇◇◇ ◇◇◇◇◇ ◇◇◇◇◇ ◇◇◇◇◇ ◇◇◇◇◇ ◇◇◇◇◇ ◇◇◇◇◇ ◇◇◇◇◇ ◇◇◇◇◇	865,500
◇◇◇◇◇ ◇◇◇◇◇ ◇◇◇◇◇ ◇◇◇◇◇ ◇◇◇◇◇ ◇◇◇◇◇ ◇◇◇◇◇ ◇◇◇◇◇ ◇◇◇◇◇ ◇◇◇◇◇	865,550
◇◇◇◇◇ ◇◇◇◇◇ ◇◇◇◇◇ ◇◇◇◇◇ ◇◇◇◇◇ ◇◇◇◇◇ ◇◇◇◇◇ ◇◇◇◇◇ ◇◇◇◇◇ ◇◇◇◇◇	865,600
◇◇◇◇◇ ◇◇◇◇◇ ◇◇◇◇◇ ◇◇◇◇◇ ◇◇◇◇◇ ◇◇◇◇◇ ◇◇◇◇◇ ◇◇◇◇◇ ◇◇◇◇◇ ◇◇◇◇◇	865,650
◇◇◇◇◇ ◇◇◇◇◇ ◇◇◇◇◇ ◇◇◇◇◇ ◇◇◇◇◇ ◇◇◇◇◇ ◇◇◇◇◇ ◇◇◇◇◇ ◇◇◇◇◇ ◇◇◇◇◇	865,700
◇◇◇◇◇ ◇◇◇◇◇ ◇◇◇◇◇ ◇◇◇◇◇ ◇◇◇◇◇ ◇◇◇◇◇ ◇◇◇◇◇ ◇◇◇◇◇ ◇◇◇◇◇ ◇◇◇◇◇	865,750
◇◇◇◇◇ ◇◇◇◇◇ ◇◇◇◇◇ ◇◇◇◇◇ ◇◇◇◇◇ ◇◇◇◇◇ ◇◇◇◇◇ ◇◇◇◇◇ ◇◇◇◇◇ ◇◇◇◇◇	865,800
◇◇◇◇◇ ◇◇◇◇◇ ◇◇◇◇◇ ◇◇◇◇◇ ◇◇◇◇◇ ◇◇◇◇◇ ◇◇◇◇◇ ◇◇◇◇◇ ◇◇◇◇◇ ◇◇◇◇◇	865,850
◇◇◇◇◇ ◇◇◇◇◇ ◇◇◇◇◇ ◇◇◇◇◇ ◇◇◇◇◇ ◇◇◇◇◇ ◇◇◇◇◇ ◇◇◇◇◇ ◇◇◇◇◇ ◇◇◇◇◇	865,900
◇◇◇◇◇ ◇◇◇◇◇ ◇◇◇◇◇ ◇◇◇◇◇ ◇◇◇◇◇ ◇◇◇◇◇ ◇◇◇◇◇ ◇◇◇◇◇ ◇◇◇◇◇ ◇◇◇◇◇	865,950
◇◇◇◇◇ ◇◇◇◇◇ ◇◇◇◇◇ ◇◇◇◇◇ ◇◇◇◇◇ ◇◇◇◇◇ ◇◇◇◇◇ ◇◇◇◇◇ ◇◇◇◇◇ ◇◇◇◇◇	**866,000**

The Number 866,001 to the Number 868,000

Start		Notes
	866,050	
	866,100	
	866,150	
	866,200	
	866,250	
	866,300	
	866,350	
	866,400	
	866,450	
	866,500	
	866,550	
	866,600	
	866,650	
	866,700	
	866,750	
	866,800	
	866,850	
	866,900	
	866,950	
	867,000	
	867,050	
	867,100	
	867,150	
	867,200	
	867,250	
	867,300	
	867,350	
	867,400	
	867,450	
	867,500	
	867,550	
	867,600	
	867,650	
	867,700	
	867,750	
	867,800	
	867,850	
	867,900	
	867,950	
	868,000	

The Number 868,001 to the Number 870,000

Start   Notes

	868,050
	868,100
	868,150
	868,200
	868,250
	868,300
	868,350
	868,400
	868,450
	868,500
	868,550
	868,600
	868,650
	868,700
	868,750
	868,800
	868,850
	868,900
	868,950
	869,000
	869,050
	869,100
	869,150
	869,200
	869,250
	869,300
	869,350
	869,400
	869,450
	869,500
	869,550
	869,600
	869,650
	869,700
	869,750
	869,800
	869,850
	869,900
	869,950
	870,000

The Number 870,001 to the Number 872,000

Start	Notes
	870,050
	870,100
	870,150
	870,200
	870,250
	870,300
	870,350
	870,400
	870,450
	870,500
	870,550
	870,600
	870,650
	870,700
	870,750
	870,800
	870,850
	870,900
	870,950
	871,000
	871,050
	871,100
	871,150
	871,200
	871,250
	871,300
	871,350
	871,400
	871,450
	871,500
	871,550
	871,600
	871,650
	871,700
	871,750
	871,800
	871,850
	871,900
	871,950
	872,000

The Number 872,001 to the Number 874,000

Start   Notes

◊◊◊◊◊ ◊◊◊◊◊ ◊◊◊◊◊ ◊◊◊◊◊ ◊◊◊◊◊ ◊◊◊◊◊ ◊◊◊◊◊ ◊◊◊◊◊ ◊◊◊◊◊ ◊◊◊◊◊	872,050
◊◊◊◊◊ ◊◊◊◊◊ ◊◊◊◊◊ ◊◊◊◊◊ ◊◊◊◊◊ ◊◊◊◊◊ ◊◊◊◊◊ ◊◊◊◊◊ ◊◊◊◊◊ ◊◊◊◊◊	872,100
◊◊◊◊◊ ◊◊◊◊◊ ◊◊◊◊◊ ◊◊◊◊◊ ◊◊◊◊◊ ◊◊◊◊◊ ◊◊◊◊◊ ◊◊◊◊◊ ◊◊◊◊◊ ◊◊◊◊◊	872,150
◊◊◊◊◊ ◊◊◊◊◊ ◊◊◊◊◊ ◊◊◊◊◊ ◊◊◊◊◊ ◊◊◊◊◊ ◊◊◊◊◊ ◊◊◊◊◊ ◊◊◊◊◊ ◊◊◊◊◊	872,200
◊◊◊◊◊ ◊◊◊◊◊ ◊◊◊◊◊ ◊◊◊◊◊ ◊◊◊◊◊ ◊◊◊◊◊ ◊◊◊◊◊ ◊◊◊◊◊ ◊◊◊◊◊ ◊◊◊◊◊	872,250
◊◊◊◊◊ ◊◊◊◊◊ ◊◊◊◊◊ ◊◊◊◊◊ ◊◊◊◊◊ ◊◊◊◊◊ ◊◊◊◊◊ ◊◊◊◊◊ ◊◊◊◊◊ ◊◊◊◊◊	872,300
◊◊◊◊◊ ◊◊◊◊◊ ◊◊◊◊◊ ◊◊◊◊◊ ◊◊◊◊◊ ◊◊◊◊◊ ◊◊◊◊◊ ◊◊◊◊◊ ◊◊◊◊◊ ◊◊◊◊◊	872,350
◊◊◊◊◊ ◊◊◊◊◊ ◊◊◊◊◊ ◊◊◊◊◊ ◊◊◊◊◊ ◊◊◊◊◊ ◊◊◊◊◊ ◊◊◊◊◊ ◊◊◊◊◊ ◊◊◊◊◊	872,400
◊◊◊◊◊ ◊◊◊◊◊ ◊◊◊◊◊ ◊◊◊◊◊ ◊◊◊◊◊ ◊◊◊◊◊ ◊◊◊◊◊ ◊◊◊◊◊ ◊◊◊◊◊ ◊◊◊◊◊	872,450
◊◊◊◊◊ ◊◊◊◊◊ ◊◊◊◊◊ ◊◊◊◊◊ ◊◊◊◊◊ ◊◊◊◊◊ ◊◊◊◊◊ ◊◊◊◊◊ ◊◊◊◊◊ ◊◊◊◊◊	872,500
◊◊◊◊◊ ◊◊◊◊◊ ◊◊◊◊◊ ◊◊◊◊◊ ◊◊◊◊◊ ◊◊◊◊◊ ◊◊◊◊◊ ◊◊◊◊◊ ◊◊◊◊◊ ◊◊◊◊◊	872,550
◊◊◊◊◊ ◊◊◊◊◊ ◊◊◊◊◊ ◊◊◊◊◊ ◊◊◊◊◊ ◊◊◊◊◊ ◊◊◊◊◊ ◊◊◊◊◊ ◊◊◊◊◊ ◊◊◊◊◊	872,600
◊◊◊◊◊ ◊◊◊◊◊ ◊◊◊◊◊ ◊◊◊◊◊ ◊◊◊◊◊ ◊◊◊◊◊ ◊◊◊◊◊ ◊◊◊◊◊ ◊◊◊◊◊ ◊◊◊◊◊	872,650
◊◊◊◊◊ ◊◊◊◊◊ ◊◊◊◊◊ ◊◊◊◊◊ ◊◊◊◊◊ ◊◊◊◊◊ ◊◊◊◊◊ ◊◊◊◊◊ ◊◊◊◊◊ ◊◊◊◊◊	872,700
◊◊◊◊◊ ◊◊◊◊◊ ◊◊◊◊◊ ◊◊◊◊◊ ◊◊◊◊◊ ◊◊◊◊◊ ◊◊◊◊◊ ◊◊◊◊◊ ◊◊◊◊◊ ◊◊◊◊◊	872,750
◊◊◊◊◊ ◊◊◊◊◊ ◊◊◊◊◊ ◊◊◊◊◊ ◊◊◊◊◊ ◊◊◊◊◊ ◊◊◊◊◊ ◊◊◊◊◊ ◊◊◊◊◊ ◊◊◊◊◊	872,800
◊◊◊◊◊ ◊◊◊◊◊ ◊◊◊◊◊ ◊◊◊◊◊ ◊◊◊◊◊ ◊◊◊◊◊ ◊◊◊◊◊ ◊◊◊◊◊ ◊◊◊◊◊ ◊◊◊◊◊	872,850
◊◊◊◊◊ ◊◊◊◊◊ ◊◊◊◊◊ ◊◊◊◊◊ ◊◊◊◊◊ ◊◊◊◊◊ ◊◊◊◊◊ ◊◊◊◊◊ ◊◊◊◊◊ ◊◊◊◊◊	872,900
◊◊◊◊◊ ◊◊◊◊◊ ◊◊◊◊◊ ◊◊◊◊◊ ◊◊◊◊◊ ◊◊◊◊◊ ◊◊◊◊◊ ◊◊◊◊◊ ◊◊◊◊◊ ◊◊◊◊◊	872,950
◊◊◊◊◊ ◊◊◊◊◊ ◊◊◊◊◊ ◊◊◊◊◊ ◊◊◊◊◊ ◊◊◊◊◊ ◊◊◊◊◊ ◊◊◊◊◊ ◊◊◊◊◊ ◊◊◊◊◊	**873,000**
◊◊◊◊◊ ◊◊◊◊◊ ◊◊◊◊◊ ◊◊◊◊◊ ◊◊◊◊◊ ◊◊◊◊◊ ◊◊◊◊◊ ◊◊◊◊◊ ◊◊◊◊◊ ◊◊◊◊◊	873,050
◊◊◊◊◊ ◊◊◊◊◊ ◊◊◊◊◊ ◊◊◊◊◊ ◊◊◊◊◊ ◊◊◊◊◊ ◊◊◊◊◊ ◊◊◊◊◊ ◊◊◊◊◊ ◊◊◊◊◊	873,100
◊◊◊◊◊ ◊◊◊◊◊ ◊◊◊◊◊ ◊◊◊◊◊ ◊◊◊◊◊ ◊◊◊◊◊ ◊◊◊◊◊ ◊◊◊◊◊ ◊◊◊◊◊ ◊◊◊◊◊	873,150
◊◊◊◊◊ ◊◊◊◊◊ ◊◊◊◊◊ ◊◊◊◊◊ ◊◊◊◊◊ ◊◊◊◊◊ ◊◊◊◊◊ ◊◊◊◊◊ ◊◊◊◊◊ ◊◊◊◊◊	873,200
◊◊◊◊◊ ◊◊◊◊◊ ◊◊◊◊◊ ◊◊◊◊◊ ◊◊◊◊◊ ◊◊◊◊◊ ◊◊◊◊◊ ◊◊◊◊◊ ◊◊◊◊◊ ◊◊◊◊◊	873,250
◊◊◊◊◊ ◊◊◊◊◊ ◊◊◊◊◊ ◊◊◊◊◊ ◊◊◊◊◊ ◊◊◊◊◊ ◊◊◊◊◊ ◊◊◊◊◊ ◊◊◊◊◊ ◊◊◊◊◊	873,300
◊◊◊◊◊ ◊◊◊◊◊ ◊◊◊◊◊ ◊◊◊◊◊ ◊◊◊◊◊ ◊◊◊◊◊ ◊◊◊◊◊ ◊◊◊◊◊ ◊◊◊◊◊ ◊◊◊◊◊	873,350
◊◊◊◊◊ ◊◊◊◊◊ ◊◊◊◊◊ ◊◊◊◊◊ ◊◊◊◊◊ ◊◊◊◊◊ ◊◊◊◊◊ ◊◊◊◊◊ ◊◊◊◊◊ ◊◊◊◊◊	873,400
◊◊◊◊◊ ◊◊◊◊◊ ◊◊◊◊◊ ◊◊◊◊◊ ◊◊◊◊◊ ◊◊◊◊◊ ◊◊◊◊◊ ◊◊◊◊◊ ◊◊◊◊◊ ◊◊◊◊◊	873,450
◊◊◊◊◊ ◊◊◊◊◊ ◊◊◊◊◊ ◊◊◊◊◊ ◊◊◊◊◊ ◊◊◊◊◊ ◊◊◊◊◊ ◊◊◊◊◊ ◊◊◊◊◊ ◊◊◊◊◊	873,500
◊◊◊◊◊ ◊◊◊◊◊ ◊◊◊◊◊ ◊◊◊◊◊ ◊◊◊◊◊ ◊◊◊◊◊ ◊◊◊◊◊ ◊◊◊◊◊ ◊◊◊◊◊ ◊◊◊◊◊	873,550
◊◊◊◊◊ ◊◊◊◊◊ ◊◊◊◊◊ ◊◊◊◊◊ ◊◊◊◊◊ ◊◊◊◊◊ ◊◊◊◊◊ ◊◊◊◊◊ ◊◊◊◊◊ ◊◊◊◊◊	873,600
◊◊◊◊◊ ◊◊◊◊◊ ◊◊◊◊◊ ◊◊◊◊◊ ◊◊◊◊◊ ◊◊◊◊◊ ◊◊◊◊◊ ◊◊◊◊◊ ◊◊◊◊◊ ◊◊◊◊◊	873,650
◊◊◊◊◊ ◊◊◊◊◊ ◊◊◊◊◊ ◊◊◊◊◊ ◊◊◊◊◊ ◊◊◊◊◊ ◊◊◊◊◊ ◊◊◊◊◊ ◊◊◊◊◊ ◊◊◊◊◊	873,700
◊◊◊◊◊ ◊◊◊◊◊ ◊◊◊◊◊ ◊◊◊◊◊ ◊◊◊◊◊ ◊◊◊◊◊ ◊◊◊◊◊ ◊◊◊◊◊ ◊◊◊◊◊ ◊◊◊◊◊	873,750
◊◊◊◊◊ ◊◊◊◊◊ ◊◊◊◊◊ ◊◊◊◊◊ ◊◊◊◊◊ ◊◊◊◊◊ ◊◊◊◊◊ ◊◊◊◊◊ ◊◊◊◊◊ ◊◊◊◊◊	873,800
◊◊◊◊◊ ◊◊◊◊◊ ◊◊◊◊◊ ◊◊◊◊◊ ◊◊◊◊◊ ◊◊◊◊◊ ◊◊◊◊◊ ◊◊◊◊◊ ◊◊◊◊◊ ◊◊◊◊◊	873,850
◊◊◊◊◊ ◊◊◊◊◊ ◊◊◊◊◊ ◊◊◊◊◊ ◊◊◊◊◊ ◊◊◊◊◊ ◊◊◊◊◊ ◊◊◊◊◊ ◊◊◊◊◊ ◊◊◊◊◊	873,900
◊◊◊◊◊ ◊◊◊◊◊ ◊◊◊◊◊ ◊◊◊◊◊ ◊◊◊◊◊ ◊◊◊◊◊ ◊◊◊◊◊ ◊◊◊◊◊ ◊◊◊◊◊ ◊◊◊◊◊	873,950
◊◊◊◊◊ ◊◊◊◊◊ ◊◊◊◊◊ ◊◊◊◊◊ ◊◊◊◊◊ ◊◊◊◊◊ ◊◊◊◊◊ ◊◊◊◊◊ ◊◊◊◊◊ ◊◊◊◊◊	**874,000**

The Number 874,001 to the Number 876,000

	Notes
874,050	
874,100	
874,150	
874,200	
874,250	
874,300	
874,350	
874,400	
874,450	
874,500	
874,550	
874,600	
874,650	
874,700	
874,750	
874,800	
874,850	
874,900	
874,950	
875,000	
875,050	
875,100	
875,150	
875,200	
875,250	
875,300	
875,350	
875,400	
875,450	
875,500	
875,550	
875,600	
875,650	
875,700	
875,750	
875,800	
875,850	
875,900	
875,950	
876,000	

The Number 876,001 to the Number 878,000

Start | | **Notes**

◊◊◊◊◊ ◊◊◊◊◊ ... ◊◊◊◊◊	876,050
◊◊◊◊◊ ◊◊◊◊◊ ... ◊◊◊◊◊	876,100
◊◊◊◊◊ ◊◊◊◊◊ ... ◊◊◊◊◊	876,150
◊◊◊◊◊ ◊◊◊◊◊ ... ◊◊◊◊◊	876,200
◊◊◊◊◊ ◊◊◊◊◊ ... ◊◊◊◊◊	876,250
◊◊◊◊◊ ◊◊◊◊◊ ... ◊◊◊◊◊	876,300
◊◊◊◊◊ ◊◊◊◊◊ ... ◊◊◊◊◊	876,350
◊◊◊◊◊ ◊◊◊◊◊ ... ◊◊◊◊◊	876,400
◊◊◊◊◊ ◊◊◊◊◊ ... ◊◊◊◊◊	876,450
◊◊◊◊◊ ◊◊◊◊◊ ... ◊◊◊◊◊	876,500
◊◊◊◊◊ ◊◊◊◊◊ ... ◊◊◊◊◊	876,550
◊◊◊◊◊ ◊◊◊◊◊ ... ◊◊◊◊◊	876,600
◊◊◊◊◊ ◊◊◊◊◊ ... ◊◊◊◊◊	876,650
◊◊◊◊◊ ◊◊◊◊◊ ... ◊◊◊◊◊	876,700
◊◊◊◊◊ ◊◊◊◊◊ ... ◊◊◊◊◊	876,750
◊◊◊◊◊ ◊◊◊◊◊ ... ◊◊◊◊◊	876,800
◊◊◊◊◊ ◊◊◊◊◊ ... ◊◊◊◊◊	876,850
◊◊◊◊◊ ◊◊◊◊◊ ... ◊◊◊◊◊	876,900
◊◊◊◊◊ ◊◊◊◊◊ ... ◊◊◊◊◊	876,950
◊◊◊◊◊ ◊◊◊◊◊ ... ◊◊◊◊◊	**877,000**
◊◊◊◊◊ ◊◊◊◊◊ ... ◊◊◊◊◊	877,050
◊◊◊◊◊ ◊◊◊◊◊ ... ◊◊◊◊◊	877,100
◊◊◊◊◊ ◊◊◊◊◊ ... ◊◊◊◊◊	877,150
◊◊◊◊◊ ◊◊◊◊◊ ... ◊◊◊◊◊	877,200
◊◊◊◊◊ ◊◊◊◊◊ ... ◊◊◊◊◊	877,250
◊◊◊◊◊ ◊◊◊◊◊ ... ◊◊◊◊◊	877,300
◊◊◊◊◊ ◊◊◊◊◊ ... ◊◊◊◊◊	877,350
◊◊◊◊◊ ◊◊◊◊◊ ... ◊◊◊◊◊	877,400
◊◊◊◊◊ ◊◊◊◊◊ ... ◊◊◊◊◊	877,450
◊◊◊◊◊ ◊◊◊◊◊ ... ◊◊◊◊◊	877,500
◊◊◊◊◊ ◊◊◊◊◊ ... ◊◊◊◊◊	877,550
◊◊◊◊◊ ◊◊◊◊◊ ... ◊◊◊◊◊	877,600
◊◊◊◊◊ ◊◊◊◊◊ ... ◊◊◊◊◊	877,650
◊◊◊◊◊ ◊◊◊◊◊ ... ◊◊◊◊◊	877,700
◊◊◊◊◊ ◊◊◊◊◊ ... ◊◊◊◊◊	877,750
◊◊◊◊◊ ◊◊◊◊◊ ... ◊◊◊◊◊	877,800
◊◊◊◊◊ ◊◊◊◊◊ ... ◊◊◊◊◊	877,850
◊◊◊◊◊ ◊◊◊◊◊ ... ◊◊◊◊◊	877,900
◊◊◊◊◊ ◊◊◊◊◊ ... ◊◊◊◊◊	877,950
◊◊◊◊◊ ◊◊◊◊◊ ... ◊◊◊◊◊	**878,000**

The Number 878,001 to the Number 880,000

Start

Notes

	878,050
	878,100
	878,150
	878,200
	878,250
	878,300
	878,350
	878,400
	878,450
	878,500
	878,550
	878,600
	878,650
	878,700
	878,750
	878,800
	878,850
	878,900
	878,950
	879,000
	879,050
	879,100
	879,150
	879,200
	879,250
	879,300
	879,350
	879,400
	879,450
	879,500
	879,550
	879,600
	879,650
	879,700
	879,750
	879,800
	879,850
	879,900
	879,950
	880,000

The Number 880,001 to the Number 882,000

Start

Notes

◊◊◊◊◊ ◊◊◊◊◊ ◊◊◊◊◊ ◊◊◊◊◊ ◊◊◊◊◊ ◊◊◊◊◊ ◊◊◊◊◊ ◊◊◊◊◊ ◊◊◊◊◊ ◊◊◊◊◊	880,050	
◊◊◊◊◊ ◊◊◊◊◊ ◊◊◊◊◊ ◊◊◊◊◊ ◊◊◊◊◊ ◊◊◊◊◊ ◊◊◊◊◊ ◊◊◊◊◊ ◊◊◊◊◊ ◊◊◊◊◊	880,100	
◊◊◊◊◊ ◊◊◊◊◊ ◊◊◊◊◊ ◊◊◊◊◊ ◊◊◊◊◊ ◊◊◊◊◊ ◊◊◊◊◊ ◊◊◊◊◊ ◊◊◊◊◊ ◊◊◊◊◊	880,150	
◊◊◊◊◊ ◊◊◊◊◊ ◊◊◊◊◊ ◊◊◊◊◊ ◊◊◊◊◊ ◊◊◊◊◊ ◊◊◊◊◊ ◊◊◊◊◊ ◊◊◊◊◊ ◊◊◊◊◊	880,200	
◊◊◊◊◊ ◊◊◊◊◊ ◊◊◊◊◊ ◊◊◊◊◊ ◊◊◊◊◊ ◊◊◊◊◊ ◊◊◊◊◊ ◊◊◊◊◊ ◊◊◊◊◊ ◊◊◊◊◊	880,250	
◊◊◊◊◊ ◊◊◊◊◊ ◊◊◊◊◊ ◊◊◊◊◊ ◊◊◊◊◊ ◊◊◊◊◊ ◊◊◊◊◊ ◊◊◊◊◊ ◊◊◊◊◊ ◊◊◊◊◊	880,300	
◊◊◊◊◊ ◊◊◊◊◊ ◊◊◊◊◊ ◊◊◊◊◊ ◊◊◊◊◊ ◊◊◊◊◊ ◊◊◊◊◊ ◊◊◊◊◊ ◊◊◊◊◊ ◊◊◊◊◊	880,350	
◊◊◊◊◊ ◊◊◊◊◊ ◊◊◊◊◊ ◊◊◊◊◊ ◊◊◊◊◊ ◊◊◊◊◊ ◊◊◊◊◊ ◊◊◊◊◊ ◊◊◊◊◊ ◊◊◊◊◊	880,400	
◊◊◊◊◊ ◊◊◊◊◊ ◊◊◊◊◊ ◊◊◊◊◊ ◊◊◊◊◊ ◊◊◊◊◊ ◊◊◊◊◊ ◊◊◊◊◊ ◊◊◊◊◊ ◊◊◊◊◊	880,450	
◊◊◊◊◊ ◊◊◊◊◊ ◊◊◊◊◊ ◊◊◊◊◊ ◊◊◊◊◊ ◊◊◊◊◊ ◊◊◊◊◊ ◊◊◊◊◊ ◊◊◊◊◊ ◊◊◊◊◊	880,500	
◊◊◊◊◊ ◊◊◊◊◊ ◊◊◊◊◊ ◊◊◊◊◊ ◊◊◊◊◊ ◊◊◊◊◊ ◊◊◊◊◊ ◊◊◊◊◊ ◊◊◊◊◊ ◊◊◊◊◊	880,550	
◊◊◊◊◊ ◊◊◊◊◊ ◊◊◊◊◊ ◊◊◊◊◊ ◊◊◊◊◊ ◊◊◊◊◊ ◊◊◊◊◊ ◊◊◊◊◊ ◊◊◊◊◊ ◊◊◊◊◊	880,600	
◊◊◊◊◊ ◊◊◊◊◊ ◊◊◊◊◊ ◊◊◊◊◊ ◊◊◊◊◊ ◊◊◊◊◊ ◊◊◊◊◊ ◊◊◊◊◊ ◊◊◊◊◊ ◊◊◊◊◊	880,650	
◊◊◊◊◊ ◊◊◊◊◊ ◊◊◊◊◊ ◊◊◊◊◊ ◊◊◊◊◊ ◊◊◊◊◊ ◊◊◊◊◊ ◊◊◊◊◊ ◊◊◊◊◊ ◊◊◊◊◊	880,700	
◊◊◊◊◊ ◊◊◊◊◊ ◊◊◊◊◊ ◊◊◊◊◊ ◊◊◊◊◊ ◊◊◊◊◊ ◊◊◊◊◊ ◊◊◊◊◊ ◊◊◊◊◊ ◊◊◊◊◊	880,750	
◊◊◊◊◊ ◊◊◊◊◊ ◊◊◊◊◊ ◊◊◊◊◊ ◊◊◊◊◊ ◊◊◊◊◊ ◊◊◊◊◊ ◊◊◊◊◊ ◊◊◊◊◊ ◊◊◊◊◊	880,800	
◊◊◊◊◊ ◊◊◊◊◊ ◊◊◊◊◊ ◊◊◊◊◊ ◊◊◊◊◊ ◊◊◊◊◊ ◊◊◊◊◊ ◊◊◊◊◊ ◊◊◊◊◊ ◊◊◊◊◊	880,850	
◊◊◊◊◊ ◊◊◊◊◊ ◊◊◊◊◊ ◊◊◊◊◊ ◊◊◊◊◊ ◊◊◊◊◊ ◊◊◊◊◊ ◊◊◊◊◊ ◊◊◊◊◊ ◊◊◊◊◊	880,900	
◊◊◊◊◊ ◊◊◊◊◊ ◊◊◊◊◊ ◊◊◊◊◊ ◊◊◊◊◊ ◊◊◊◊◊ ◊◊◊◊◊ ◊◊◊◊◊ ◊◊◊◊◊ ◊◊◊◊◊	880,950	
◊◊◊◊◊ ◊◊◊◊◊ ◊◊◊◊◊ ◊◊◊◊◊ ◊◊◊◊◊ ◊◊◊◊◊ ◊◊◊◊◊ ◊◊◊◊◊ ◊◊◊◊◊ ◊◊◊◊◊	**881,000**	
◊◊◊◊◊ ◊◊◊◊◊ ◊◊◊◊◊ ◊◊◊◊◊ ◊◊◊◊◊ ◊◊◊◊◊ ◊◊◊◊◊ ◊◊◊◊◊ ◊◊◊◊◊ ◊◊◊◊◊	881,050	
◊◊◊◊◊ ◊◊◊◊◊ ◊◊◊◊◊ ◊◊◊◊◊ ◊◊◊◊◊ ◊◊◊◊◊ ◊◊◊◊◊ ◊◊◊◊◊ ◊◊◊◊◊ ◊◊◊◊◊	881,100	
◊◊◊◊◊ ◊◊◊◊◊ ◊◊◊◊◊ ◊◊◊◊◊ ◊◊◊◊◊ ◊◊◊◊◊ ◊◊◊◊◊ ◊◊◊◊◊ ◊◊◊◊◊ ◊◊◊◊◊	881,150	
◊◊◊◊◊ ◊◊◊◊◊ ◊◊◊◊◊ ◊◊◊◊◊ ◊◊◊◊◊ ◊◊◊◊◊ ◊◊◊◊◊ ◊◊◊◊◊ ◊◊◊◊◊ ◊◊◊◊◊	881,200	
◊◊◊◊◊ ◊◊◊◊◊ ◊◊◊◊◊ ◊◊◊◊◊ ◊◊◊◊◊ ◊◊◊◊◊ ◊◊◊◊◊ ◊◊◊◊◊ ◊◊◊◊◊ ◊◊◊◊◊	881,250	
◊◊◊◊◊ ◊◊◊◊◊ ◊◊◊◊◊ ◊◊◊◊◊ ◊◊◊◊◊ ◊◊◊◊◊ ◊◊◊◊◊ ◊◊◊◊◊ ◊◊◊◊◊ ◊◊◊◊◊	881,300	
◊◊◊◊◊ ◊◊◊◊◊ ◊◊◊◊◊ ◊◊◊◊◊ ◊◊◊◊◊ ◊◊◊◊◊ ◊◊◊◊◊ ◊◊◊◊◊ ◊◊◊◊◊ ◊◊◊◊◊	881,350	
◊◊◊◊◊ ◊◊◊◊◊ ◊◊◊◊◊ ◊◊◊◊◊ ◊◊◊◊◊ ◊◊◊◊◊ ◊◊◊◊◊ ◊◊◊◊◊ ◊◊◊◊◊ ◊◊◊◊◊	881,400	
◊◊◊◊◊ ◊◊◊◊◊ ◊◊◊◊◊ ◊◊◊◊◊ ◊◊◊◊◊ ◊◊◊◊◊ ◊◊◊◊◊ ◊◊◊◊◊ ◊◊◊◊◊ ◊◊◊◊◊	881,450	
◊◊◊◊◊ ◊◊◊◊◊ ◊◊◊◊◊ ◊◊◊◊◊ ◊◊◊◊◊ ◊◊◊◊◊ ◊◊◊◊◊ ◊◊◊◊◊ ◊◊◊◊◊ ◊◊◊◊◊	881,500	
◊◊◊◊◊ ◊◊◊◊◊ ◊◊◊◊◊ ◊◊◊◊◊ ◊◊◊◊◊ ◊◊◊◊◊ ◊◊◊◊◊ ◊◊◊◊◊ ◊◊◊◊◊ ◊◊◊◊◊	881,550	
◊◊◊◊◊ ◊◊◊◊◊ ◊◊◊◊◊ ◊◊◊◊◊ ◊◊◊◊◊ ◊◊◊◊◊ ◊◊◊◊◊ ◊◊◊◊◊ ◊◊◊◊◊ ◊◊◊◊◊	881,600	
◊◊◊◊◊ ◊◊◊◊◊ ◊◊◊◊◊ ◊◊◊◊◊ ◊◊◊◊◊ ◊◊◊◊◊ ◊◊◊◊◊ ◊◊◊◊◊ ◊◊◊◊◊ ◊◊◊◊◊	881,650	
◊◊◊◊◊ ◊◊◊◊◊ ◊◊◊◊◊ ◊◊◊◊◊ ◊◊◊◊◊ ◊◊◊◊◊ ◊◊◊◊◊ ◊◊◊◊◊ ◊◊◊◊◊ ◊◊◊◊◊	881,700	
◊◊◊◊◊ ◊◊◊◊◊ ◊◊◊◊◊ ◊◊◊◊◊ ◊◊◊◊◊ ◊◊◊◊◊ ◊◊◊◊◊ ◊◊◊◊◊ ◊◊◊◊◊ ◊◊◊◊◊	881,750	
◊◊◊◊◊ ◊◊◊◊◊ ◊◊◊◊◊ ◊◊◊◊◊ ◊◊◊◊◊ ◊◊◊◊◊ ◊◊◊◊◊ ◊◊◊◊◊ ◊◊◊◊◊ ◊◊◊◊◊	881,800	
◊◊◊◊◊ ◊◊◊◊◊ ◊◊◊◊◊ ◊◊◊◊◊ ◊◊◊◊◊ ◊◊◊◊◊ ◊◊◊◊◊ ◊◊◊◊◊ ◊◊◊◊◊ ◊◊◊◊◊	881,850	
◊◊◊◊◊ ◊◊◊◊◊ ◊◊◊◊◊ ◊◊◊◊◊ ◊◊◊◊◊ ◊◊◊◊◊ ◊◊◊◊◊ ◊◊◊◊◊ ◊◊◊◊◊ ◊◊◊◊◊	881,900	
◊◊◊◊◊ ◊◊◊◊◊ ◊◊◊◊◊ ◊◊◊◊◊ ◊◊◊◊◊ ◊◊◊◊◊ ◊◊◊◊◊ ◊◊◊◊◊ ◊◊◊◊◊ ◊◊◊◊◊	881,950	
◊◊◊◊◊ ◊◊◊◊◊ ◊◊◊◊◊ ◊◊◊◊◊ ◊◊◊◊◊ ◊◊◊◊◊ ◊◊◊◊◊ ◊◊◊◊◊ ◊◊◊◊◊ ◊◊◊◊◊	**882,000**	

The Number 882,001 to the Number 884,000

Start	Notes
	882,050
	882,100
	882,150
	882,200
	882,250
	882,300
	882,350
	882,400
	882,450
	882,500
	882,550
	882,600
	882,650
	882,700
	882,750
	882,800
	882,850
	882,900
	882,950
	883,000
	883,050
	883,100
	883,150
	883,200
	883,250
	883,300
	883,350
	883,400
	883,450
	883,500
	883,550
	883,600
	883,650
	883,700
	883,750
	883,800
	883,850
	883,900
	883,950
	884,000

The Number 884,001 to the Number 886,000

Start

	884,050
	884,100
	884,150
	884,200
	884,250
	884,300
	884,350
	884,400
	884,450
	884,500
	884,550
	884,600
	884,650
	884,700
	884,750
	884,800
	884,850
	884,900
	884,950
	885,000
	885,050
	885,100
	885,150
	885,200
	885,250
	885,300
	885,350
	885,400
	885,450
	885,500
	885,550
	885,600
	885,650
	885,700
	885,750
	885,800
	885,850
	885,900
	885,950
	886,000

The Number 886,001 to the Number 888,000

Start | | Notes

	886,050
	886,100
	886,150
	886,200
	886,250
	886,300
	886,350
	886,400
	886,450
	886,500
	886,550
	886,600
	886,650
	886,700
	886,750
	886,800
	886,850
	886,900
	886,950
	887,000
	887,050
	887,100
	887,150
	887,200
	887,250
	887,300
	887,350
	887,400
	887,450
	887,500
	887,550
	887,600
	887,650
	887,700
	887,750
	887,800
	887,850
	887,900
	887,950
	888,000

The Number 888,001 to the Number 890,000

Start		Notes
888,050
888,100
888,150
888,200
888,250
888,300
888,350
888,400
888,450
888,500
888,550
888,600
888,650
888,700
888,750
888,800
888,850
888,900
888,950
889,000
889,050
889,100
889,150
889,200
889,250
889,300
889,350
889,400
889,450
889,500
889,550
889,600
889,650
889,700
889,750
889,800
889,850
889,900
889,950
890,000

The Number 890,001 to the Number 892,000

Start

	890,050
	890,100
	890,150
	890,200
	890,250
	890,300
	890,350
	890,400
	890,450
	890,500
	890,550
	890,600
	890,650
	890,700
	890,750
	890,800
	890,850
	890,900
	890,950
	891,000
	891,050
	891,100
	891,150
	891,200
	891,250
	891,300
	891,350
	891,400
	891,450
	891,500
	891,550
	891,600
	891,650
	891,700
	891,750
	891,800
	891,850
	891,900
	891,950
	892,000

The Number 892,001 to the Number 894,000

Start | | Notes

	Number	Notes
	892,050	
	892,100	
	892,150	
	892,200	
	892,250	
	892,300	
	892,350	
	892,400	
	892,450	
	892,500	
	892,550	
	892,600	
	892,650	
	892,700	
	892,750	
	892,800	
	892,850	
	892,900	
	892,950	
	893,000	
	893,050	
	893,100	
	893,150	
	893,200	
	893,250	
	893,300	
	893,350	
	893,400	
	893,450	
	893,500	
	893,550	
	893,600	
	893,650	
	893,700	
	893,750	
	893,800	
	893,850	
	893,900	
	893,950	
	894,000	

The Number 894,001 to the Number 896,000

Start	Notes
◊◊◊◊◊ ◊◊◊◊◊ ◊◊◊◊◊ ◊◊◊◊◊ ◊◊◊◊◊ ◊◊◊◊◊ ◊◊◊◊◊ ◊◊◊◊◊ ◊◊◊◊◊ ◊◊◊◊◊	894,050
◊◊◊◊◊ ◊◊◊◊◊ ◊◊◊◊◊ ◊◊◊◊◊ ◊◊◊◊◊ ◊◊◊◊◊ ◊◊◊◊◊ ◊◊◊◊◊ ◊◊◊◊◊ ◊◊◊◊◊	894,100
◊◊◊◊◊ ◊◊◊◊◊ ◊◊◊◊◊ ◊◊◊◊◊ ◊◊◊◊◊ ◊◊◊◊◊ ◊◊◊◊◊ ◊◊◊◊◊ ◊◊◊◊◊ ◊◊◊◊◊	894,150
◊◊◊◊◊ ◊◊◊◊◊ ◊◊◊◊◊ ◊◊◊◊◊ ◊◊◊◊◊ ◊◊◊◊◊ ◊◊◊◊◊ ◊◊◊◊◊ ◊◊◊◊◊ ◊◊◊◊◊	894,200
◊◊◊◊◊ ◊◊◊◊◊ ◊◊◊◊◊ ◊◊◊◊◊ ◊◊◊◊◊ ◊◊◊◊◊ ◊◊◊◊◊ ◊◊◊◊◊ ◊◊◊◊◊ ◊◊◊◊◊	894,250
◊◊◊◊◊ ◊◊◊◊◊ ◊◊◊◊◊ ◊◊◊◊◊ ◊◊◊◊◊ ◊◊◊◊◊ ◊◊◊◊◊ ◊◊◊◊◊ ◊◊◊◊◊ ◊◊◊◊◊	894,300
◊◊◊◊◊ ◊◊◊◊◊ ◊◊◊◊◊ ◊◊◊◊◊ ◊◊◊◊◊ ◊◊◊◊◊ ◊◊◊◊◊ ◊◊◊◊◊ ◊◊◊◊◊ ◊◊◊◊◊	894,350
◊◊◊◊◊ ◊◊◊◊◊ ◊◊◊◊◊ ◊◊◊◊◊ ◊◊◊◊◊ ◊◊◊◊◊ ◊◊◊◊◊ ◊◊◊◊◊ ◊◊◊◊◊ ◊◊◊◊◊	894,400
◊◊◊◊◊ ◊◊◊◊◊ ◊◊◊◊◊ ◊◊◊◊◊ ◊◊◊◊◊ ◊◊◊◊◊ ◊◊◊◊◊ ◊◊◊◊◊ ◊◊◊◊◊ ◊◊◊◊◊	894,450
◊◊◊◊◊ ◊◊◊◊◊ ◊◊◊◊◊ ◊◊◊◊◊ ◊◊◊◊◊ ◊◊◊◊◊ ◊◊◊◊◊ ◊◊◊◊◊ ◊◊◊◊◊ ◊◊◊◊◊	894,500
◊◊◊◊◊ ◊◊◊◊◊ ◊◊◊◊◊ ◊◊◊◊◊ ◊◊◊◊◊ ◊◊◊◊◊ ◊◊◊◊◊ ◊◊◊◊◊ ◊◊◊◊◊ ◊◊◊◊◊	894,550
◊◊◊◊◊ ◊◊◊◊◊ ◊◊◊◊◊ ◊◊◊◊◊ ◊◊◊◊◊ ◊◊◊◊◊ ◊◊◊◊◊ ◊◊◊◊◊ ◊◊◊◊◊ ◊◊◊◊◊	894,600
◊◊◊◊◊ ◊◊◊◊◊ ◊◊◊◊◊ ◊◊◊◊◊ ◊◊◊◊◊ ◊◊◊◊◊ ◊◊◊◊◊ ◊◊◊◊◊ ◊◊◊◊◊ ◊◊◊◊◊	894,650
◊◊◊◊◊ ◊◊◊◊◊ ◊◊◊◊◊ ◊◊◊◊◊ ◊◊◊◊◊ ◊◊◊◊◊ ◊◊◊◊◊ ◊◊◊◊◊ ◊◊◊◊◊ ◊◊◊◊◊	894,700
◊◊◊◊◊ ◊◊◊◊◊ ◊◊◊◊◊ ◊◊◊◊◊ ◊◊◊◊◊ ◊◊◊◊◊ ◊◊◊◊◊ ◊◊◊◊◊ ◊◊◊◊◊ ◊◊◊◊◊	894,750
◊◊◊◊◊ ◊◊◊◊◊ ◊◊◊◊◊ ◊◊◊◊◊ ◊◊◊◊◊ ◊◊◊◊◊ ◊◊◊◊◊ ◊◊◊◊◊ ◊◊◊◊◊ ◊◊◊◊◊	894,800
◊◊◊◊◊ ◊◊◊◊◊ ◊◊◊◊◊ ◊◊◊◊◊ ◊◊◊◊◊ ◊◊◊◊◊ ◊◊◊◊◊ ◊◊◊◊◊ ◊◊◊◊◊ ◊◊◊◊◊	894,850
◊◊◊◊◊ ◊◊◊◊◊ ◊◊◊◊◊ ◊◊◊◊◊ ◊◊◊◊◊ ◊◊◊◊◊ ◊◊◊◊◊ ◊◊◊◊◊ ◊◊◊◊◊ ◊◊◊◊◊	894,900
◊◊◊◊◊ ◊◊◊◊◊ ◊◊◊◊◊ ◊◊◊◊◊ ◊◊◊◊◊ ◊◊◊◊◊ ◊◊◊◊◊ ◊◊◊◊◊ ◊◊◊◊◊ ◊◊◊◊◊	894,950
◊◊◊◊◊ ◊◊◊◊◊ ◊◊◊◊◊ ◊◊◊◊◊ ◊◊◊◊◊ ◊◊◊◊◊ ◊◊◊◊◊ ◊◊◊◊◊ ◊◊◊◊◊ ◊◊◊◊◊	**895,000**
◊◊◊◊◊ ◊◊◊◊◊ ◊◊◊◊◊ ◊◊◊◊◊ ◊◊◊◊◊ ◊◊◊◊◊ ◊◊◊◊◊ ◊◊◊◊◊ ◊◊◊◊◊ ◊◊◊◊◊	895,050
◊◊◊◊◊ ◊◊◊◊◊ ◊◊◊◊◊ ◊◊◊◊◊ ◊◊◊◊◊ ◊◊◊◊◊ ◊◊◊◊◊ ◊◊◊◊◊ ◊◊◊◊◊ ◊◊◊◊◊	895,100
◊◊◊◊◊ ◊◊◊◊◊ ◊◊◊◊◊ ◊◊◊◊◊ ◊◊◊◊◊ ◊◊◊◊◊ ◊◊◊◊◊ ◊◊◊◊◊ ◊◊◊◊◊ ◊◊◊◊◊	895,150
◊◊◊◊◊ ◊◊◊◊◊ ◊◊◊◊◊ ◊◊◊◊◊ ◊◊◊◊◊ ◊◊◊◊◊ ◊◊◊◊◊ ◊◊◊◊◊ ◊◊◊◊◊ ◊◊◊◊◊	895,200
◊◊◊◊◊ ◊◊◊◊◊ ◊◊◊◊◊ ◊◊◊◊◊ ◊◊◊◊◊ ◊◊◊◊◊ ◊◊◊◊◊ ◊◊◊◊◊ ◊◊◊◊◊ ◊◊◊◊◊	895,250
◊◊◊◊◊ ◊◊◊◊◊ ◊◊◊◊◊ ◊◊◊◊◊ ◊◊◊◊◊ ◊◊◊◊◊ ◊◊◊◊◊ ◊◊◊◊◊ ◊◊◊◊◊ ◊◊◊◊◊	895,300
◊◊◊◊◊ ◊◊◊◊◊ ◊◊◊◊◊ ◊◊◊◊◊ ◊◊◊◊◊ ◊◊◊◊◊ ◊◊◊◊◊ ◊◊◊◊◊ ◊◊◊◊◊ ◊◊◊◊◊	895,350
◊◊◊◊◊ ◊◊◊◊◊ ◊◊◊◊◊ ◊◊◊◊◊ ◊◊◊◊◊ ◊◊◊◊◊ ◊◊◊◊◊ ◊◊◊◊◊ ◊◊◊◊◊ ◊◊◊◊◊	895,400
◊◊◊◊◊ ◊◊◊◊◊ ◊◊◊◊◊ ◊◊◊◊◊ ◊◊◊◊◊ ◊◊◊◊◊ ◊◊◊◊◊ ◊◊◊◊◊ ◊◊◊◊◊ ◊◊◊◊◊	895,450
◊◊◊◊◊ ◊◊◊◊◊ ◊◊◊◊◊ ◊◊◊◊◊ ◊◊◊◊◊ ◊◊◊◊◊ ◊◊◊◊◊ ◊◊◊◊◊ ◊◊◊◊◊ ◊◊◊◊◊	895,500
◊◊◊◊◊ ◊◊◊◊◊ ◊◊◊◊◊ ◊◊◊◊◊ ◊◊◊◊◊ ◊◊◊◊◊ ◊◊◊◊◊ ◊◊◊◊◊ ◊◊◊◊◊ ◊◊◊◊◊	895,550
◊◊◊◊◊ ◊◊◊◊◊ ◊◊◊◊◊ ◊◊◊◊◊ ◊◊◊◊◊ ◊◊◊◊◊ ◊◊◊◊◊ ◊◊◊◊◊ ◊◊◊◊◊ ◊◊◊◊◊	895,600
◊◊◊◊◊ ◊◊◊◊◊ ◊◊◊◊◊ ◊◊◊◊◊ ◊◊◊◊◊ ◊◊◊◊◊ ◊◊◊◊◊ ◊◊◊◊◊ ◊◊◊◊◊ ◊◊◊◊◊	895,650
◊◊◊◊◊ ◊◊◊◊◊ ◊◊◊◊◊ ◊◊◊◊◊ ◊◊◊◊◊ ◊◊◊◊◊ ◊◊◊◊◊ ◊◊◊◊◊ ◊◊◊◊◊ ◊◊◊◊◊	895,700
◊◊◊◊◊ ◊◊◊◊◊ ◊◊◊◊◊ ◊◊◊◊◊ ◊◊◊◊◊ ◊◊◊◊◊ ◊◊◊◊◊ ◊◊◊◊◊ ◊◊◊◊◊ ◊◊◊◊◊	895,750
◊◊◊◊◊ ◊◊◊◊◊ ◊◊◊◊◊ ◊◊◊◊◊ ◊◊◊◊◊ ◊◊◊◊◊ ◊◊◊◊◊ ◊◊◊◊◊ ◊◊◊◊◊ ◊◊◊◊◊	895,800
◊◊◊◊◊ ◊◊◊◊◊ ◊◊◊◊◊ ◊◊◊◊◊ ◊◊◊◊◊ ◊◊◊◊◊ ◊◊◊◊◊ ◊◊◊◊◊ ◊◊◊◊◊ ◊◊◊◊◊	895,850
◊◊◊◊◊ ◊◊◊◊◊ ◊◊◊◊◊ ◊◊◊◊◊ ◊◊◊◊◊ ◊◊◊◊◊ ◊◊◊◊◊ ◊◊◊◊◊ ◊◊◊◊◊ ◊◊◊◊◊	895,900
◊◊◊◊◊ ◊◊◊◊◊ ◊◊◊◊◊ ◊◊◊◊◊ ◊◊◊◊◊ ◊◊◊◊◊ ◊◊◊◊◊ ◊◊◊◊◊ ◊◊◊◊◊ ◊◊◊◊◊	895,950
◊◊◊◊◊ ◊◊◊◊◊ ◊◊◊◊◊ ◊◊◊◊◊ ◊◊◊◊◊ ◊◊◊◊◊ ◊◊◊◊◊ ◊◊◊◊◊ ◊◊◊◊◊ ◊◊◊◊◊	**896,000**

◇ The Number 896,001 to the Number 898,000

Start   **Notes**

	896,050
	896,100
	896,150
	896,200
	896,250
	896,300
	896,350
	896,400
	896,450
	896,500
	896,550
	896,600
	896,650
	896,700
	896,750
	896,800
	896,850
	896,900
	896,950
	897,000
	897,050
	897,100
	897,150
	897,200
	897,250
	897,300
	897,350
	897,400
	897,450
	897,500
	897,550
	897,600
	897,650
	897,700
	897,750
	897,800
	897,850
	897,900
	897,950
	898,000

The Number 898,001 to the Number 900,000

Start		Notes
	898,050	
	898,100	
	898,150	
	898,200	
	898,250	
	898,300	
	898,350	
	898,400	
	898,450	
	898,500	
	898,550	
	898,600	
	898,650	
	898,700	
	898,750	
	898,800	
	898,850	
	898,900	
	898,950	
	899,000	
	899,050	
	899,100	
	899,150	
	899,200	
	899,250	
	899,300	
	899,350	
	899,400	
	899,450	
	899,500	
	899,550	
	899,600	
	899,650	
	899,700	
	899,750	
	899,800	
	899,850	
	899,900	
	899,950	
	900,000	

The Number 900,001 to the Number 902,000

Start **Notes**

	900,050
	900,100
	900,150
	900,200
	900,250
	900,300
	900,350
	900,400
	900,450
	900,500
	900,550
	900,600
	900,650
	900,700
	900,750
	900,800
	900,850
	900,900
	900,950
	901,000
	901,050
	901,100
	901,150
	901,200
	901,250
	901,300
	901,350
	901,400
	901,450
	901,500
	901,550
	901,600
	901,650
	901,700
	901,750
	901,800
	901,850
	901,900
	901,950
	902,000

The Number 902,001 to the Number 904,000

Start   Notes

	902,050
	902,100
	902,150
	902,200
	902,250
	902,300
	902,350
	902,400
	902,450
	902,500
	902,550
	902,600
	902,650
	902,700
	902,750
	902,800
	902,850
	902,900
	902,950
	903,000
	903,050
	903,100
	903,150
	903,200
	903,250
	903,300
	903,350
	903,400
	903,450
	903,500
	903,550
	903,600
	903,650
	903,700
	903,750
	903,800
	903,850
	903,900
	903,950
	904,000

The Number 904,001 to the Number 906,000

Start   Notes

	Number
◊◊◊◊◊ ◊◊◊◊◊ ◊◊◊◊◊ ◊◊◊◊◊ ◊◊◊◊◊ ◊◊◊◊◊ ◊◊◊◊◊ ◊◊◊◊◊ ◊◊◊◊◊ ◊◊◊◊◊	904,050
◊◊◊◊◊ ◊◊◊◊◊ ◊◊◊◊◊ ◊◊◊◊◊ ◊◊◊◊◊ ◊◊◊◊◊ ◊◊◊◊◊ ◊◊◊◊◊ ◊◊◊◊◊ ◊◊◊◊◊	904,100
◊◊◊◊◊ ◊◊◊◊◊ ◊◊◊◊◊ ◊◊◊◊◊ ◊◊◊◊◊ ◊◊◊◊◊ ◊◊◊◊◊ ◊◊◊◊◊ ◊◊◊◊◊ ◊◊◊◊◊	904,150
◊◊◊◊◊ ◊◊◊◊◊ ◊◊◊◊◊ ◊◊◊◊◊ ◊◊◊◊◊ ◊◊◊◊◊ ◊◊◊◊◊ ◊◊◊◊◊ ◊◊◊◊◊ ◊◊◊◊◊	904,200
◊◊◊◊◊ ◊◊◊◊◊ ◊◊◊◊◊ ◊◊◊◊◊ ◊◊◊◊◊ ◊◊◊◊◊ ◊◊◊◊◊ ◊◊◊◊◊ ◊◊◊◊◊ ◊◊◊◊◊	904,250
◊◊◊◊◊ ◊◊◊◊◊ ◊◊◊◊◊ ◊◊◊◊◊ ◊◊◊◊◊ ◊◊◊◊◊ ◊◊◊◊◊ ◊◊◊◊◊ ◊◊◊◊◊ ◊◊◊◊◊	904,300
◊◊◊◊◊ ◊◊◊◊◊ ◊◊◊◊◊ ◊◊◊◊◊ ◊◊◊◊◊ ◊◊◊◊◊ ◊◊◊◊◊ ◊◊◊◊◊ ◊◊◊◊◊ ◊◊◊◊◊	904,350
◊◊◊◊◊ ◊◊◊◊◊ ◊◊◊◊◊ ◊◊◊◊◊ ◊◊◊◊◊ ◊◊◊◊◊ ◊◊◊◊◊ ◊◊◊◊◊ ◊◊◊◊◊ ◊◊◊◊◊	904,400
◊◊◊◊◊ ◊◊◊◊◊ ◊◊◊◊◊ ◊◊◊◊◊ ◊◊◊◊◊ ◊◊◊◊◊ ◊◊◊◊◊ ◊◊◊◊◊ ◊◊◊◊◊ ◊◊◊◊◊	904,450
◊◊◊◊◊ ◊◊◊◊◊ ◊◊◊◊◊ ◊◊◊◊◊ ◊◊◊◊◊ ◊◊◊◊◊ ◊◊◊◊◊ ◊◊◊◊◊ ◊◊◊◊◊ ◊◊◊◊◊	904,500
◊◊◊◊◊ ◊◊◊◊◊ ◊◊◊◊◊ ◊◊◊◊◊ ◊◊◊◊◊ ◊◊◊◊◊ ◊◊◊◊◊ ◊◊◊◊◊ ◊◊◊◊◊ ◊◊◊◊◊	904,550
◊◊◊◊◊ ◊◊◊◊◊ ◊◊◊◊◊ ◊◊◊◊◊ ◊◊◊◊◊ ◊◊◊◊◊ ◊◊◊◊◊ ◊◊◊◊◊ ◊◊◊◊◊ ◊◊◊◊◊	904,600
◊◊◊◊◊ ◊◊◊◊◊ ◊◊◊◊◊ ◊◊◊◊◊ ◊◊◊◊◊ ◊◊◊◊◊ ◊◊◊◊◊ ◊◊◊◊◊ ◊◊◊◊◊ ◊◊◊◊◊	904,650
◊◊◊◊◊ ◊◊◊◊◊ ◊◊◊◊◊ ◊◊◊◊◊ ◊◊◊◊◊ ◊◊◊◊◊ ◊◊◊◊◊ ◊◊◊◊◊ ◊◊◊◊◊ ◊◊◊◊◊	904,700
◊◊◊◊◊ ◊◊◊◊◊ ◊◊◊◊◊ ◊◊◊◊◊ ◊◊◊◊◊ ◊◊◊◊◊ ◊◊◊◊◊ ◊◊◊◊◊ ◊◊◊◊◊ ◊◊◊◊◊	904,750
◊◊◊◊◊ ◊◊◊◊◊ ◊◊◊◊◊ ◊◊◊◊◊ ◊◊◊◊◊ ◊◊◊◊◊ ◊◊◊◊◊ ◊◊◊◊◊ ◊◊◊◊◊ ◊◊◊◊◊	904,800
◊◊◊◊◊ ◊◊◊◊◊ ◊◊◊◊◊ ◊◊◊◊◊ ◊◊◊◊◊ ◊◊◊◊◊ ◊◊◊◊◊ ◊◊◊◊◊ ◊◊◊◊◊ ◊◊◊◊◊	904,850
◊◊◊◊◊ ◊◊◊◊◊ ◊◊◊◊◊ ◊◊◊◊◊ ◊◊◊◊◊ ◊◊◊◊◊ ◊◊◊◊◊ ◊◊◊◊◊ ◊◊◊◊◊ ◊◊◊◊◊	904,900
◊◊◊◊◊ ◊◊◊◊◊ ◊◊◊◊◊ ◊◊◊◊◊ ◊◊◊◊◊ ◊◊◊◊◊ ◊◊◊◊◊ ◊◊◊◊◊ ◊◊◊◊◊ ◊◊◊◊◊	904,950
◊◊◊◊◊ ◊◊◊◊◊ ◊◊◊◊◊ ◊◊◊◊◊ ◊◊◊◊◊ ◊◊◊◊◊ ◊◊◊◊◊ ◊◊◊◊◊ ◊◊◊◊◊ ◊◊◊◊◊	**905,000**
◊◊◊◊◊ ◊◊◊◊◊ ◊◊◊◊◊ ◊◊◊◊◊ ◊◊◊◊◊ ◊◊◊◊◊ ◊◊◊◊◊ ◊◊◊◊◊ ◊◊◊◊◊ ◊◊◊◊◊	905,050
◊◊◊◊◊ ◊◊◊◊◊ ◊◊◊◊◊ ◊◊◊◊◊ ◊◊◊◊◊ ◊◊◊◊◊ ◊◊◊◊◊ ◊◊◊◊◊ ◊◊◊◊◊ ◊◊◊◊◊	905,100
◊◊◊◊◊ ◊◊◊◊◊ ◊◊◊◊◊ ◊◊◊◊◊ ◊◊◊◊◊ ◊◊◊◊◊ ◊◊◊◊◊ ◊◊◊◊◊ ◊◊◊◊◊ ◊◊◊◊◊	905,150
◊◊◊◊◊ ◊◊◊◊◊ ◊◊◊◊◊ ◊◊◊◊◊ ◊◊◊◊◊ ◊◊◊◊◊ ◊◊◊◊◊ ◊◊◊◊◊ ◊◊◊◊◊ ◊◊◊◊◊	905,200
◊◊◊◊◊ ◊◊◊◊◊ ◊◊◊◊◊ ◊◊◊◊◊ ◊◊◊◊◊ ◊◊◊◊◊ ◊◊◊◊◊ ◊◊◊◊◊ ◊◊◊◊◊ ◊◊◊◊◊	905,250
◊◊◊◊◊ ◊◊◊◊◊ ◊◊◊◊◊ ◊◊◊◊◊ ◊◊◊◊◊ ◊◊◊◊◊ ◊◊◊◊◊ ◊◊◊◊◊ ◊◊◊◊◊ ◊◊◊◊◊	905,300
◊◊◊◊◊ ◊◊◊◊◊ ◊◊◊◊◊ ◊◊◊◊◊ ◊◊◊◊◊ ◊◊◊◊◊ ◊◊◊◊◊ ◊◊◊◊◊ ◊◊◊◊◊ ◊◊◊◊◊	905,350
◊◊◊◊◊ ◊◊◊◊◊ ◊◊◊◊◊ ◊◊◊◊◊ ◊◊◊◊◊ ◊◊◊◊◊ ◊◊◊◊◊ ◊◊◊◊◊ ◊◊◊◊◊ ◊◊◊◊◊	905,400
◊◊◊◊◊ ◊◊◊◊◊ ◊◊◊◊◊ ◊◊◊◊◊ ◊◊◊◊◊ ◊◊◊◊◊ ◊◊◊◊◊ ◊◊◊◊◊ ◊◊◊◊◊ ◊◊◊◊◊	905,450
◊◊◊◊◊ ◊◊◊◊◊ ◊◊◊◊◊ ◊◊◊◊◊ ◊◊◊◊◊ ◊◊◊◊◊ ◊◊◊◊◊ ◊◊◊◊◊ ◊◊◊◊◊ ◊◊◊◊◊	905,500
◊◊◊◊◊ ◊◊◊◊◊ ◊◊◊◊◊ ◊◊◊◊◊ ◊◊◊◊◊ ◊◊◊◊◊ ◊◊◊◊◊ ◊◊◊◊◊ ◊◊◊◊◊ ◊◊◊◊◊	905,550
◊◊◊◊◊ ◊◊◊◊◊ ◊◊◊◊◊ ◊◊◊◊◊ ◊◊◊◊◊ ◊◊◊◊◊ ◊◊◊◊◊ ◊◊◊◊◊ ◊◊◊◊◊ ◊◊◊◊◊	905,600
◊◊◊◊◊ ◊◊◊◊◊ ◊◊◊◊◊ ◊◊◊◊◊ ◊◊◊◊◊ ◊◊◊◊◊ ◊◊◊◊◊ ◊◊◊◊◊ ◊◊◊◊◊ ◊◊◊◊◊	905,650
◊◊◊◊◊ ◊◊◊◊◊ ◊◊◊◊◊ ◊◊◊◊◊ ◊◊◊◊◊ ◊◊◊◊◊ ◊◊◊◊◊ ◊◊◊◊◊ ◊◊◊◊◊ ◊◊◊◊◊	905,700
◊◊◊◊◊ ◊◊◊◊◊ ◊◊◊◊◊ ◊◊◊◊◊ ◊◊◊◊◊ ◊◊◊◊◊ ◊◊◊◊◊ ◊◊◊◊◊ ◊◊◊◊◊ ◊◊◊◊◊	905,750
◊◊◊◊◊ ◊◊◊◊◊ ◊◊◊◊◊ ◊◊◊◊◊ ◊◊◊◊◊ ◊◊◊◊◊ ◊◊◊◊◊ ◊◊◊◊◊ ◊◊◊◊◊ ◊◊◊◊◊	905,800
◊◊◊◊◊ ◊◊◊◊◊ ◊◊◊◊◊ ◊◊◊◊◊ ◊◊◊◊◊ ◊◊◊◊◊ ◊◊◊◊◊ ◊◊◊◊◊ ◊◊◊◊◊ ◊◊◊◊◊	905,850
◊◊◊◊◊ ◊◊◊◊◊ ◊◊◊◊◊ ◊◊◊◊◊ ◊◊◊◊◊ ◊◊◊◊◊ ◊◊◊◊◊ ◊◊◊◊◊ ◊◊◊◊◊ ◊◊◊◊◊	905,900
◊◊◊◊◊ ◊◊◊◊◊ ◊◊◊◊◊ ◊◊◊◊◊ ◊◊◊◊◊ ◊◊◊◊◊ ◊◊◊◊◊ ◊◊◊◊◊ ◊◊◊◊◊ ◊◊◊◊◊	905,950
◊◊◊◊◊ ◊◊◊◊◊ ◊◊◊◊◊ ◊◊◊◊◊ ◊◊◊◊◊ ◊◊◊◊◊ ◊◊◊◊◊ ◊◊◊◊◊ ◊◊◊◊◊ ◊◊◊◊◊	**906,000**

The Number 906,001 to the Number 908,000

Start

Notes

906,050
906,100
906,150
906,200
906,250
906,300
906,350
906,400
906,450
906,500
906,550
906,600
906,650
906,700
906,750
906,800
906,850
906,900
906,950
907,000
907,050
907,100
907,150
907,200
907,250
907,300
907,350
907,400
907,450
907,500
907,550
907,600
907,650
907,700
907,750
907,800
907,850
907,900
907,950
908,000

The Number 908,001 to the Number 910,000

Start   Notes

	908,050
	908,100
	908,150
	908,200
	908,250
	908,300
	908,350
	908,400
	908,450
	908,500
	908,550
	908,600
	908,650
	908,700
	908,750
	908,800
	908,850
	908,900
	908,950
	909,000
	909,050
	909,100
	909,150
	909,200
	909,250
	909,300
	909,350
	909,400
	909,450
	909,500
	909,550
	909,600
	909,650
	909,700
	909,750
	909,800
	909,850
	909,900
	909,950
	910,000

The Number 910,001 to the Number 912,000

	Notes
910,050	
910,100	
910,150	
910,200	
910,250	
910,300	
910,350	
910,400	
910,450	
910,500	
910,550	
910,600	
910,650	
910,700	
910,750	
910,800	
910,850	
910,900	
910,950	
911,000	
911,050	
911,100	
911,150	
911,200	
911,250	
911,300	
911,350	
911,400	
911,450	
911,500	
911,550	
911,600	
911,650	
911,700	
911,750	
911,800	
911,850	
911,900	
911,950	
912,000	

The Number 912,001 to the Number 914,000

	Start	Notes
	912,050	
	912,100	
	912,150	
	912,200	
	912,250	
	912,300	
	912,350	
	912,400	
	912,450	
	912,500	
	912,550	
	912,600	
	912,650	
	912,700	
	912,750	
	912,800	
	912,850	
	912,900	
	912,950	
	913,000	
	913,050	
	913,100	
	913,150	
	913,200	
	913,250	
	913,300	
	913,350	
	913,400	
	913,450	
	913,500	
	913,550	
	913,600	
	913,650	
	913,700	
	913,750	
	913,800	
	913,850	
	913,900	
	913,950	
	914,000	

The Number 914,001 to the Number 916,000

Start											Notes
										914,050	
										914,100	
										914,150	
										914,200	
										914,250	
										914,300	
										914,350	
										914,400	
										914,450	
										914,500	
										914,550	
										914,600	
										914,650	
										914,700	
										914,750	
										914,800	
										914,850	
										914,900	
										914,950	
										915,000	
										915,050	
										915,100	
										915,150	
										915,200	
										915,250	
										915,300	
										915,350	
										915,400	
										915,450	
										915,500	
										915,550	
										915,600	
										915,650	
										915,700	
										915,750	
										915,800	
										915,850	
										915,900	
										915,950	
										916,000	

The Number 916,001 to the Number 918,000

Start

Notes

	916,050
	916,100
	916,150
	916,200
	916,250
	916,300
	916,350
	916,400
	916,450
	916,500
	916,550
	916,600
	916,650
	916,700
	916,750
	916,800
	916,850
	916,900
	916,950
	917,000
	917,050
	917,100
	917,150
	917,200
	917,250
	917,300
	917,350
	917,400
	917,450
	917,500
	917,550
	917,600
	917,650
	917,700
	917,750
	917,800
	917,850
	917,900
	917,950
	918,000

The Number 918,001 to the Number 920,000

Start	Notes
	918,050
	918,100
	918,150
	918,200
	918,250
	918,300
	918,350
	918,400
	918,450
	918,500
	918,550
	918,600
	918,650
	918,700
	918,750
	918,800
	918,850
	918,900
	918,950
	919,000
	919,050
	919,100
	919,150
	919,200
	919,250
	919,300
	919,350
	919,400
	919,450
	919,500
	919,550
	919,600
	919,650
	919,700
	919,750
	919,800
	919,850
	919,900
	919,950
	920,000

The Number 920,001 to the Number 922,000

Start

Notes

◊◊◊◊◊ ◊◊◊◊◊ ◊◊◊◊◊ ◊◊◊◊◊ ◊◊◊◊◊ ◊◊◊◊◊ ◊◊◊◊◊ ◊◊◊◊◊ ◊◊◊◊◊ ◊◊◊◊◊	920,050
◊◊◊◊◊ ◊◊◊◊◊ ◊◊◊◊◊ ◊◊◊◊◊ ◊◊◊◊◊ ◊◊◊◊◊ ◊◊◊◊◊ ◊◊◊◊◊ ◊◊◊◊◊ ◊◊◊◊◊	920,100
◊◊◊◊◊ ◊◊◊◊◊ ◊◊◊◊◊ ◊◊◊◊◊ ◊◊◊◊◊ ◊◊◊◊◊ ◊◊◊◊◊ ◊◊◊◊◊ ◊◊◊◊◊ ◊◊◊◊◊	920,150
◊◊◊◊◊ ◊◊◊◊◊ ◊◊◊◊◊ ◊◊◊◊◊ ◊◊◊◊◊ ◊◊◊◊◊ ◊◊◊◊◊ ◊◊◊◊◊ ◊◊◊◊◊ ◊◊◊◊◊	920,200
◊◊◊◊◊ ◊◊◊◊◊ ◊◊◊◊◊ ◊◊◊◊◊ ◊◊◊◊◊ ◊◊◊◊◊ ◊◊◊◊◊ ◊◊◊◊◊ ◊◊◊◊◊ ◊◊◊◊◊	920,250
◊◊◊◊◊ ◊◊◊◊◊ ◊◊◊◊◊ ◊◊◊◊◊ ◊◊◊◊◊ ◊◊◊◊◊ ◊◊◊◊◊ ◊◊◊◊◊ ◊◊◊◊◊ ◊◊◊◊◊	920,300
◊◊◊◊◊ ◊◊◊◊◊ ◊◊◊◊◊ ◊◊◊◊◊ ◊◊◊◊◊ ◊◊◊◊◊ ◊◊◊◊◊ ◊◊◊◊◊ ◊◊◊◊◊ ◊◊◊◊◊	920,350
◊◊◊◊◊ ◊◊◊◊◊ ◊◊◊◊◊ ◊◊◊◊◊ ◊◊◊◊◊ ◊◊◊◊◊ ◊◊◊◊◊ ◊◊◊◊◊ ◊◊◊◊◊ ◊◊◊◊◊	920,400
◊◊◊◊◊ ◊◊◊◊◊ ◊◊◊◊◊ ◊◊◊◊◊ ◊◊◊◊◊ ◊◊◊◊◊ ◊◊◊◊◊ ◊◊◊◊◊ ◊◊◊◊◊ ◊◊◊◊◊	920,450
◊◊◊◊◊ ◊◊◊◊◊ ◊◊◊◊◊ ◊◊◊◊◊ ◊◊◊◊◊ ◊◊◊◊◊ ◊◊◊◊◊ ◊◊◊◊◊ ◊◊◊◊◊ ◊◊◊◊◊	920,500
◊◊◊◊◊ ◊◊◊◊◊ ◊◊◊◊◊ ◊◊◊◊◊ ◊◊◊◊◊ ◊◊◊◊◊ ◊◊◊◊◊ ◊◊◊◊◊ ◊◊◊◊◊ ◊◊◊◊◊	920,550
◊◊◊◊◊ ◊◊◊◊◊ ◊◊◊◊◊ ◊◊◊◊◊ ◊◊◊◊◊ ◊◊◊◊◊ ◊◊◊◊◊ ◊◊◊◊◊ ◊◊◊◊◊ ◊◊◊◊◊	920,600
◊◊◊◊◊ ◊◊◊◊◊ ◊◊◊◊◊ ◊◊◊◊◊ ◊◊◊◊◊ ◊◊◊◊◊ ◊◊◊◊◊ ◊◊◊◊◊ ◊◊◊◊◊ ◊◊◊◊◊	920,650
◊◊◊◊◊ ◊◊◊◊◊ ◊◊◊◊◊ ◊◊◊◊◊ ◊◊◊◊◊ ◊◊◊◊◊ ◊◊◊◊◊ ◊◊◊◊◊ ◊◊◊◊◊ ◊◊◊◊◊	920,700
◊◊◊◊◊ ◊◊◊◊◊ ◊◊◊◊◊ ◊◊◊◊◊ ◊◊◊◊◊ ◊◊◊◊◊ ◊◊◊◊◊ ◊◊◊◊◊ ◊◊◊◊◊ ◊◊◊◊◊	920,750
◊◊◊◊◊ ◊◊◊◊◊ ◊◊◊◊◊ ◊◊◊◊◊ ◊◊◊◊◊ ◊◊◊◊◊ ◊◊◊◊◊ ◊◊◊◊◊ ◊◊◊◊◊ ◊◊◊◊◊	920,800
◊◊◊◊◊ ◊◊◊◊◊ ◊◊◊◊◊ ◊◊◊◊◊ ◊◊◊◊◊ ◊◊◊◊◊ ◊◊◊◊◊ ◊◊◊◊◊ ◊◊◊◊◊ ◊◊◊◊◊	920,850
◊◊◊◊◊ ◊◊◊◊◊ ◊◊◊◊◊ ◊◊◊◊◊ ◊◊◊◊◊ ◊◊◊◊◊ ◊◊◊◊◊ ◊◊◊◊◊ ◊◊◊◊◊ ◊◊◊◊◊	920,900
◊◊◊◊◊ ◊◊◊◊◊ ◊◊◊◊◊ ◊◊◊◊◊ ◊◊◊◊◊ ◊◊◊◊◊ ◊◊◊◊◊ ◊◊◊◊◊ ◊◊◊◊◊ ◊◊◊◊◊	920,950
◊◊◊◊◊ ◊◊◊◊◊ ◊◊◊◊◊ ◊◊◊◊◊ ◊◊◊◊◊ ◊◊◊◊◊ ◊◊◊◊◊ ◊◊◊◊◊ ◊◊◊◊◊ ◊◊◊◊◊	**921,000**
◊◊◊◊◊ ◊◊◊◊◊ ◊◊◊◊◊ ◊◊◊◊◊ ◊◊◊◊◊ ◊◊◊◊◊ ◊◊◊◊◊ ◊◊◊◊◊ ◊◊◊◊◊ ◊◊◊◊◊	921,050
◊◊◊◊◊ ◊◊◊◊◊ ◊◊◊◊◊ ◊◊◊◊◊ ◊◊◊◊◊ ◊◊◊◊◊ ◊◊◊◊◊ ◊◊◊◊◊ ◊◊◊◊◊ ◊◊◊◊◊	921,100
◊◊◊◊◊ ◊◊◊◊◊ ◊◊◊◊◊ ◊◊◊◊◊ ◊◊◊◊◊ ◊◊◊◊◊ ◊◊◊◊◊ ◊◊◊◊◊ ◊◊◊◊◊ ◊◊◊◊◊	921,150
◊◊◊◊◊ ◊◊◊◊◊ ◊◊◊◊◊ ◊◊◊◊◊ ◊◊◊◊◊ ◊◊◊◊◊ ◊◊◊◊◊ ◊◊◊◊◊ ◊◊◊◊◊ ◊◊◊◊◊	921,200
◊◊◊◊◊ ◊◊◊◊◊ ◊◊◊◊◊ ◊◊◊◊◊ ◊◊◊◊◊ ◊◊◊◊◊ ◊◊◊◊◊ ◊◊◊◊◊ ◊◊◊◊◊ ◊◊◊◊◊	921,250
◊◊◊◊◊ ◊◊◊◊◊ ◊◊◊◊◊ ◊◊◊◊◊ ◊◊◊◊◊ ◊◊◊◊◊ ◊◊◊◊◊ ◊◊◊◊◊ ◊◊◊◊◊ ◊◊◊◊◊	921,300
◊◊◊◊◊ ◊◊◊◊◊ ◊◊◊◊◊ ◊◊◊◊◊ ◊◊◊◊◊ ◊◊◊◊◊ ◊◊◊◊◊ ◊◊◊◊◊ ◊◊◊◊◊ ◊◊◊◊◊	921,350
◊◊◊◊◊ ◊◊◊◊◊ ◊◊◊◊◊ ◊◊◊◊◊ ◊◊◊◊◊ ◊◊◊◊◊ ◊◊◊◊◊ ◊◊◊◊◊ ◊◊◊◊◊ ◊◊◊◊◊	921,400
◊◊◊◊◊ ◊◊◊◊◊ ◊◊◊◊◊ ◊◊◊◊◊ ◊◊◊◊◊ ◊◊◊◊◊ ◊◊◊◊◊ ◊◊◊◊◊ ◊◊◊◊◊ ◊◊◊◊◊	921,450
◊◊◊◊◊ ◊◊◊◊◊ ◊◊◊◊◊ ◊◊◊◊◊ ◊◊◊◊◊ ◊◊◊◊◊ ◊◊◊◊◊ ◊◊◊◊◊ ◊◊◊◊◊ ◊◊◊◊◊	921,500
◊◊◊◊◊ ◊◊◊◊◊ ◊◊◊◊◊ ◊◊◊◊◊ ◊◊◊◊◊ ◊◊◊◊◊ ◊◊◊◊◊ ◊◊◊◊◊ ◊◊◊◊◊ ◊◊◊◊◊	921,550
◊◊◊◊◊ ◊◊◊◊◊ ◊◊◊◊◊ ◊◊◊◊◊ ◊◊◊◊◊ ◊◊◊◊◊ ◊◊◊◊◊ ◊◊◊◊◊ ◊◊◊◊◊ ◊◊◊◊◊	921,600
◊◊◊◊◊ ◊◊◊◊◊ ◊◊◊◊◊ ◊◊◊◊◊ ◊◊◊◊◊ ◊◊◊◊◊ ◊◊◊◊◊ ◊◊◊◊◊ ◊◊◊◊◊ ◊◊◊◊◊	921,650
◊◊◊◊◊ ◊◊◊◊◊ ◊◊◊◊◊ ◊◊◊◊◊ ◊◊◊◊◊ ◊◊◊◊◊ ◊◊◊◊◊ ◊◊◊◊◊ ◊◊◊◊◊ ◊◊◊◊◊	921,700
◊◊◊◊◊ ◊◊◊◊◊ ◊◊◊◊◊ ◊◊◊◊◊ ◊◊◊◊◊ ◊◊◊◊◊ ◊◊◊◊◊ ◊◊◊◊◊ ◊◊◊◊◊ ◊◊◊◊◊	921,750
◊◊◊◊◊ ◊◊◊◊◊ ◊◊◊◊◊ ◊◊◊◊◊ ◊◊◊◊◊ ◊◊◊◊◊ ◊◊◊◊◊ ◊◊◊◊◊ ◊◊◊◊◊ ◊◊◊◊◊	921,800
◊◊◊◊◊ ◊◊◊◊◊ ◊◊◊◊◊ ◊◊◊◊◊ ◊◊◊◊◊ ◊◊◊◊◊ ◊◊◊◊◊ ◊◊◊◊◊ ◊◊◊◊◊ ◊◊◊◊◊	921,850
◊◊◊◊◊ ◊◊◊◊◊ ◊◊◊◊◊ ◊◊◊◊◊ ◊◊◊◊◊ ◊◊◊◊◊ ◊◊◊◊◊ ◊◊◊◊◊ ◊◊◊◊◊ ◊◊◊◊◊	921,900
◊◊◊◊◊ ◊◊◊◊◊ ◊◊◊◊◊ ◊◊◊◊◊ ◊◊◊◊◊ ◊◊◊◊◊ ◊◊◊◊◊ ◊◊◊◊◊ ◊◊◊◊◊ ◊◊◊◊◊	921,950
◊◊◊◊◊ ◊◊◊◊◊ ◊◊◊◊◊ ◊◊◊◊◊ ◊◊◊◊◊ ◊◊◊◊◊ ◊◊◊◊◊ ◊◊◊◊◊ ◊◊◊◊◊ ◊◊◊◊◊	**922,000**

The Number 922,001 to the Number 924,000

Start   Notes

	922,050
	922,100
	922,150
	922,200
	922,250
	922,300
	922,350
	922,400
	922,450
	922,500
	922,550
	922,600
	922,650
	922,700
	922,750
	922,800
	922,850
	922,900
	922,950
	923,000
	923,050
	923,100
	923,150
	923,200
	923,250
	923,300
	923,350
	923,400
	923,450
	923,500
	923,550
	923,600
	923,650
	923,700
	923,750
	923,800
	923,850
	923,900
	923,950
	924,000

The Number 924,001 to the Number 926,000

Start | Notes

	924,050
	924,100
	924,150
	924,200
	924,250
	924,300
	924,350
	924,400
	924,450
	924,500
	924,550
	924,600
	924,650
	924,700
	924,750
	924,800
	924,850
	924,900
	924,950
	925,000
	925,050
	925,100
	925,150
	925,200
	925,250
	925,300
	925,350
	925,400
	925,450
	925,500
	925,550
	925,600
	925,650
	925,700
	925,750
	925,800
	925,850
	925,900
	925,950
	926,000

The Number 926,001 to the Number 928,000

Start		Notes
	926,050
	926,100
	926,150
	926,200
	926,250
	926,300
	926,350
	926,400
	926,450
	926,500
	926,550
	926,600
	926,650
	926,700
	926,750
	926,800
	926,850
	926,900
	926,950
	927,000
	927,050
	927,100
	927,150
	927,200
	927,250
	927,300
	927,350
	927,400
	927,450
	927,500
	927,550
	927,600
	927,650
	927,700
	927,750
	927,800
	927,850
	927,900
	927,950
	928,000

◇ The Number 928,001 to the Number 930,000

Start   Notes

	928,050
	928,100
	928,150
	928,200
	928,250
	928,300
	928,350
	928,400
	928,450
	928,500
	928,550
	928,600
	928,650
	928,700
	928,750
	928,800
	928,850
	928,900
	928,950
	929,000
	929,050
	929,100
	929,150
	929,200
	929,250
	929,300
	929,350
	929,400
	929,450
	929,500
	929,550
	929,600
	929,650
	929,700
	929,750
	929,800
	929,850
	929,900
	929,950
	930,000

The Number 930,001 to the Number 932,000

Start		Notes

Number	Notes
930,050	
930,100	
930,150	
930,200	
930,250	
930,300	
930,350	
930,400	
930,450	
930,500	
930,550	
930,600	
930,650	
930,700	
930,750	
930,800	
930,850	
930,900	
930,950	
931,000	
931,050	
931,100	
931,150	
931,200	
931,250	
931,300	
931,350	
931,400	
931,450	
931,500	
931,550	
931,600	
931,650	
931,700	
931,750	
931,800	
931,850	
931,900	
931,950	
932,000	

The Number 932,001 to the Number 934,000

Start   Notes

	932,050
	932,100
	932,150
	932,200
	932,250
	932,300
	932,350
	932,400
	932,450
	932,500
	932,550
	932,600
	932,650
	932,700
	932,750
	932,800
	932,850
	932,900
	932,950
	933,000
	933,050
	933,100
	933,150
	933,200
	933,250
	933,300
	933,350
	933,400
	933,450
	933,500
	933,550
	933,600
	933,650
	933,700
	933,750
	933,800
	933,850
	933,900
	933,950
	934,000

The Number 934,001 to the Number 936,000

Start	Notes
	934,050
	934,100
	934,150
	934,200
	934,250
	934,300
	934,350
	934,400
	934,450
	934,500
	934,550
	934,600
	934,650
	934,700
	934,750
	934,800
	934,850
	934,900
	934,950
	935,000
	935,050
	935,100
	935,150
	935,200
	935,250
	935,300
	935,350
	935,400
	935,450
	935,500
	935,550
	935,600
	935,650
	935,700
	935,750
	935,800
	935,850
	935,900
	935,950
	936,000

The Number 936,001 to the Number 938,000

Start   Notes

936,050
936,100
936,150
936,200
936,250
936,300
936,350
936,400
936,450
936,500
936,550
936,600
936,650
936,700
936,750
936,800
936,850
936,900
936,950
937,000
937,050
937,100
937,150
937,200
937,250
937,300
937,350
937,400
937,450
937,500
937,550
937,600
937,650
937,700
937,750
937,800
937,850
937,900
937,950
938,000

The Number 938,001 to the Number 940,000

Start

Notes

	938,050
	938,100
	938,150
	938,200
	938,250
	938,300
	938,350
	938,400
	938,450
	938,500
	938,550
	938,600
	938,650
	938,700
	938,750
	938,800
	938,850
	938,900
	938,950
	939,000
	939,050
	939,100
	939,150
	939,200
	939,250
	939,300
	939,350
	939,400
	939,450
	939,500
	939,550
	939,600
	939,650
	939,700
	939,750
	939,800
	939,850
	939,900
	939,950
	940,000

The Number 940,001 to the Number 942,000

Start   Notes

	940,050
	940,100
	940,150
	940,200
	940,250
	940,300
	940,350
	940,400
	940,450
	940,500
	940,550
	940,600
	940,650
	940,700
	940,750
	940,800
	940,850
	940,900
	940,950
	941,000
	941,050
	941,100
	941,150
	941,200
	941,250
	941,300
	941,350
	941,400
	941,450
	941,500
	941,550
	941,600
	941,650
	941,700
	941,750
	941,800
	941,850
	941,900
	941,950
	942,000

The Number 942,001 to the Number 944,000

Start		Notes
	942,050
	942,100
	942,150
	942,200
	942,250
	942,300
	942,350
	942,400
	942,450
	942,500
	942,550
	942,600
	942,650
	942,700
	942,750
	942,800
	942,850
	942,900
	942,950
	943,000
	943,050
	943,100
	943,150
	943,200
	943,250
	943,300
	943,350
	943,400
	943,450
	943,500
	943,550
	943,600
	943,650
	943,700
	943,750
	943,800
	943,850
	943,900
	943,950
	944,000

The Number 944,001 to the Number 946,000

Start   **Notes**

	944,050
	944,100
	944,150
	944,200
	944,250
	944,300
	944,350
	944,400
	944,450
	944,500
	944,550
	944,600
	944,650
	944,700
	944,750
	944,800
	944,850
	944,900
	944,950
	945,000
	945,050
	945,100
	945,150
	945,200
	945,250
	945,300
	945,350
	945,400
	945,450
	945,500
	945,550
	945,600
	945,650
	945,700
	945,750
	945,800
	945,850
	945,900
	945,950
	946,000

The Number 946,001 to the Number 948,000

Start
Notes

	946,050
	946,100
	946,150
	946,200
	946,250
	946,300
	946,350
	946,400
	946,450
	946,500
	946,550
	946,600
	946,650
	946,700
	946,750
	946,800
	946,850
	946,900
	946,950
	947,000
	947,050
	947,100
	947,150
	947,200
	947,250
	947,300
	947,350
	947,400
	947,450
	947,500
	947,550
	947,600
	947,650
	947,700
	947,750
	947,800
	947,850
	947,900
	947,950
	948,000

The Number 948,001 to the Number 950,000

Start   **Notes**

◊◊◊◊◊ ◊◊◊◊◊ ◊◊◊◊◊ ◊◊◊◊◊ ◊◊◊◊◊ ◊◊◊◊◊ ◊◊◊◊◊ ◊◊◊◊◊ ◊◊◊◊◊ ◊◊◊◊◊	948,050
◊◊◊◊◊ ◊◊◊◊◊ ◊◊◊◊◊ ◊◊◊◊◊ ◊◊◊◊◊ ◊◊◊◊◊ ◊◊◊◊◊ ◊◊◊◊◊ ◊◊◊◊◊ ◊◊◊◊◊	948,100
◊◊◊◊◊ ◊◊◊◊◊ ◊◊◊◊◊ ◊◊◊◊◊ ◊◊◊◊◊ ◊◊◊◊◊ ◊◊◊◊◊ ◊◊◊◊◊ ◊◊◊◊◊ ◊◊◊◊◊	948,150
◊◊◊◊◊ ◊◊◊◊◊ ◊◊◊◊◊ ◊◊◊◊◊ ◊◊◊◊◊ ◊◊◊◊◊ ◊◊◊◊◊ ◊◊◊◊◊ ◊◊◊◊◊ ◊◊◊◊◊	948,200
◊◊◊◊◊ ◊◊◊◊◊ ◊◊◊◊◊ ◊◊◊◊◊ ◊◊◊◊◊ ◊◊◊◊◊ ◊◊◊◊◊ ◊◊◊◊◊ ◊◊◊◊◊ ◊◊◊◊◊	948,250
◊◊◊◊◊ ◊◊◊◊◊ ◊◊◊◊◊ ◊◊◊◊◊ ◊◊◊◊◊ ◊◊◊◊◊ ◊◊◊◊◊ ◊◊◊◊◊ ◊◊◊◊◊ ◊◊◊◊◊	948,300
◊◊◊◊◊ ◊◊◊◊◊ ◊◊◊◊◊ ◊◊◊◊◊ ◊◊◊◊◊ ◊◊◊◊◊ ◊◊◊◊◊ ◊◊◊◊◊ ◊◊◊◊◊ ◊◊◊◊◊	948,350
◊◊◊◊◊ ◊◊◊◊◊ ◊◊◊◊◊ ◊◊◊◊◊ ◊◊◊◊◊ ◊◊◊◊◊ ◊◊◊◊◊ ◊◊◊◊◊ ◊◊◊◊◊ ◊◊◊◊◊	948,400
◊◊◊◊◊ ◊◊◊◊◊ ◊◊◊◊◊ ◊◊◊◊◊ ◊◊◊◊◊ ◊◊◊◊◊ ◊◊◊◊◊ ◊◊◊◊◊ ◊◊◊◊◊ ◊◊◊◊◊	948,450
◊◊◊◊◊ ◊◊◊◊◊ ◊◊◊◊◊ ◊◊◊◊◊ ◊◊◊◊◊ ◊◊◊◊◊ ◊◊◊◊◊ ◊◊◊◊◊ ◊◊◊◊◊ ◊◊◊◊◊	948,500
◊◊◊◊◊ ◊◊◊◊◊ ◊◊◊◊◊ ◊◊◊◊◊ ◊◊◊◊◊ ◊◊◊◊◊ ◊◊◊◊◊ ◊◊◊◊◊ ◊◊◊◊◊ ◊◊◊◊◊	948,550
◊◊◊◊◊ ◊◊◊◊◊ ◊◊◊◊◊ ◊◊◊◊◊ ◊◊◊◊◊ ◊◊◊◊◊ ◊◊◊◊◊ ◊◊◊◊◊ ◊◊◊◊◊ ◊◊◊◊◊	948,600
◊◊◊◊◊ ◊◊◊◊◊ ◊◊◊◊◊ ◊◊◊◊◊ ◊◊◊◊◊ ◊◊◊◊◊ ◊◊◊◊◊ ◊◊◊◊◊ ◊◊◊◊◊ ◊◊◊◊◊	948,650
◊◊◊◊◊ ◊◊◊◊◊ ◊◊◊◊◊ ◊◊◊◊◊ ◊◊◊◊◊ ◊◊◊◊◊ ◊◊◊◊◊ ◊◊◊◊◊ ◊◊◊◊◊ ◊◊◊◊◊	948,700
◊◊◊◊◊ ◊◊◊◊◊ ◊◊◊◊◊ ◊◊◊◊◊ ◊◊◊◊◊ ◊◊◊◊◊ ◊◊◊◊◊ ◊◊◊◊◊ ◊◊◊◊◊ ◊◊◊◊◊	948,750
◊◊◊◊◊ ◊◊◊◊◊ ◊◊◊◊◊ ◊◊◊◊◊ ◊◊◊◊◊ ◊◊◊◊◊ ◊◊◊◊◊ ◊◊◊◊◊ ◊◊◊◊◊ ◊◊◊◊◊	948,800
◊◊◊◊◊ ◊◊◊◊◊ ◊◊◊◊◊ ◊◊◊◊◊ ◊◊◊◊◊ ◊◊◊◊◊ ◊◊◊◊◊ ◊◊◊◊◊ ◊◊◊◊◊ ◊◊◊◊◊	948,850
◊◊◊◊◊ ◊◊◊◊◊ ◊◊◊◊◊ ◊◊◊◊◊ ◊◊◊◊◊ ◊◊◊◊◊ ◊◊◊◊◊ ◊◊◊◊◊ ◊◊◊◊◊ ◊◊◊◊◊	948,900
◊◊◊◊◊ ◊◊◊◊◊ ◊◊◊◊◊ ◊◊◊◊◊ ◊◊◊◊◊ ◊◊◊◊◊ ◊◊◊◊◊ ◊◊◊◊◊ ◊◊◊◊◊ ◊◊◊◊◊	948,950
◊◊◊◊◊ ◊◊◊◊◊ ◊◊◊◊◊ ◊◊◊◊◊ ◊◊◊◊◊ ◊◊◊◊◊ ◊◊◊◊◊ ◊◊◊◊◊ ◊◊◊◊◊ ◊◊◊◊◊	**949,000**
◊◊◊◊◊ ◊◊◊◊◊ ◊◊◊◊◊ ◊◊◊◊◊ ◊◊◊◊◊ ◊◊◊◊◊ ◊◊◊◊◊ ◊◊◊◊◊ ◊◊◊◊◊ ◊◊◊◊◊	949,050
◊◊◊◊◊ ◊◊◊◊◊ ◊◊◊◊◊ ◊◊◊◊◊ ◊◊◊◊◊ ◊◊◊◊◊ ◊◊◊◊◊ ◊◊◊◊◊ ◊◊◊◊◊ ◊◊◊◊◊	949,100
◊◊◊◊◊ ◊◊◊◊◊ ◊◊◊◊◊ ◊◊◊◊◊ ◊◊◊◊◊ ◊◊◊◊◊ ◊◊◊◊◊ ◊◊◊◊◊ ◊◊◊◊◊ ◊◊◊◊◊	949,150
◊◊◊◊◊ ◊◊◊◊◊ ◊◊◊◊◊ ◊◊◊◊◊ ◊◊◊◊◊ ◊◊◊◊◊ ◊◊◊◊◊ ◊◊◊◊◊ ◊◊◊◊◊ ◊◊◊◊◊	949,200
◊◊◊◊◊ ◊◊◊◊◊ ◊◊◊◊◊ ◊◊◊◊◊ ◊◊◊◊◊ ◊◊◊◊◊ ◊◊◊◊◊ ◊◊◊◊◊ ◊◊◊◊◊ ◊◊◊◊◊	949,250
◊◊◊◊◊ ◊◊◊◊◊ ◊◊◊◊◊ ◊◊◊◊◊ ◊◊◊◊◊ ◊◊◊◊◊ ◊◊◊◊◊ ◊◊◊◊◊ ◊◊◊◊◊ ◊◊◊◊◊	949,300
◊◊◊◊◊ ◊◊◊◊◊ ◊◊◊◊◊ ◊◊◊◊◊ ◊◊◊◊◊ ◊◊◊◊◊ ◊◊◊◊◊ ◊◊◊◊◊ ◊◊◊◊◊ ◊◊◊◊◊	949,350
◊◊◊◊◊ ◊◊◊◊◊ ◊◊◊◊◊ ◊◊◊◊◊ ◊◊◊◊◊ ◊◊◊◊◊ ◊◊◊◊◊ ◊◊◊◊◊ ◊◊◊◊◊ ◊◊◊◊◊	949,400
◊◊◊◊◊ ◊◊◊◊◊ ◊◊◊◊◊ ◊◊◊◊◊ ◊◊◊◊◊ ◊◊◊◊◊ ◊◊◊◊◊ ◊◊◊◊◊ ◊◊◊◊◊ ◊◊◊◊◊	949,450
◊◊◊◊◊ ◊◊◊◊◊ ◊◊◊◊◊ ◊◊◊◊◊ ◊◊◊◊◊ ◊◊◊◊◊ ◊◊◊◊◊ ◊◊◊◊◊ ◊◊◊◊◊ ◊◊◊◊◊	949,500
◊◊◊◊◊ ◊◊◊◊◊ ◊◊◊◊◊ ◊◊◊◊◊ ◊◊◊◊◊ ◊◊◊◊◊ ◊◊◊◊◊ ◊◊◊◊◊ ◊◊◊◊◊ ◊◊◊◊◊	949,550
◊◊◊◊◊ ◊◊◊◊◊ ◊◊◊◊◊ ◊◊◊◊◊ ◊◊◊◊◊ ◊◊◊◊◊ ◊◊◊◊◊ ◊◊◊◊◊ ◊◊◊◊◊ ◊◊◊◊◊	949,600
◊◊◊◊◊ ◊◊◊◊◊ ◊◊◊◊◊ ◊◊◊◊◊ ◊◊◊◊◊ ◊◊◊◊◊ ◊◊◊◊◊ ◊◊◊◊◊ ◊◊◊◊◊ ◊◊◊◊◊	949,650
◊◊◊◊◊ ◊◊◊◊◊ ◊◊◊◊◊ ◊◊◊◊◊ ◊◊◊◊◊ ◊◊◊◊◊ ◊◊◊◊◊ ◊◊◊◊◊ ◊◊◊◊◊ ◊◊◊◊◊	949,700
◊◊◊◊◊ ◊◊◊◊◊ ◊◊◊◊◊ ◊◊◊◊◊ ◊◊◊◊◊ ◊◊◊◊◊ ◊◊◊◊◊ ◊◊◊◊◊ ◊◊◊◊◊ ◊◊◊◊◊	949,750
◊◊◊◊◊ ◊◊◊◊◊ ◊◊◊◊◊ ◊◊◊◊◊ ◊◊◊◊◊ ◊◊◊◊◊ ◊◊◊◊◊ ◊◊◊◊◊ ◊◊◊◊◊ ◊◊◊◊◊	949,800
◊◊◊◊◊ ◊◊◊◊◊ ◊◊◊◊◊ ◊◊◊◊◊ ◊◊◊◊◊ ◊◊◊◊◊ ◊◊◊◊◊ ◊◊◊◊◊ ◊◊◊◊◊ ◊◊◊◊◊	949,850
◊◊◊◊◊ ◊◊◊◊◊ ◊◊◊◊◊ ◊◊◊◊◊ ◊◊◊◊◊ ◊◊◊◊◊ ◊◊◊◊◊ ◊◊◊◊◊ ◊◊◊◊◊ ◊◊◊◊◊	949,900
◊◊◊◊◊ ◊◊◊◊◊ ◊◊◊◊◊ ◊◊◊◊◊ ◊◊◊◊◊ ◊◊◊◊◊ ◊◊◊◊◊ ◊◊◊◊◊ ◊◊◊◊◊ ◊◊◊◊◊	949,950
◊◊◊◊◊ ◊◊◊◊◊ ◊◊◊◊◊ ◊◊◊◊◊ ◊◊◊◊◊ ◊◊◊◊◊ ◊◊◊◊◊ ◊◊◊◊◊ ◊◊◊◊◊ ◊◊◊◊◊	**950,000**

The Number 950,001 to the Number 952,000

Start		Notes
	950,050
	950,100
	950,150
	950,200
	950,250
	950,300
	950,350
	950,400
	950,450
	950,500
	950,550
	950,600
	950,650
	950,700
	950,750
	950,800
	950,850
	950,900
	950,950
	951,000
	951,050
	951,100
	951,150
	951,200
	951,250
	951,300
	951,350
	951,400
	951,450
	951,500
	951,550
	951,600
	951,650
	951,700
	951,750
	951,800
	951,850
	951,900
	951,950
	952,000

◇ The Number 952,001 to the Number 954,000

Start	Notes
	952,050
	952,100
	952,150
	952,200
	952,250
	952,300
	952,350
	952,400
	952,450
	952,500
	952,550
	952,600
	952,650
	952,700
	952,750
	952,800
	952,850
	952,900
	952,950
	953,000
	953,050
	953,100
	953,150
	953,200
	953,250
	953,300
	953,350
	953,400
	953,450
	953,500
	953,550
	953,600
	953,650
	953,700
	953,750
	953,800
	953,850
	953,900
	953,950
	954,000

The Number 954,001 to the Number 956,000

Start | | Notes

	Number	Notes
	954,050	
	954,100	
	954,150	
	954,200	
	954,250	
	954,300	
	954,350	
	954,400	
	954,450	
	954,500	
	954,550	
	954,600	
	954,650	
	954,700	
	954,750	
	954,800	
	954,850	
	954,900	
	954,950	
	955,000	
	955,050	
	955,100	
	955,150	
	955,200	
	955,250	
	955,300	
	955,350	
	955,400	
	955,450	
	955,500	
	955,550	
	955,600	
	955,650	
	955,700	
	955,750	
	955,800	
	955,850	
	955,900	
	955,950	
	956,000	

◊ The Number 956,001 to the Number 958,000

Start   **Notes**

	Number
	956,050
	956,100
	956,150
	956,200
	956,250
	956,300
	956,350
	956,400
	956,450
	956,500
	956,550
	956,600
	956,650
	956,700
	956,750
	956,800
	956,850
	956,900
	956,950
	957,000
	957,050
	957,100
	957,150
	957,200
	957,250
	957,300
	957,350
	957,400
	957,450
	957,500
	957,550
	957,600
	957,650
	957,700
	957,750
	957,800
	957,850
	957,900
	957,950
	958,000

The Number 958,001 to the Number 960,000

Start	Notes
	958,050
	958,100
	958,150
	958,200
	958,250
	958,300
	958,350
	958,400
	958,450
	958,500
	958,550
	958,600
	958,650
	958,700
	958,750
	958,800
	958,850
	958,900
	958,950
	959,000
	959,050
	959,100
	959,150
	959,200
	959,250
	959,300
	959,350
	959,400
	959,450
	959,500
	959,550
	959,600
	959,650
	959,700
	959,750
	959,800
	959,850
	959,900
	959,950
	960,000

The Number 960,001 to the Number 962,000

Start		Notes

	Notes
960,050	
960,100	
960,150	
960,200	
960,250	
960,300	
960,350	
960,400	
960,450	
960,500	
960,550	
960,600	
960,650	
960,700	
960,750	
960,800	
960,850	
960,900	
960,950	
961,000	
961,050	
961,100	
961,150	
961,200	
961,250	
961,300	
961,350	
961,400	
961,450	
961,500	
961,550	
961,600	
961,650	
961,700	
961,750	
961,800	
961,850	
961,900	
961,950	
962,000	

The Number 962,001 to the Number 964,000

Start

Notes

962,050
962,100
962,150
962,200
962,250
962,300
962,350
962,400
962,450
962,500
962,550
962,600
962,650
962,700
962,750
962,800
962,850
962,900
962,950
963,000
963,050
963,100
963,150
963,200
963,250
963,300
963,350
963,400
963,450
963,500
963,550
963,600
963,650
963,700
963,750
963,800
963,850
963,900
963,950
964,000

◇ The Number 964,001 to the Number 966,000

Start   Notes

	964,050
	964,100
	964,150
	964,200
	964,250
	964,300
	964,350
	964,400
	964,450
	964,500
	964,550
	964,600
	964,650
	964,700
	964,750
	964,800
	964,850
	964,900
	964,950
	965,000
	965,050
	965,100
	965,150
	965,200
	965,250
	965,300
	965,350
	965,400
	965,450
	965,500
	965,550
	965,600
	965,650
	965,700
	965,750
	965,800
	965,850
	965,900
	965,950
	966,000

The Number 966,001 to the Number 968,000

Start

Notes

966,050	
966,100	
966,150	
966,200	
966,250	
966,300	
966,350	
966,400	
966,450	
966,500	
966,550	
966,600	
966,650	
966,700	
966,750	
966,800	
966,850	
966,900	
966,950	
967,000	
967,050	
967,100	
967,150	
967,200	
967,250	
967,300	
967,350	
967,400	
967,450	
967,500	
967,550	
967,600	
967,650	
967,700	
967,750	
967,800	
967,850	
967,900	
967,950	
968,000	

◇ The Number 968,001 to the Number 970,000

Start   Notes

Number	Notes
968,050	
968,100	
968,150	
968,200	
968,250	
968,300	
968,350	
968,400	
968,450	
968,500	
968,550	
968,600	
968,650	
968,700	
968,750	
968,800	
968,850	
968,900	
968,950	
969,000	
969,050	
969,100	
969,150	
969,200	
969,250	
969,300	
969,350	
969,400	
969,450	
969,500	
969,550	
969,600	
969,650	
969,700	
969,750	
969,800	
969,850	
969,900	
969,950	
970,000	

The Number 970,001 to the Number 972,000

◇ Start

Notes

	970,050
	970,100
	970,150
	970,200
	970,250
	970,300
	970,350
	970,400
	970,450
	970,500
	970,550
	970,600
	970,650
	970,700
	970,750
	970,800
	970,850
	970,900
	970,950
	971,000
	971,050
	971,100
	971,150
	971,200
	971,250
	971,300
	971,350
	971,400
	971,450
	971,500
	971,550
	971,600
	971,650
	971,700
	971,750
	971,800
	971,850
	971,900
	971,950
	972,000

The Number 972,001 to the Number 974,000

Start   Notes

	972,050
	972,100
	972,150
	972,200
	972,250
	972,300
	972,350
	972,400
	972,450
	972,500
	972,550
	972,600
	972,650
	972,700
	972,750
	972,800
	972,850
	972,900
	972,950
	973,000
	973,050
	973,100
	973,150
	973,200
	973,250
	973,300
	973,350
	973,400
	973,450
	973,500
	973,550
	973,600
	973,650
	973,700
	973,750
	973,800
	973,850
	973,900
	973,950
	974,000

The Number 974,001 to the Number 976,000

Start

Notes

	974,050
	974,100
	974,150
	974,200
	974,250
	974,300
	974,350
	974,400
	974,450
	974,500
	974,550
	974,600
	974,650
	974,700
	974,750
	974,800
	974,850
	974,900
	974,950
	975,000
	975,050
	975,100
	975,150
	975,200
	975,250
	975,300
	975,350
	975,400
	975,450
	975,500
	975,550
	975,600
	975,650
	975,700
	975,750
	975,800
	975,850
	975,900
	975,950
	976,000

◊ The Number 976,001 to the Number 978,000

Start   **Notes**

	976,050
	976,100
	976,150
	976,200
	976,250
	976,300
	976,350
	976,400
	976,450
	976,500
	976,550
	976,600
	976,650
	976,700
	976,750
	976,800
	976,850
	976,900
	976,950
	977,000
	977,050
	977,100
	977,150
	977,200
	977,250
	977,300
	977,350
	977,400
	977,450
	977,500
	977,550
	977,600
	977,650
	977,700
	977,750
	977,800
	977,850
	977,900
	977,950
	978,000

The Number 978,001 to the Number 980,000

Start

Notes

	978,050
	978,100
	978,150
	978,200
	978,250
	978,300
	978,350
	978,400
	978,450
	978,500
	978,550
	978,600
	978,650
	978,700
	978,750
	978,800
	978,850
	978,900
	978,950
	979,000
	979,050
	979,100
	979,150
	979,200
	979,250
	979,300
	979,350
	979,400
	979,450
	979,500
	979,550
	979,600
	979,650
	979,700
	979,750
	979,800
	979,850
	979,900
	979,950
	980,000

The Number 980,001 to the Number 982,000

Start   Notes

◇◇◇◇◇ ◇◇◇◇◇ ◇◇◇◇◇ ◇◇◇◇◇ ◇◇◇◇◇ ◇◇◇◇◇ ◇◇◇◇◇ ◇◇◇◇◇ ◇◇◇◇◇ ◇◇◇◇◇	980,050
◇◇◇◇◇ ◇◇◇◇◇ ◇◇◇◇◇ ◇◇◇◇◇ ◇◇◇◇◇ ◇◇◇◇◇ ◇◇◇◇◇ ◇◇◇◇◇ ◇◇◇◇◇ ◇◇◇◇◇	980,100
◇◇◇◇◇ ◇◇◇◇◇ ◇◇◇◇◇ ◇◇◇◇◇ ◇◇◇◇◇ ◇◇◇◇◇ ◇◇◇◇◇ ◇◇◇◇◇ ◇◇◇◇◇ ◇◇◇◇◇	980,150
◇◇◇◇◇ ◇◇◇◇◇ ◇◇◇◇◇ ◇◇◇◇◇ ◇◇◇◇◇ ◇◇◇◇◇ ◇◇◇◇◇ ◇◇◇◇◇ ◇◇◇◇◇ ◇◇◇◇◇	980,200
◇◇◇◇◇ ◇◇◇◇◇ ◇◇◇◇◇ ◇◇◇◇◇ ◇◇◇◇◇ ◇◇◇◇◇ ◇◇◇◇◇ ◇◇◇◇◇ ◇◇◇◇◇ ◇◇◇◇◇	980,250
◇◇◇◇◇ ◇◇◇◇◇ ◇◇◇◇◇ ◇◇◇◇◇ ◇◇◇◇◇ ◇◇◇◇◇ ◇◇◇◇◇ ◇◇◇◇◇ ◇◇◇◇◇ ◇◇◇◇◇	980,300
◇◇◇◇◇ ◇◇◇◇◇ ◇◇◇◇◇ ◇◇◇◇◇ ◇◇◇◇◇ ◇◇◇◇◇ ◇◇◇◇◇ ◇◇◇◇◇ ◇◇◇◇◇ ◇◇◇◇◇	980,350
◇◇◇◇◇ ◇◇◇◇◇ ◇◇◇◇◇ ◇◇◇◇◇ ◇◇◇◇◇ ◇◇◇◇◇ ◇◇◇◇◇ ◇◇◇◇◇ ◇◇◇◇◇ ◇◇◇◇◇	980,400
◇◇◇◇◇ ◇◇◇◇◇ ◇◇◇◇◇ ◇◇◇◇◇ ◇◇◇◇◇ ◇◇◇◇◇ ◇◇◇◇◇ ◇◇◇◇◇ ◇◇◇◇◇ ◇◇◇◇◇	980,450
◇◇◇◇◇ ◇◇◇◇◇ ◇◇◇◇◇ ◇◇◇◇◇ ◇◇◇◇◇ ◇◇◇◇◇ ◇◇◇◇◇ ◇◇◇◇◇ ◇◇◇◇◇ ◇◇◇◇◇	980,500
◇◇◇◇◇ ◇◇◇◇◇ ◇◇◇◇◇ ◇◇◇◇◇ ◇◇◇◇◇ ◇◇◇◇◇ ◇◇◇◇◇ ◇◇◇◇◇ ◇◇◇◇◇ ◇◇◇◇◇	980,550
◇◇◇◇◇ ◇◇◇◇◇ ◇◇◇◇◇ ◇◇◇◇◇ ◇◇◇◇◇ ◇◇◇◇◇ ◇◇◇◇◇ ◇◇◇◇◇ ◇◇◇◇◇ ◇◇◇◇◇	980,600
◇◇◇◇◇ ◇◇◇◇◇ ◇◇◇◇◇ ◇◇◇◇◇ ◇◇◇◇◇ ◇◇◇◇◇ ◇◇◇◇◇ ◇◇◇◇◇ ◇◇◇◇◇ ◇◇◇◇◇	980,650
◇◇◇◇◇ ◇◇◇◇◇ ◇◇◇◇◇ ◇◇◇◇◇ ◇◇◇◇◇ ◇◇◇◇◇ ◇◇◇◇◇ ◇◇◇◇◇ ◇◇◇◇◇ ◇◇◇◇◇	980,700
◇◇◇◇◇ ◇◇◇◇◇ ◇◇◇◇◇ ◇◇◇◇◇ ◇◇◇◇◇ ◇◇◇◇◇ ◇◇◇◇◇ ◇◇◇◇◇ ◇◇◇◇◇ ◇◇◇◇◇	980,750
◇◇◇◇◇ ◇◇◇◇◇ ◇◇◇◇◇ ◇◇◇◇◇ ◇◇◇◇◇ ◇◇◇◇◇ ◇◇◇◇◇ ◇◇◇◇◇ ◇◇◇◇◇ ◇◇◇◇◇	980,800
◇◇◇◇◇ ◇◇◇◇◇ ◇◇◇◇◇ ◇◇◇◇◇ ◇◇◇◇◇ ◇◇◇◇◇ ◇◇◇◇◇ ◇◇◇◇◇ ◇◇◇◇◇ ◇◇◇◇◇	980,850
◇◇◇◇◇ ◇◇◇◇◇ ◇◇◇◇◇ ◇◇◇◇◇ ◇◇◇◇◇ ◇◇◇◇◇ ◇◇◇◇◇ ◇◇◇◇◇ ◇◇◇◇◇ ◇◇◇◇◇	980,900
◇◇◇◇◇ ◇◇◇◇◇ ◇◇◇◇◇ ◇◇◇◇◇ ◇◇◇◇◇ ◇◇◇◇◇ ◇◇◇◇◇ ◇◇◇◇◇ ◇◇◇◇◇ ◇◇◇◇◇	980,950
◇◇◇◇◇ ◇◇◇◇◇ ◇◇◇◇◇ ◇◇◇◇◇ ◇◇◇◇◇ ◇◇◇◇◇ ◇◇◇◇◇ ◇◇◇◇◇ ◇◇◇◇◇ ◇◇◇◇◇	**981,000**
◇◇◇◇◇ ◇◇◇◇◇ ◇◇◇◇◇ ◇◇◇◇◇ ◇◇◇◇◇ ◇◇◇◇◇ ◇◇◇◇◇ ◇◇◇◇◇ ◇◇◇◇◇ ◇◇◇◇◇	981,050
◇◇◇◇◇ ◇◇◇◇◇ ◇◇◇◇◇ ◇◇◇◇◇ ◇◇◇◇◇ ◇◇◇◇◇ ◇◇◇◇◇ ◇◇◇◇◇ ◇◇◇◇◇ ◇◇◇◇◇	981,100
◇◇◇◇◇ ◇◇◇◇◇ ◇◇◇◇◇ ◇◇◇◇◇ ◇◇◇◇◇ ◇◇◇◇◇ ◇◇◇◇◇ ◇◇◇◇◇ ◇◇◇◇◇ ◇◇◇◇◇	981,150
◇◇◇◇◇ ◇◇◇◇◇ ◇◇◇◇◇ ◇◇◇◇◇ ◇◇◇◇◇ ◇◇◇◇◇ ◇◇◇◇◇ ◇◇◇◇◇ ◇◇◇◇◇ ◇◇◇◇◇	981,200
◇◇◇◇◇ ◇◇◇◇◇ ◇◇◇◇◇ ◇◇◇◇◇ ◇◇◇◇◇ ◇◇◇◇◇ ◇◇◇◇◇ ◇◇◇◇◇ ◇◇◇◇◇ ◇◇◇◇◇	981,250
◇◇◇◇◇ ◇◇◇◇◇ ◇◇◇◇◇ ◇◇◇◇◇ ◇◇◇◇◇ ◇◇◇◇◇ ◇◇◇◇◇ ◇◇◇◇◇ ◇◇◇◇◇ ◇◇◇◇◇	981,300
◇◇◇◇◇ ◇◇◇◇◇ ◇◇◇◇◇ ◇◇◇◇◇ ◇◇◇◇◇ ◇◇◇◇◇ ◇◇◇◇◇ ◇◇◇◇◇ ◇◇◇◇◇ ◇◇◇◇◇	981,350
◇◇◇◇◇ ◇◇◇◇◇ ◇◇◇◇◇ ◇◇◇◇◇ ◇◇◇◇◇ ◇◇◇◇◇ ◇◇◇◇◇ ◇◇◇◇◇ ◇◇◇◇◇ ◇◇◇◇◇	981,400
◇◇◇◇◇ ◇◇◇◇◇ ◇◇◇◇◇ ◇◇◇◇◇ ◇◇◇◇◇ ◇◇◇◇◇ ◇◇◇◇◇ ◇◇◇◇◇ ◇◇◇◇◇ ◇◇◇◇◇	981,450
◇◇◇◇◇ ◇◇◇◇◇ ◇◇◇◇◇ ◇◇◇◇◇ ◇◇◇◇◇ ◇◇◇◇◇ ◇◇◇◇◇ ◇◇◇◇◇ ◇◇◇◇◇ ◇◇◇◇◇	981,500
◇◇◇◇◇ ◇◇◇◇◇ ◇◇◇◇◇ ◇◇◇◇◇ ◇◇◇◇◇ ◇◇◇◇◇ ◇◇◇◇◇ ◇◇◇◇◇ ◇◇◇◇◇ ◇◇◇◇◇	981,550
◇◇◇◇◇ ◇◇◇◇◇ ◇◇◇◇◇ ◇◇◇◇◇ ◇◇◇◇◇ ◇◇◇◇◇ ◇◇◇◇◇ ◇◇◇◇◇ ◇◇◇◇◇ ◇◇◇◇◇	981,600
◇◇◇◇◇ ◇◇◇◇◇ ◇◇◇◇◇ ◇◇◇◇◇ ◇◇◇◇◇ ◇◇◇◇◇ ◇◇◇◇◇ ◇◇◇◇◇ ◇◇◇◇◇ ◇◇◇◇◇	981,650
◇◇◇◇◇ ◇◇◇◇◇ ◇◇◇◇◇ ◇◇◇◇◇ ◇◇◇◇◇ ◇◇◇◇◇ ◇◇◇◇◇ ◇◇◇◇◇ ◇◇◇◇◇ ◇◇◇◇◇	981,700
◇◇◇◇◇ ◇◇◇◇◇ ◇◇◇◇◇ ◇◇◇◇◇ ◇◇◇◇◇ ◇◇◇◇◇ ◇◇◇◇◇ ◇◇◇◇◇ ◇◇◇◇◇ ◇◇◇◇◇	981,750
◇◇◇◇◇ ◇◇◇◇◇ ◇◇◇◇◇ ◇◇◇◇◇ ◇◇◇◇◇ ◇◇◇◇◇ ◇◇◇◇◇ ◇◇◇◇◇ ◇◇◇◇◇ ◇◇◇◇◇	981,800
◇◇◇◇◇ ◇◇◇◇◇ ◇◇◇◇◇ ◇◇◇◇◇ ◇◇◇◇◇ ◇◇◇◇◇ ◇◇◇◇◇ ◇◇◇◇◇ ◇◇◇◇◇ ◇◇◇◇◇	981,850
◇◇◇◇◇ ◇◇◇◇◇ ◇◇◇◇◇ ◇◇◇◇◇ ◇◇◇◇◇ ◇◇◇◇◇ ◇◇◇◇◇ ◇◇◇◇◇ ◇◇◇◇◇ ◇◇◇◇◇	981,900
◇◇◇◇◇ ◇◇◇◇◇ ◇◇◇◇◇ ◇◇◇◇◇ ◇◇◇◇◇ ◇◇◇◇◇ ◇◇◇◇◇ ◇◇◇◇◇ ◇◇◇◇◇ ◇◇◇◇◇	981,950
◇◇◇◇◇ ◇◇◇◇◇ ◇◇◇◇◇ ◇◇◇◇◇ ◇◇◇◇◇ ◇◇◇◇◇ ◇◇◇◇◇ ◇◇◇◇◇ ◇◇◇◇◇ ◇◇◇◇◇	**982,000**

The Number 982,001 to the Number 984,000

Start   Notes

	982,050
	982,100
	982,150
	982,200
	982,250
	982,300
	982,350
	982,400
	982,450
	982,500
	982,550
	982,600
	982,650
	982,700
	982,750
	982,800
	982,850
	982,900
	982,950
	983,000
	983,050
	983,100
	983,150
	983,200
	983,250
	983,300
	983,350
	983,400
	983,450
	983,500
	983,550
	983,600
	983,650
	983,700
	983,750
	983,800
	983,850
	983,900
	983,950
	984,000

The Number 984,001 to the Number 986,000

Start | | Notes

	984,050
	984,100
	984,150
	984,200
	984,250
	984,300
	984,350
	984,400
	984,450
	984,500
	984,550
	984,600
	984,650
	984,700
	984,750
	984,800
	984,850
	984,900
	984,950
	985,000
	985,050
	985,100
	985,150
	985,200
	985,250
	985,300
	985,350
	985,400
	985,450
	985,500
	985,550
	985,600
	985,650
	985,700
	985,750
	985,800
	985,850
	985,900
	985,950
	986,000

The Number 986,001 to the Number 988,000

Start

Notes

	986,050
	986,100
	986,150
	986,200
	986,250
	986,300
	986,350
	986,400
	986,450
	986,500
	986,550
	986,600
	986,650
	986,700
	986,750
	986,800
	986,850
	986,900
	986,950
	987,000
	987,050
	987,100
	987,150
	987,200
	987,250
	987,300
	987,350
	987,400
	987,450
	987,500
	987,550
	987,600
	987,650
	987,700
	987,750
	987,800
	987,850
	987,900
	987,950
	988,000

The Number 988,001 to the Number 990,000

Start

	Notes
	988,050
	988,100
	988,150
	988,200
	988,250
	988,300
	988,350
	988,400
	988,450
	988,500
	988,550
	988,600
	988,650
	988,700
	988,750
	988,800
	988,850
	988,900
	988,950
	989,000
	989,050
	989,100
	989,150
	989,200
	989,250
	989,300
	989,350
	989,400
	989,450
	989,500
	989,550
	989,600
	989,650
	989,700
	989,750
	989,800
	989,850
	989,900
	989,950
	990,000

The Number 990,001 to the Number 992,000

Start

	990,050
	990,100
	990,150
	990,200
	990,250
	990,300
	990,350
	990,400
	990,450
	990,500
	990,550
	990,600
	990,650
	990,700
	990,750
	990,800
	990,850
	990,900
	990,950
	991,000
	991,050
	991,100
	991,150
	991,200
	991,250
	991,300
	991,350
	991,400
	991,450
	991,500
	991,550
	991,600
	991,650
	991,700
	991,750
	991,800
	991,850
	991,900
	991,950
	992,000

The Number 992,001 to the Number 994,000

Start   Notes

Number
992,050
992,100
992,150
992,200
992,250
992,300
992,350
992,400
992,450
992,500
992,550
992,600
992,650
992,700
992,750
992,800
992,850
992,900
992,950
993,000
993,050
993,100
993,150
993,200
993,250
993,300
993,350
993,400
993,450
993,500
993,550
993,600
993,650
993,700
993,750
993,800
993,850
993,900
993,950
994,000

The Number 994,001 to the Number 996,000

Start	Notes
	994,050
	994,100
	994,150
	994,200
	994,250
	994,300
	994,350
	994,400
	994,450
	994,500
	994,550
	994,600
	994,650
	994,700
	994,750
	994,800
	994,850
	994,900
	994,950
	995,000
	995,050
	995,100
	995,150
	995,200
	995,250
	995,300
	995,350
	995,400
	995,450
	995,500
	995,550
	995,600
	995,650
	995,700
	995,750
	995,800
	995,850
	995,900
	995,950
	996,000

The Number 996,001 to the Number 998,000

Start   Notes

◊◊◊◊◊ ◊◊◊◊◊ ◊◊◊◊◊ ◊◊◊◊◊ ◊◊◊◊◊ ◊◊◊◊◊ ◊◊◊◊◊ ◊◊◊◊◊ ◊◊◊◊◊ ◊◊◊◊◊	996,050
◊◊◊◊◊ ◊◊◊◊◊ ◊◊◊◊◊ ◊◊◊◊◊ ◊◊◊◊◊ ◊◊◊◊◊ ◊◊◊◊◊ ◊◊◊◊◊ ◊◊◊◊◊ ◊◊◊◊◊	996,100
◊◊◊◊◊ ◊◊◊◊◊ ◊◊◊◊◊ ◊◊◊◊◊ ◊◊◊◊◊ ◊◊◊◊◊ ◊◊◊◊◊ ◊◊◊◊◊ ◊◊◊◊◊ ◊◊◊◊◊	996,150
◊◊◊◊◊ ◊◊◊◊◊ ◊◊◊◊◊ ◊◊◊◊◊ ◊◊◊◊◊ ◊◊◊◊◊ ◊◊◊◊◊ ◊◊◊◊◊ ◊◊◊◊◊ ◊◊◊◊◊	996,200
◊◊◊◊◊ ◊◊◊◊◊ ◊◊◊◊◊ ◊◊◊◊◊ ◊◊◊◊◊ ◊◊◊◊◊ ◊◊◊◊◊ ◊◊◊◊◊ ◊◊◊◊◊ ◊◊◊◊◊	996,250
◊◊◊◊◊ ◊◊◊◊◊ ◊◊◊◊◊ ◊◊◊◊◊ ◊◊◊◊◊ ◊◊◊◊◊ ◊◊◊◊◊ ◊◊◊◊◊ ◊◊◊◊◊ ◊◊◊◊◊	996,300
◊◊◊◊◊ ◊◊◊◊◊ ◊◊◊◊◊ ◊◊◊◊◊ ◊◊◊◊◊ ◊◊◊◊◊ ◊◊◊◊◊ ◊◊◊◊◊ ◊◊◊◊◊ ◊◊◊◊◊	996,350
◊◊◊◊◊ ◊◊◊◊◊ ◊◊◊◊◊ ◊◊◊◊◊ ◊◊◊◊◊ ◊◊◊◊◊ ◊◊◊◊◊ ◊◊◊◊◊ ◊◊◊◊◊ ◊◊◊◊◊	996,400
◊◊◊◊◊ ◊◊◊◊◊ ◊◊◊◊◊ ◊◊◊◊◊ ◊◊◊◊◊ ◊◊◊◊◊ ◊◊◊◊◊ ◊◊◊◊◊ ◊◊◊◊◊ ◊◊◊◊◊	996,450
◊◊◊◊◊ ◊◊◊◊◊ ◊◊◊◊◊ ◊◊◊◊◊ ◊◊◊◊◊ ◊◊◊◊◊ ◊◊◊◊◊ ◊◊◊◊◊ ◊◊◊◊◊ ◊◊◊◊◊	996,500
◊◊◊◊◊ ◊◊◊◊◊ ◊◊◊◊◊ ◊◊◊◊◊ ◊◊◊◊◊ ◊◊◊◊◊ ◊◊◊◊◊ ◊◊◊◊◊ ◊◊◊◊◊ ◊◊◊◊◊	996,550
◊◊◊◊◊ ◊◊◊◊◊ ◊◊◊◊◊ ◊◊◊◊◊ ◊◊◊◊◊ ◊◊◊◊◊ ◊◊◊◊◊ ◊◊◊◊◊ ◊◊◊◊◊ ◊◊◊◊◊	996,600
◊◊◊◊◊ ◊◊◊◊◊ ◊◊◊◊◊ ◊◊◊◊◊ ◊◊◊◊◊ ◊◊◊◊◊ ◊◊◊◊◊ ◊◊◊◊◊ ◊◊◊◊◊ ◊◊◊◊◊	996,650
◊◊◊◊◊ ◊◊◊◊◊ ◊◊◊◊◊ ◊◊◊◊◊ ◊◊◊◊◊ ◊◊◊◊◊ ◊◊◊◊◊ ◊◊◊◊◊ ◊◊◊◊◊ ◊◊◊◊◊	996,700
◊◊◊◊◊ ◊◊◊◊◊ ◊◊◊◊◊ ◊◊◊◊◊ ◊◊◊◊◊ ◊◊◊◊◊ ◊◊◊◊◊ ◊◊◊◊◊ ◊◊◊◊◊ ◊◊◊◊◊	996,750
◊◊◊◊◊ ◊◊◊◊◊ ◊◊◊◊◊ ◊◊◊◊◊ ◊◊◊◊◊ ◊◊◊◊◊ ◊◊◊◊◊ ◊◊◊◊◊ ◊◊◊◊◊ ◊◊◊◊◊	996,800
◊◊◊◊◊ ◊◊◊◊◊ ◊◊◊◊◊ ◊◊◊◊◊ ◊◊◊◊◊ ◊◊◊◊◊ ◊◊◊◊◊ ◊◊◊◊◊ ◊◊◊◊◊ ◊◊◊◊◊	996,850
◊◊◊◊◊ ◊◊◊◊◊ ◊◊◊◊◊ ◊◊◊◊◊ ◊◊◊◊◊ ◊◊◊◊◊ ◊◊◊◊◊ ◊◊◊◊◊ ◊◊◊◊◊ ◊◊◊◊◊	996,900
◊◊◊◊◊ ◊◊◊◊◊ ◊◊◊◊◊ ◊◊◊◊◊ ◊◊◊◊◊ ◊◊◊◊◊ ◊◊◊◊◊ ◊◊◊◊◊ ◊◊◊◊◊ ◊◊◊◊◊	996,950
◊◊◊◊◊ ◊◊◊◊◊ ◊◊◊◊◊ ◊◊◊◊◊ ◊◊◊◊◊ ◊◊◊◊◊ ◊◊◊◊◊ ◊◊◊◊◊ ◊◊◊◊◊ ◊◊◊◊◊	**997,000**
◊◊◊◊◊ ◊◊◊◊◊ ◊◊◊◊◊ ◊◊◊◊◊ ◊◊◊◊◊ ◊◊◊◊◊ ◊◊◊◊◊ ◊◊◊◊◊ ◊◊◊◊◊ ◊◊◊◊◊	997,050
◊◊◊◊◊ ◊◊◊◊◊ ◊◊◊◊◊ ◊◊◊◊◊ ◊◊◊◊◊ ◊◊◊◊◊ ◊◊◊◊◊ ◊◊◊◊◊ ◊◊◊◊◊ ◊◊◊◊◊	997,100
◊◊◊◊◊ ◊◊◊◊◊ ◊◊◊◊◊ ◊◊◊◊◊ ◊◊◊◊◊ ◊◊◊◊◊ ◊◊◊◊◊ ◊◊◊◊◊ ◊◊◊◊◊ ◊◊◊◊◊	997,150
◊◊◊◊◊ ◊◊◊◊◊ ◊◊◊◊◊ ◊◊◊◊◊ ◊◊◊◊◊ ◊◊◊◊◊ ◊◊◊◊◊ ◊◊◊◊◊ ◊◊◊◊◊ ◊◊◊◊◊	997,200
◊◊◊◊◊ ◊◊◊◊◊ ◊◊◊◊◊ ◊◊◊◊◊ ◊◊◊◊◊ ◊◊◊◊◊ ◊◊◊◊◊ ◊◊◊◊◊ ◊◊◊◊◊ ◊◊◊◊◊	997,250
◊◊◊◊◊ ◊◊◊◊◊ ◊◊◊◊◊ ◊◊◊◊◊ ◊◊◊◊◊ ◊◊◊◊◊ ◊◊◊◊◊ ◊◊◊◊◊ ◊◊◊◊◊ ◊◊◊◊◊	997,300
◊◊◊◊◊ ◊◊◊◊◊ ◊◊◊◊◊ ◊◊◊◊◊ ◊◊◊◊◊ ◊◊◊◊◊ ◊◊◊◊◊ ◊◊◊◊◊ ◊◊◊◊◊ ◊◊◊◊◊	997,350
◊◊◊◊◊ ◊◊◊◊◊ ◊◊◊◊◊ ◊◊◊◊◊ ◊◊◊◊◊ ◊◊◊◊◊ ◊◊◊◊◊ ◊◊◊◊◊ ◊◊◊◊◊ ◊◊◊◊◊	997,400
◊◊◊◊◊ ◊◊◊◊◊ ◊◊◊◊◊ ◊◊◊◊◊ ◊◊◊◊◊ ◊◊◊◊◊ ◊◊◊◊◊ ◊◊◊◊◊ ◊◊◊◊◊ ◊◊◊◊◊	997,450
◊◊◊◊◊ ◊◊◊◊◊ ◊◊◊◊◊ ◊◊◊◊◊ ◊◊◊◊◊ ◊◊◊◊◊ ◊◊◊◊◊ ◊◊◊◊◊ ◊◊◊◊◊ ◊◊◊◊◊	997,500
◊◊◊◊◊ ◊◊◊◊◊ ◊◊◊◊◊ ◊◊◊◊◊ ◊◊◊◊◊ ◊◊◊◊◊ ◊◊◊◊◊ ◊◊◊◊◊ ◊◊◊◊◊ ◊◊◊◊◊	997,550
◊◊◊◊◊ ◊◊◊◊◊ ◊◊◊◊◊ ◊◊◊◊◊ ◊◊◊◊◊ ◊◊◊◊◊ ◊◊◊◊◊ ◊◊◊◊◊ ◊◊◊◊◊ ◊◊◊◊◊	997,600
◊◊◊◊◊ ◊◊◊◊◊ ◊◊◊◊◊ ◊◊◊◊◊ ◊◊◊◊◊ ◊◊◊◊◊ ◊◊◊◊◊ ◊◊◊◊◊ ◊◊◊◊◊ ◊◊◊◊◊	997,650
◊◊◊◊◊ ◊◊◊◊◊ ◊◊◊◊◊ ◊◊◊◊◊ ◊◊◊◊◊ ◊◊◊◊◊ ◊◊◊◊◊ ◊◊◊◊◊ ◊◊◊◊◊ ◊◊◊◊◊	997,700
◊◊◊◊◊ ◊◊◊◊◊ ◊◊◊◊◊ ◊◊◊◊◊ ◊◊◊◊◊ ◊◊◊◊◊ ◊◊◊◊◊ ◊◊◊◊◊ ◊◊◊◊◊ ◊◊◊◊◊	997,750
◊◊◊◊◊ ◊◊◊◊◊ ◊◊◊◊◊ ◊◊◊◊◊ ◊◊◊◊◊ ◊◊◊◊◊ ◊◊◊◊◊ ◊◊◊◊◊ ◊◊◊◊◊ ◊◊◊◊◊	997,800
◊◊◊◊◊ ◊◊◊◊◊ ◊◊◊◊◊ ◊◊◊◊◊ ◊◊◊◊◊ ◊◊◊◊◊ ◊◊◊◊◊ ◊◊◊◊◊ ◊◊◊◊◊ ◊◊◊◊◊	997,850
◊◊◊◊◊ ◊◊◊◊◊ ◊◊◊◊◊ ◊◊◊◊◊ ◊◊◊◊◊ ◊◊◊◊◊ ◊◊◊◊◊ ◊◊◊◊◊ ◊◊◊◊◊ ◊◊◊◊◊	997,900
◊◊◊◊◊ ◊◊◊◊◊ ◊◊◊◊◊ ◊◊◊◊◊ ◊◊◊◊◊ ◊◊◊◊◊ ◊◊◊◊◊ ◊◊◊◊◊ ◊◊◊◊◊ ◊◊◊◊◊	997,950
◊◊◊◊◊ ◊◊◊◊◊ ◊◊◊◊◊ ◊◊◊◊◊ ◊◊◊◊◊ ◊◊◊◊◊ ◊◊◊◊◊ ◊◊◊◊◊ ◊◊◊◊◊ ◊◊◊◊◊	**998,000**

The Number 998,001 to the Number 1,000,000

Start		Notes
	998,050	
	998,100	
	998,150	
	998,200	
	998,250	
	998,300	
	998,350	
	998,400	
	998,450	
	998,500	
	998,550	
	998,600	
	998,650	
	998,700	
	998,750	
	998,800	
	998,850	
	998,900	
	998,950	
	999,000	
	999,050	
	999,100	
	999,150	
	999,200	
	999,250	
	999,300	
	999,350	
	999,400	
	999,450	
	999,500	
	999,550	
	999,600	
	999,650	
	999,700	
	999,750	
	999,800	
	999,850	
	999,900	
	999,950	
	1,000,000	

Acknowledgements

"Feeling gratitude and not expressing it is like wrapping a present and not giving it."

William Arthur Ward

Sadly, I do not have a boatload of people to thank for the contents of this book or for its inspiration—at least any who are not already thanked on this work's *Dedication. Page.* The idea of doing some sort of a million-themed book has been with me since I was a teenager. I have not been a teenager for some time now.

My sources, published or otherwise, have been very helpful, especially my internet ones.

My computer software must also be allowed to ride with me in the triumphal chariot.

Russ Rowlett of the University of North Carolina at Chapel Hill also gets an individual high five for devising his most elegant solution to the competing name systems for big numbers. It is better than any I might have come up with. I am all for these kinds of small revolutions. I hope it catches on.

Thus, there is only one multitude I need to thank, the unknowable number of my Readers that showed me the kindness to buy this book. May I tender one million thanks to all of you. (Please remember to divide these up evenly among yourselves.)

As always, any and all errors of commission or omission between the covers of this book are mine alone.

Sources

"We must find time to stop and thank the people who made a difference in our lives."

John F. Kennedy

There are just not going to be a lot of references here. Generating one million images of something is actually pretty simple to do on a computer. All the below helped in some way and are excellent resources for further reading or research. All my sources have helped me understand *scale*; which, in-turn, helped me understand that particular and most special scale: one million.

- Gullberg, Jan, *Mathematics: From the Birth of Numbers*, 1997, W. W. Norton & Co., 500 Fifth Avenue, New York, NY, USA, 10110
- Levy, Joel, *A Curious History of Mathematics: The Big Ideas from Early Number Concepts to Chaos Theory*, 2013, André Deutsch, an Imprint of Carlton Publishing Company, 20 Mortimer Street, London, W1T 3JW
- *Livescience.com*, an online resource
- *Mathematics,com*, an online resource
- Midgley, Ruth Managing Editor, *Comparisons of distance, size, area, volume, mass, weight, density, energy, temperature, time, speed and number throughout the universe*, 1980, The Diagram Group, Diagram Visual Information Ltd, St. Martin's Press, Inc., 175 Fifth Avenue, New York, NY, USA, 10010
- Morrison, Philip, Morrison, Phyllis & the Office of Charles & Ray Eames, *Powers of Ten: About the Relative Size of Things in the Universe and the Effect of Adding Another Zero*, 1982, 1994, Scientific American Library, a Division of HPHLP, Distributed by W. H. Freeman & Co., 41 Madison Avenue, New York, NY, USA, 10010
- Packard, Edward, *Imagining the Universe: A Visual Journey*, 1994, Perigee Books, Published by the Berkley Publishing Co., 200 Madison Avenue, New York, NY, USA, 10016
- Pickover, Clifford A., *The Math Book: From Pythagoras to the 57th Dimension, 250 Milestones in the History of Mathematics*, 2009, Sterling Publishing Co., Inc., 387 Park Avenue South, New York, NY, USA, 10016
- *Sites.google.com/ Russ Rowlett's Greek Based -illions*, an internet resource.
- Twocan Publishing, *Would It Help if I Said I'm Sorry a Million Times?*, 1987, Twocan Publishing, San Francisco, CA, USA
- Weeks, Marcus, *How Many Elephants in a Blue Whale: Measuring What You Don't Know in Terms of What You Do*, 2010, Puzzlewright Press, an Imprint of Sterling Publishing Company, Inc., 387 Park Avenue South, New York, NY, USA, 10016
- *Wikipedia*, an internet resource

About the Author

The author was born in Riverside, California, USA, on the day the Korean War ended. Sadly, Dale Mitchell has little memory of that most famous day in his life. Things have seldom been boring, though. Fame in human society may have been fleeting, but Nature is steadfast. At this point, the author has over 50 years of personal nature study, much of it learned alone in the days when having a *scientific* orientation was not yet 'cool' in a teenager or young man. He feels himself to be a production of the Peterson Field Guide System, as he learned about the natural world book-by-book over the years as each 'Peterson' was published.

When he is not engaged with the natural world, the author likes to prowl libraries and bookstores. His interests have been expanded greatly by chance discoveries, such as classic science fiction (where he first heard of *ecology* from the book *Dune,* by Frank Herbert), high fantasy (through J. R. R. Tolkien and Stephen Donaldson), history (from a book tossed in the trash, *The Longest Day June 6, 1944,* by Cornelius Ryan) and Andrew Lloyd Weber / Tim Rice musicals. Moreover, in the annuals of how to be a good person, perhaps his greatest example came from the very interesting folks of the Society for Creative Anachronism.

The author's heart-felt goal is to change the world in little ways. Not scary upheavals like religion, politics or . . . etiquette, but more along the lines of replacing QWERTY typewriters, spelling reform or base 12 counting. Sadly, it seems unlikely he will *master* any of these goals, but he hopes pushing the best of these keeps utilitarian change visible to the future. *Amphibians of the World: The Nature Lover's Life List* was the author's first book. Over a span of seven years, that work included the unexpected task of coining, or modifying, thousands of English names to 'solve' the problem of "What is a frog or a toad?"—an almost unique experience, and one resulting in a strongly iconoclastic work. This book on big numbers is a bit of a departure from his core interests. It is not especially iconoclastic . . .

Now living with an insanely beautiful, nature-loving woman, whom he adores, the author is happy and knows he has been most deeply fortunate. He shares his life with a plotting jealous cat, a big friendly dog of destructive, otter-like, tail and a bulging library in Corvallis, Oregon.

Epilogue:

"Rabbits can count up to four. Any number above four is <u>hrair</u>—"a lot" . . ."

Richard Adams, *Watership Down*

One, Two, Three, Four . . . *Lots:* The Fascination of Big Numbers

Numbers Matter. Indeed, you have likely been savvy to numerical importance since you were about, say, seven or eight, if not earlier. One can look at numbers as constituting a full third of the intellectual part of one's life. (Reading and social intelligence being other realms.) That said, *perhaps* you do not know this to the depths that I do. No, I am not being conceited and it is certainly not something to crow about. I simply do not have numbers, or other *quantities*, in most areas of my life. Numbers might be briefly learned, sometimes; but then swiftly 'melt away' from me, time and time again. Thus, I, the author of *The Visible One Million*, am profoundly innumerate.

So, why does this book exist, if its author is so mathematically-declined? The answer is that I, and many of my fellow innumerates, still see numbers in other ways than just quantity and 'figuring'. For example, prime numbers are within my bailiwick. Though I cannot calculate any, the shocking fact that simple one-by-one counting nevertheless manages to generate *built-in* chaos troubles my sense of logic and reality. Like a flourish of 'magic', Fibonacci numbers produce their Golden Means, and grave sunflower heads into pretty spirals, for me just as they do for you, Dear Reader. *Flatland*, Edwin Abbott's little book on dimensions (and its modern spin-offs), is an endless delight. Logarithmic charts are both beautiful and useful to me. It can even be explained to me why one should not put undying trust in statistics, winning streaks or 'luck'. (Sadly, a noisy minority of innumerates actually flaunt their ignorance of things numerical and even contend that this is something others should admire or acquire. I have no truck with that sort; because their aim to shut your mind to many of the songs through which beauty and wonder are expressed.) We all need some level of mathematics to find mental fulfillment.

In short, when quantities are explained in creative prose, illustrated, graphed, or otherwise made 'transparent', most innumerates are given a real chance to show that they do, indeed, possess a true species of math within them. We could try to expand upon this nascent mathematical potential by teaching, say, algebra or geometry *with calculators*, as a work-around for what is often the real stumbling block of most innumerates: Arithmetic. To quote Gullberg's, *Mathematics: From the Birth of Numbers*, "The most common fallacy, even among otherwise well-informed people, is . . . to confuse mathematics with elementary arithmetic, and thus to suppose that progress in mathematics consists in performing ever more complicated calculations with speed, dexterity, and accuracy." His contention, and mine, is that the beautiful creature that is 'real' mathematics is not summoned by making change at the local hamburger foodie. Of course, I am not advocating we drop teaching arithmetic, or the life skills that result from it (goodness knows, I know this); but, in the same way that not all of us *can ever* learn the necessary talent to, say, catch a bag of peanuts tossed by a baseball stadium vender; so, *sometimes it must be this way* with a few of us regarding many arithmetic skills. We, as a society, might be surprised at what can happen if we take the non-intuitive step of jumping innumerates to higher math. Why not?

Thus, it was my fascination with that strange parallel world of 'numbers', so poignant in my peripheral vision, that was father to this book on a particular big number and its subsets. I wanted to teach myself something. I did, indeed, find that something. I hope you, too, take from here your own information and delight in my little sally as a stranger in a strange land.

"In China when you're one in a million, there are 1,300 people just like you."

Bill Gates

◊

Other Books by Dale W. Mitchell

Amphibians of the World: The Nature Lover's Life List
Including Many New and More Systematic English Names